URANIUM FRENZY

Saga of the Nuclear West

D1534234

Uranium Frenzy

Saga of the Nuclear West

Revised and Expanded Edition

by

Raye C. Ringholz

UTAH STATE UNIVERSITY PRESS
Logan, Utah

Utah State University Press
Logan, Utah 84322-7800

Manufactured in the United States of America
Printed on acid-free paper

Cover design by K. C. Muscolino

Front Cover: Military personnel observing one of the bomb tests; *U. S. Department of Energy photograph*. Back Cover: Penny stock certificate; *Western Mining and Railroad Museum*.

Library of Congress Cataloging-in-Publication Data

Ringholz, Raye Carleson.
   Uranium frenzy : saga of the nuclear west / by Raye C. Ringholz.
Rev. and expanded ed.
      p. cm.
Includes bibliographical references and index.
   ISBN 0-87421-432-7 (pbk. : alk. paper)
   1. Uranium industry—Four Corners Region—History. I. Title.
   HD9539.U72 U5366 2002
   338.2'74932'097913
                                 2002009421

For Carol Houck Smith

CONTENTS

INTRODUCTION

IN 1952, A CURRENTLY unemployed geologist from Texas, Charlie Steen, succeeded in locating the largest deposit of high-grade uranium ore in the United States. His Mi Vida Mine, struck in an area thought to be unpromising, started a stampede of prospecting that was like the westward movement and the great gold rush rolled into one.

It was the first and only mineral rush triggered by the U.S. government. America, on the threshold of the nuclear age, was desperate for a domestic source of uranium. The Atomic Energy Commission was the only buyer of the ore. But it was ordinary citizens who engaged in a massive treasure hunt to satisfy the nation's needs.

Real estate salesmen, schoolteachers, hash-slingers and lawyers competed with trained mining engineers on a near insane quest to discover uranium in the redrock deserts of the Colorado Plateau, where Utah, Arizona, Colorado and New Mexico conjoin. Hard on the heels of the ore-seekers, promoters snapped up the claims—with or without any indication of ore—and went after venture capital to get the mines operating. Some of these "wheeler dealers" were financial wizards who contributed to the buildup of domestic uranium reserves. Others were rogues, with flimsy get-rich-quick schemes. Overnight, most of them became millionaires, and many lost it all because they couldn't resist the next deal.

Self-proclaimed securities brokers created an over-the-counter frenzy. Staid Salt Lake City became "The Wall Street of Uranium Stocks." Market "experts," most of them fresh out of college, were like Super Bowl rookies who had never played the big game. Motivated by quick money and the lure of action, they were the forerunners of the risk-takers we have often seen on Wall Street.

But there was a dark side to the uranium dream. An invisible, odorless and intangible menace lurked underground, where radon daughters—the decay products of uranium—were destroying the lungs of American miners. As early as 1950, industrial hygienists like Duncan Holaday warned of tragedy unless radon levels in the mines were substantially lowered through better ventilation. But few listened or took responsibility for the miners' safety.

I wanted to write this book for twenty years, since the days I lived through the uranium frenzy in Salt Lake City, and young marrieds like myself talked as much about uranium stock as we did about Dr. Spock. In the late 1960s, I began interviewing friends who had been in the forefront of the action as stockbrokers, promoters, and even prospectors (and were involved again in the second uranium boom of the sixties). Then I branched out for further interviews with my tape recorder. Circumstances necessitated that I set the project aside for a number of years, but I was able to return to it in the mid-eighties. I wanted to preserve this colorful period of Western American history with its myriad paradoxes before it was too late, before the older principals were gone. And by then, the story of the miners and radiation had become all too clear.

Now, in the beginning of a new century, the continuing and controversial aftermath of the uranium era calls for further examination. In this edition I will more fully examine the Nevada nuclear bomb tests and the effect of their radioactive fallout on military personnel involved and people living downwind of the test site. I will focus on the plight of Native American uranium workers and the lengthy battle to obtain compassionate compensation for a broad spectrum of millers, ore-haulers and other radiation victims. The gnawing problem of radioactive waste disposal will also be discussed.

Some who lived through this period may feel that the "best incident" or "the most interesting person" was left out. I have been selective in my telling of this story. By focusing on particular persons or events, I have attempted to take a broad look at the play of elements that constituted the rush for uranium and its variant consequences. It was a vivid and sometimes crazy time, perhaps the last time that an ordinary individual could wrest riches from the earth with his own two hands. It was a time dominated by expediency and a real—or exaggerated—national crisis.

The twenty-first century evokes a new chapter of development on the Colorado Plateau. Rural areas are shrinking as they are overtaken by retirees, vacationers and those seeking "second homes." There is talk of revival of the uranium industry as energy needs proliferate and terrorists fan international tensions. Yesterday we were ignorant, or maybe just plain cavalier, about destroying the environment and diminishing our quality of life. May a new conscience ensure that the past mistakes chronicled in these pages will not play a part in the future.

ACKNOWLEDGMENTS

The bulk of my research for this book came from interviews with people who were involved in the era, and material from their personal memorabilia. I am deeply indebted to all who helped me.

Mitchell Melich spent many hours talking with me about the Charlie Steen story and happenings in Moab, and generously shared his records and photographs. Charles and M. L. Steen, and Floyd Odlum gave me personal insight into their experiences during a number of interviews.

Jack Coombs was a tremendous help with information about the penny stock boom and more recent market action. Duncan Holaday and Dr. Victor Archer were constant in their willingness to inform me about their uranium miner study and explain the complexities of radiation.

Stewart Udall and his associate, Kenley Brunsdale, generously opened their files to me and saved me untold hours of digging up court testimonies, correspondence and scores of inaccessible government documents relating to the radiation tragedy. Trial proceedings for this new edition were furnished by Judge Bruce S. Jenkins and Dan Bushnell, who provided additional information in an interview. My appreciation to Ray Brim for loaning a congressional record concerning the bomb tests, and providing photographs.

Probably more than anyone else, Eola Garner gave of herself in relating her frustration and hardships during her battle for workman's compensation after the death of her miner husband. The letters, court records and other materials she let me use were invaluable.

The moving stories of the downwinders and activists who have campaigned long and hard to gain compensation for radiation victims and fight dangerous nuclear waste disposal has added depth to this edition. My thanks to Janet Gordon, Dennis and Denise Nelson, Lori Goodman, Steve Erickson, Ken Sleight, Claudia Peterson, Melton Martinez, McRae Bulloch, Marilyn Cox, John Weisheit, David Orr, Bob Pattison, Marilyn Cox and Rebecca Rockwell.

Space will not allow me to detail the contributions of all who helped me in compiling material for this book, but I do want to acknowledge them here. Information about personalities and events involved in uranium prospecting and mining was given by Maxine Steen Boyd, Charles A. Steen, Jr., Vernon Pick, Dan O'Laurie, Bob Barrett, Estelee Silver, Jim Dufford, Mel Swanson, Sam Taylor, Maxine Newell, Pete Souvall, Dr. Paul Mayberry, Al Dearth, Barney Stalcup, Dan Wegner, Gordon Babbel, Howard Balsley, Pete Byrd, J. W. Corbin, Jack Turner, Reo Hunt, Harold Ekker and Nate Knight.

Stories relating to the Manhattan Project and the Atomic Energy Commission were furnished by D. L. Edward Ellinwood, Elmer Adams, Isabelle Wooden, Ruth Ann Catt, Robert Nininger, Ralph Batie, Bud Franz, E. V. "Mike" Reinhardt, Donald L. Everhart, George Grandbouche, G. A. Franz, Elton Youngberg, Don Hill, Phil Leahy, Sheldon Wimpfen, Holger Albreathsen, Jr., Earl Land and Philip Merritt. Jeff Gordon assisted me in finding material on the atom bomb tests in the files of the Department of Energy Nevada Public Reading Facility in Las Vegas.

R. G. Beverly, Ron Evans, Edward R. Farley, Jr., and Nels Stalheim gave me insight into the background and operation of the mining companies. William Deal gave me a personal tour of the White Mesa Mill and Diane Nielson, of the Utah Department of Environmental Quality explained waste disposal processes and regulations.

Dr. Geno Saccamanno and Henry Doyle discussed the miner radiation study with me, and former Secretary of Labor Willard Wirtz and his then assistant secretary, Esther Peterson, told of establishing mine radiation standards. Thanks are due also to Philip Bierbaum and Bob Roscoe of the Department of Health and Human Services, Dave and Ron Garner, Cliff Heitt, J. V. Reistrup, Charles J. Traylor and Joan Daniels, who gave further information about the Tex Garner case.

The story of the colorful over-the-counter uranium stock boom was greatly enhanced by the memories of those who made it happen. My thanks to Dick and Frank Whitney, Dick Muir, Ralph, Ray and Jack Bowman, Dewey Anderson, Zeke Dumke, A. Payne Kibbe, Bob Bernick, Max Lewis, Wallace R. Bennett, Dave Hughes, Robert Cranmer, Noland Schneider, Frank Notti, Alex Walker, Jr., Harold Bennett, Reed Brinton, G. W. Anderson, Al Bain, Paul Barracco, Sam Bemstein, Hal Cameron, Harry Metos, Plato Christopulos, Marilyn Coon, Cherry Ridges and Gail Weggeland.

My research at libraries and museums was made easier by the gracious staff members who assisted me in finding information and photographs. Karen Carver, archivist at the University of Utah Marriott Library Special Collections, delivered cartons of my former research material to my home so that I could access them for this new edition. Judy Prosser, at the Museum of Western Colorado; Sam Quigley, Madge Tomsic and Lori Perez, Helper Western Mining and Railroad Museum; Jean McDowell, Dan O'Laurie Museum in Moab; Dr. Jack M. Hall, chief historian, Department of Energy; Barbara James and Gloria Denny, Grand County Library; Dr. Gary Shumway and Shirley Stevenson, Oral History Department, University of California, Fullerton; and staff members at the Salt Lake City Public Library, Utah Historical Society, and the Mesa County Library in Grand Junction are greatly appreciated.

My thanks to Marianne Cone for the beautiful maps.

Special appreciation goes to K.C. Muscolino, who has again created a special cover for my book.

It was Carol Houck Smith who encouraged me to write this book, and it is in thanks for her exacting editing and warm friendship that I dedicate *Uranium Frenzy* to her. Thanks to John Alley and the Utah State University Press for giving me the opportunity to complete the story of the uranium era.

Finally, thanks to my husband, Joe, who was supportive and understanding throughout the lengthy writing process.

THE NUCLEAR WEST

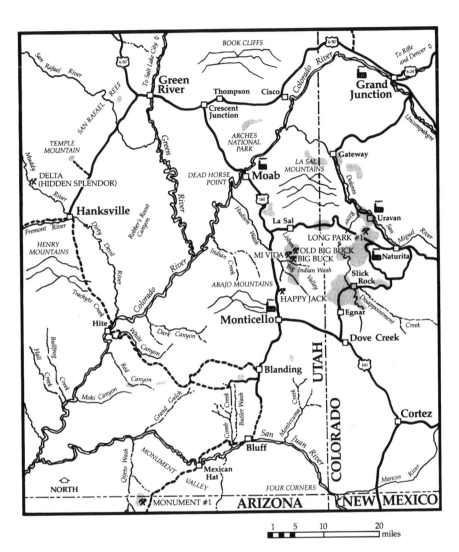

San Rafael River

8-50

To Salt Lake City

BOOK CLIFFS

Colorado River

8-50

To Rifle
and Denver

8-24

**Green
River**

Thompson   Cisco

**Grand
Junction**

SAN RAFAEL REEF

Crescent
Junction

ARCHES
NATIONAL
PARK

Uncompahgre

TEMPLE
MOUNTAIN

Gateway

Green River

DEAD HORSE
POINT

Moab

LA SAL
MOUNTAINS

Dolores River

DELTA
(HIDDEN SPLENDOR)

160

Muddy River

Hanksville

La Sal

Uravan

San Miguel River

Fremont River

Robber's Roost Canyon

Hudson Wash

LONG PARK #1

MI VIDA
OLD BIG BUCK
BIG BUCK

Naturita

HENRY
MOUNTAINS

Dirty Devil River

Colorado River

Indian Creek

Lisbon Valley

Big Indian Wash

Slick
Rock

Trachyte Creek

ABAJO MOUNTAINS

HAPPY JACK

Egnar

Disappointment Creek

Hite

White Canyon

Dark Canyon

Monticello

Dove Creek

Bullfrog Creek

Hall Creek

Red Canyon

Blanding

160

Cortez

Moki Canyon

Grand Gulch

Comb Creek

Butler Wash

San Juan River

Montezuma Creek

UTAH

COLORADO

Qjeto Wash

MONUMENT
VALLEY

Bluff

Mancos River

NORTH

**Mexican
Hat**

FOUR CORNERS

MONUMENT #1

ARIZONA

NEW MEXICO

1   5   10   20
                                    miles

 MINE   MILL   MAJOR URANIUM MINING AREA

1   THE SIREN CALL

CHARLIE DIDN'T QUITE KNOW how to tell M. L. There she was, her body all swelled up with a baby due in a couple of weeks. Their cramped rear apartment already teemed with three high-decibel kids under four years of age, crawling all over each other and on the few rickety pieces of furniture. There was barely enough money coming in to stock the fridge. Charlie felt guilty as hell but he knew he had to say that he was heading for the Colorado Plateau in a few days.

M. L. understood. It wasn't unexpected. He had read the article to her, and said it was the only way out. Life with Charlie Steen had never been dull.

It was the winter of 1949. Houston, Texas. Charlie was twenty-eight years old. He was working as a carpenter—adding a bathroom here, remodeling a kitchen there—a job he tolerated out of necessity.

His real love was geology. That was his training. He had a B.A. in geology from the Texas School of Mines and Metallurgy in El Paso. He had started a promising career as a geologist with the Standard Oil Company of Indiana. They even gave him a fifty dollar raise after his first six weeks. He spent two years with them doing field work, locating potential oil deposits. It was in the field that he was at his best.

But the job required paper work, as well. Reports. Analyses. Financial statements. Most of them nonsense as far as Charlie was concerned. Finally, he balked. He refused to submit a report he considered unnecessary. The boss thought otherwise. Charlie shot off his mouth and was fired for "rebellion against authority."[1] To make matters worse, he was blacklisted as a geologist throughout the entire oil industry.

Charlie figured that carpentry was just a stop-gap until he could get back into the field and do a bit of wildcatting for oil on his own. Maybe make a fortune. But it took a lot of cash to work a claim. More than he would ever make pounding nails and it didn't look as if anyone was willing to back him. It seemed hopeless.

Then he picked up that December 1949 copy of the *Engineering and Mining Journal* and heard the siren call.

"CAN URANIUM MINING PAY?" a headline challenged.

The accompanying story answered the question: "Yes." And it told of a potential uranium boom evolving in the ragged badlands of the Colorado Plateau.

". . . risks are being taken and profits are being made," the article said. "This speaks well for the independent producers there because many of them are newcomers to the mining field, having been farmers, ranchers, fruit growers, etc. It should also be a challenge to experienced mining men to come into the area and do as well or better."[2]

Charlie saw the article as a personal invitation. If a bunch of farmers and ranchers could locate ore certainly he, a professional geologist, could succeed. His mind soared like an eagle over the vast, rocky stretches of the Colorado Plateau.

The timing was far from ideal but Charlie wasn't one to waffle over decisions. He knew this was his only chance. Sometimes a man has to take the high dive and hope he hits the teacup of water.

Those who knew him wouldn't have said that Charles Augustus Steen looked like your classic adventurer. Prematurely balding, he was a stick-figure of a man who seemed skinny enough to slither through a keyhole. Thick round glasses veiled eyes like faded denim in a face that carried a wide grin or a black frown, depending on his mood. He liked to wear suntan drill shirts and pants that blended with his hair and complexion in a kind of protective coloration for the sere desert backcountry where he loved to roam.

But you couldn't exactly say that Charlie blended into the background. He had presence. Whether people considered his personal traits strengths or weaknesses, he was noticed. Some thought him bullheaded. Others said he had the courage of his convictions. Some considered him adventuresome. Detractors called him rash. He was labeled impulsive and patient, lucky and methodical. Certainly always in high gear; nothing in moderation. A man of extremes.

It hadn't been easy for Charlie. In his early years in Caddo, Texas, he was caught in a war zone between his mother, Rose, and his father, Charles Augustus Steen, Sr. His father had made and squandered $100,000 wildcatting for oil before he was twenty-two years old. He was thirty when he married Rose. At fifteen, she was too young to realize that he wasn't the most reliable of husbands. They broke up in 1923 when Charlie was four and his baby sister, Maxine, was only six months old.

Then came a succession of new "daddies." Lisle, in 1924; he died in a construction accident. Howell, 1928; Rose divorced him after a few months. Then came Nixon, in 1929. Rose shunted the kids from town to town with each new start. During one term, Charlie attended school in eight different states.

Another divorce brought the fatherless family back to Houston. Charlie started scrambling for 65-cent-an hour jobs to work his way through high school. He sold papers, jerked sodas, clerked groceries and did clean up work in a bookshop. At one time he supervised the mail room of a Federal Reserve Bank. When he enrolled in John Tarleton Agricultural College in Stephenville, Texas, he worked summers as a waterboy and timekeeper for Chicago Bridge and Iron Company, where his first stepfather had been employed. The company, feeling responsibility for Lisle's death, helped finance Charlie's education as partial compensation to the family.

Perhaps it was a long-held dream that decided him on a geology major during his freshman year. The idea of prospecting had always fascinated Charlie. His father had been a prospector. He had made it big, but then he lost it. Charlie figured if he studied hard and got the right kind of experience he could make his own mark in the mining world. But he would do it scientifically. He would keep his winnings. And he concentrated on his studies at Tarleton so hard that, when he met Minnie Lee Holland, a popular, dimpled brunette who preferred to be called "M. L.," he "barely lifted his eyes from his geology book to acknowledge the introduction."[3]

The next year he transferred to the Texas School of Mines and Metallurgy at El Paso to start his major in earnest. By then, he had run into M. L. again and had been more attentive the second time. And thereafter. It was a full-blown romance in 1943 when he received a B.A. in geology.

World War II put the courtship on hold, however. Following his graduation, Charlie discovered that he was ineligible for the draft, classified 4-F due to poor eyesight. Disappointed that he couldn't serve his country, he decided to further his career and signed up as a geologist for a Bolivian tin mine. Since he would be gone, M. L. enlisted in the W.A.V.E.S., the newly-created women's branch of the U.S. Navy.

Following a few months in Bolivia, Charlie moved into the Peruvian Amazon to conduct field reconnaissance for the Socony Vacuum Oil Company. While one of Peru's richest gas fields was discovered and drilled there some twenty years later, Socony's efforts did not bear fruit. The experience, however, strengthened Charlie's resolve to strike paydirt on his own. He vowed to find his bonanza "if it takes me the rest of my life."[4]

After the war, Charlie returned from South America and M. L. received a medical discharge from the W.A.V.E.S, due to a back injury. They got married and moved to Chicago, where Charlie completed one year of graduate school at the University of Chicago and M. L. got pregnant. It was the beginning of rapid-fire parenthood that would bring them four sons within the first five years of marriage.

When family responsibility put a damper on further education, Charlie got the job with Standard Oil that didn't pan out. "In retrospect I had it coming to me," he later admitted. "It taught me a number of lessons, including that the boss is not always right, but he is always the boss."[5]

Henceforth, resolving to be his own boss, he took up carpentry and formed his own contracting business, intending to pinch pennies until he could strike out again for oil. Instead, he read the article about uranium in the mining journal.

Uranium was suddenly the most critical material the United States had ever known. It fueled atomic weapons. It promised environmentally-clean electrical power, gas-free operation of cars, planes and locomotives, preservation of meat, distillation of sea water. Uncle Sam was desperate for the mineral and willing to pay for it. By law, the federal government was the only buyer.

America needed a domestic supply of uranium to keep a nuclear edge in the cold war that was developing with Russia. To encourage prospectors like Charlie, the Atomic Energy Commission (AEC) established minimum prices for the ore, guaranteeing the rates for ten years. An additional incentive was a $10,000 bonus for each separate discovery and production of high-grade uranium from new domestic deposits. What's more, a man could stake a uranium claim on public land for a dollar.[6]

Charlie couldn't help himself. One part of him said to stay put in Houston and the other pushed him toward the vast unknown where anything was possible. He asked himself why he would want to leave a steady paycheck and his family for the exhaustion, hunger, thirst, danger and loneliness of prospecting. Deep down he knew the answer. He wanted that red Lincoln car. A house and jewels for M. L. College degrees for his sons.

Once formed, the idea consumed him. He pored over geology books describing the Colorado Plateau and lost himself in the infinity of rock and sand sculpted by the winds and waters of geologic ages. He heard the silences. He burned with the heat. He relished the solitude and being dwarfed by massive arcs and pillars of sandstone.

In the *U.S. Army Corps of Engineers Report Upon the Colorado River of the West, 1861,* J. S. Newberry, the first geologist to visit the region, said of it:

"Perhaps no portion of the earth's surface is more irredeemably sterile, none more hopelessly lost to human occupation. . . ."

But to Charlie it was the ultimate challenge. The prehistoric seas and riverine marshes, as extinct as the fish and dinosaurs that once inhabited them, masked treasures that had lain petrified over hundreds of millions of years. Nothing was as it appeared. What the eye measured as ten miles turned out to be a hundred miles of flat desert slashed by deep canyons. What looked to be a lake dissolved into a sea of sand. A city skyline became a cluster of standing rocks carved from frozen dunes.

And, perhaps, here was the last American Dream. The gold rush and the western movement rolled into one. The final opportunity for a man to bet on himself and stake his own knowledge, instincts and sweat against the hidden riches of a formidable land.

Charlie knew he had to beat the flood of fortune hunters that would surely rise to the AEC's bait. He added his meagre savings to the $1,000 his mother pressed into his hand after mortgaging her home. It was enough to buy a second-hand jeep and a broken-down drill rig. M. L. assured him that she and the kids could get along fine for the time being on the monthly thirty-eight dollar disability check she received from the navy. Rose would help her when the baby came. She and the children would join Charlie later.

Charlie was well prepared. He knew the uranium mining history of the area by heart. He had followed the long story of boom and bust that had repeated itself over the centuries. It was a strange progression that found the waste products of one era becoming the sought-after prize of the next.

The Ute and Navajo Indians were the first to inhabit the Colorado Plateau during historic times. Theirs was not a search for riches, however. They dug up the red vanadium and yellow uranium-bearing rocks of their homeland unaware of any value or hidden powers in the ore. They pounded the stones into dust and mixed it with animal fat to make war paint for their bodies.

The actual mining of uranium started by accident in the mid-nineteenth century. Gold miners in the Central City district near Denver, Colorado, were exasperated by a thick, black substance that clung like tar to their mining and processing tools. The annoying material was called pitchblende. The miners tossed it into their tailings dumps where they discarded waste products.

Some years later, yellow uranium oxide became valuable as a coloring agent for porcelains and ceramics. The piles of gold mine tailings took on an importance of their own. People started to mine the cast-off mineral out of the dumps, and for over a decade the ore was shipped to London china factories.

Shortly after the turn of the century, a vital use for pitchblende developed when Madame Marie Curie and her husband Pierre discovered the "miracle element" radium. They isolated the radioactive material that seemed to promise a possible cure for cancer from pitchblende.

Most of the ore that the Curies used came from mines in the Erz Mountains of Germany and Czechoslovakia. But after a few years, those deposits were depleted. Luckily, a French chemist discovered a new source of radium in carnotite, the yellow, red and black uranium-vanadium-bearing rock that the Indians on the Colorado Plateau had used for their war paint.

The news triggered a rush of prospectors into the backcountry. When compared to the hardships entailed by most mining, the search for carnotite was a lark. Men could find a deposit within a few hours. Often outcrops showed right on the surface or in petrified logs. In the month of March 1912, alone, ten claims were staked. From then on, the Curie's primary source of radium came from carnotite on the Colorado Plateau. Mining thrived there until 1923, when discovery of high-grade ore in the Belgian Congo led to closing of most of the mines.

During the radium boom there were few known uses for uranium and vanadium was considered a waste product of carnotite. But not for long. French chemists, studying the effects of vanadium on steel, had found that if a compound of vanadium oxide and iron was dropped into molten steel just before pouring, it would greatly increase the tensile strength, wearability and elasticity of the steel. Prior to World War I, the mineral was mined in Peru and in roscoelite deposits of Colorado. With the declaration of war, exporters persuaded the British to use vanadium steel as armor for their warships. But by 1934 the mines had played out. American manufacturers of automobiles, planes, locomotives and industrial machinery forced the steel industry to seek vanadium on the Colorado Plateau. The abandoned radium dumps were mined again and a vanadium run resulted with prospectors returning to redrock country for a second mining boom.

History was to repeat itself twenty-five years later. This time uranium was the prized element. Uranium—the waste material that had been dumped in the tailings during the vanadium era.

In the early 1940s, the Manhattan Project of the U.S. Army Corps of Engineers was created to develop an atomic bomb. The key element for the bomb was uranium. But the metal was in short supply. The only known deposits of the ore were in what was then called the Belgian Congo and

Canada, but these mines were nearing depletion. Thus, the Manhattan Project began to conduct a covert campaign to locate domestic sources of uranium ore.

Shortly after the highly-classified operation got underway, a secret shipment of uranium was escorted to a Canadian refinery by Phil Merritt, director of Raw Materials for the Manhattan Project. One thousand tons of uranium oxide from the Shinkolobwe mines in the Belgian Congo had been diverted to the United States when Germany invaded Belgium. The shipment, ordered by Edgar Sengier, chief executive of the Union Minière de Haute Katanga—because "America might just need it"—had been stored in a Staten Island warehouse since 1940. This was the uranium used to build the original atom bombs.

But one thousand tons was a mere pittance in the overall uranium picture. The United States was desperate for its own stockpile of raw materials. Washington turned its attention to the Four Corners area where Utah, Colorado, Arizona and New Mexico conjoin. It was here that radium and vanadium had been mined. A modest stockpile of uranium that had been discarded in the tailings of those operations was already available. There must be more ore underground, they reasoned. So an engineering unit of the Manhattan Project moved into the Colorado Plateau. Lt. Phil Leahy was transferred to Grand Junction, Colorado, to oversee the division. He had no idea of the project's scope or its connection with an atomic bomb. All he knew was that he must squeeze all of the uranium he could out of those old vanadium tailings.

Leahy set up headquarters in a deserted log cabin at an abandoned lumberyard. He oversaw the conversion of vanadium mills in the Colorado towns of Durango, Rifle, Naturita and Uravan, and in Monticello, Utah, for extraction of uranium from their dumps. An experimental uranium-processing mill was constructed at the Manhattan Project compound. Fendoll A. Sitton, a mining man from Dove Creek, Colorado, and Blair Burwell, manager of the Uravan mill, secretly advised him about working out existing uranium mines and acquiring additional amounts of ore from independent miners. Manhattan Project geologists even went out to look for undiscovered deposits themselves.

Charlie Steen was particularly interested in reports about government prospectors who ranged over the Plateau searching for new uranium lodes. When security restrictions were lifted after the war, he read articles in newspapers and magazines that told how scores of young men in the Army Corps of Engineers disappeared into the desert to fan metal detection devices,

The Manhattan Project Grand Junction office was in an abandoned lumberyard cabin. *Museum of Western Colorado.*

called Geiger counters, over the sandstone rims of the canyons and chart any anomalies that indicated possible deposits. They were the ones who pioneered the method for seeking the elusive ore. Carrying out their highly classified wartime mission, they ushered in a brand new field of mineralogy.

Charlie was fascinated with this revolutionary era of mining. He watched it develop further when Russia entered the nuclear race and prospecting for uranium became even more urgent. It was peacetime then. The U.S. Army Corps of Engineers had disbanded the Manhattan Project and turned the uranium business over to a newly-appointed civilian body called the Atomic Energy Commission. It was the AEC that launched the massive procurement program to develop a full-fledged uranium industry in America. In 1948 they announced the first federally-controlled, federally-promoted and federally-supported mineral rush in the nation's history.

Charlie analyzed the locations of existing mines, and figured that a triangle drawn from Moab, Utah, on the west, Dove Creek, Colorado, near the Utah-Colorado border in the southeast, and Grand Junction to the north contained the heart of the action. He revved up his rattletrap jeep, hitched up the trailer and headed for Dove Creek.

In 1950, Dove Creek was little more than a clearing in the sagebrush, accommodating a few pinto-bean farms, some cattle, a bank, a general store and a scattering of houses. News of the uranium hunt had stirred the drowsy little community to subdued excitement. Normally conservative farmers took to the hills on weekend prospecting trips, and strangers suddenly began to converge upon the quiet settlement.

When Charlie arrived, he parked his trailer in town and then steered his jeep through the roadless tracts of desert country to scan the sandstone strata for signs of uranium. He marveled at the geologic record that sprang to life around him: volcanic cones, folds, terraced faults and scalloped anticlines and synclines that spoke of prehistoric tectonic forces; fossilized remnants of ancient seas; the burnished vermilion, amber, russet, salmon and tawny tones of shales, sandstones, limestones and conglomerates layered in random patterns over eons of time.

It was an inhospitable land. Yet it possessed an eerie, other-worldly beauty. Brilliant bursts of color tinted the cliffs at dawn and sunset. The midday glare muted the garish tones to pastel hues. Then came a silvered nighttime sky so clustered with stars that mere pinpoints of darkness were insinuated between them.

Time lapsed into slow motion. Charlie scarcely noted its passing, but he must have been prospecting for several days when the engine of the old jeep boiled over and died. He walked up a hill to a farmhouse to get some water. Bob Barrett, a tall, burly bean farmer with tanned skin and rough, blunt hands answered the door and invited Charlie inside.

"We got to jagging about uranium," Barrett recalls, "and he stayed there at my place for three days. He'd read about Sitton. [The vanadium miner dubbed "the Uranium King" who sold ore to the Manhattan Project during the war and advised Lt. Phil Leahy.] *Collier's* magazine had given him [Sitton] a big writeup. Charlie, when he first come to Dove Creek, was hunting Sitton. Wanted to get close to Sitton. We got to arguing and I told him, 'Sitton don't know anymore about uranium than anyone else. It's scattered all over the basin.' I loaded his jeep up on the truck and went around and showed him the whole country and back to Big Indian [a mining area where some uranium had been found]. I took him out there the first time. He was looking for a big bed of low-grade ore to drill behind."

Barrett wasn't impressed with Charlie's provisions for the hard work of prospecting.

"He didn't know a thing about getting along in that country or providing for himself," he said. "His mother sent him a little money over there to get himself something to eat and he'd buy bananas and potato chips. The first time I went out with him, he said he'd get the groceries. He liked canned peaches. Well, if you go out prospecting you ought to go out once with canned peaches!"

"Charlie, he's a hustler! When he was out, he might not have known what he was doing, but he sure covered a lot of ground. He followed his trade, all right."[7]

About two months after Charlie arrived in Dove Creek, M. L. bundled up four-year-old Johnny, Charles, Jr., year-old Andy and her brand new baby Mark and joined him. By that time, Charlie was almost broke, but he and M. L. were determined for the family to stay together. Dove Creek locals offered little sympathy, looking askance at this young, able-bodied male who chased after uranium instead of supporting his family "with a decent job."

Charlie knew he'd have to get a grubstake or quit. Finally, he wrangled an appointment with Fendoll Sitton. Sitton sent him to Bill McCormick, owner of the Dove Creek Mercantile Store, who had dabbled in uranium mining during the war. Charlie asked McCormick if he would be interested in grubstaking his search.

"We had some discussion there and I said I wasn't looking for a drug-store geologist," McCormick later remembered. "And Charlie gave me to understand he wasn't a drugstore geologist and went out to his car and brought in some papers and showed me that he had been down in South America and worked out in the jungles.

"He said he was willing to go and stay as long as I was willing to stay. We shook hands on that, and that same day we went over to Big Indian. I was working on that property then. We never made up a contract. It was simply that we met, we went out and prospected together, and the one that had the dollar paid it."[8]

McCormick doled out the supplies and Charlie did the footwork that meant long days in the searing dust with blistered metatarsals and a mouth that felt like sheep's wool. Sharp rocks bruised his feet inside his high, laced boots. The sun burned through his cotton clothing.

M. L. stayed in the trailer at Spud Patch with the kids. She stewed wild rabbit and cooked the venison that Charlie bagged out of season. Sometimes when there wasn't enough money for canned milk, she fed the baby tea and sugar.

Despite the hardships, they were happy, "happier than I have been for years," Charlie wrote his friend Edna W. Miner in Houston. "There is a good chance that I can make a huge amount of money, but at the very least, it will be an exciting adventure. We are living according to the new theme as stated by that college prexy, namely, 'Stop striving for the mythical illusion of security in the atomic age. You will live dangerously every day of your

At Yellow Cat Wash the Steens lived in a trailer linked to an 8 x 16-foot shack. *Charles A. Steen, Jr.*

life, if you live at all.' Or, as they say in the vernacular of this country, 'You don't have to be crazy to mine uranium, but it helps.' I feel that I qualify."[9]

As winter approached, there was a good chance of being snowed in at Spud Patch, a risk Charlie couldn't afford with four little ones twenty-five miles from the nearest town. Besides, the cost of living there was too much for them. Hitching up the trailer, he joined a few other squatters on a dusty plot of ground at a forsaken place called Yellow Cat Wash in southeastern Utah. Someone had hit a little low-grade ore near there. They connected the trailer to an eight-by-sixteen-foot shack that boasted a second-hand wood stove and two built-in bunks. Their home was at the face of a huge cliff with a distant view of the La Sal Mountains.

"The scenery is magnificent," M. L. wrote a friend. "Of course our conveniences here are no better than Dove Creek or the poor Mexican peon, but it's been like a prolonged vacation. There's no watching the clock (as we don't have one) . . . just eat, sleep, and do everything else if and when we feel like it. I work like a slave when I want to and go two or three days at a time without doing anything, such as labor, I mean."[10]

Charlie spent his time prospecting and trying to find a backer. He would be away from home for several weeks.

"I miss him," M. L. wrote, "but get along same as usual so far as everything else is concerned. He's really wonderful, and if we never have money, I'm rich with him."

By that time Charlie had established himself throughout the Plateau as "that nutty Texan" who tried to find uranium without a Geiger counter.

A prospector checks the radiation reading on his Geiger counter. *Museum of Western Colorado.*

What was worse, he persisted in carrying his search into areas the AEC had already labeled "barren of possibilities." Everyone laughed over "Steen's Folly."

Most uranium hunters knew little and cared less about geology. They just followed the lead of the Manhattan Project and AEC geologists and moved over the canyon ridges of the backcountry like a blind man's fingers reading a relief map. They concentrated on exposed outcroppings on the canyon rims. They looked for tell-tale red, light brown or gray sandstone that they were told would indicate something called the Salt Wash member of the Morrison formation. They didn't know what that was but it was where they were supposed to find uranium.

They put a few dollars down for a radiation counter named after the German physicist Hans Geiger. Then they hiked until the thing started clicking, swinging a detection wand that looked something like a handheld microphone over promising terrain. When the clicks chattered or the needle on a dial whirled, they checked the surrounding area more carefully. The number of pulses per minute or the distance the needle moved measured the intensity of the radioactivity.

Charlie couldn't afford a Geiger counter. He didn't want one. After all, he was a graduate geologist. He would base his explorations on that knowledge.

Geologic time inscribes itself on the Colorado Plateau in vari-colored sandstones layered like a gigantic sandwich. Each stratum of rock represents a portion of the region's genealogy. Those strata born of a vast, restless sea that covered the land hundreds of millions of years ago contain fossilized shells and fishes. Others tell of sediments gradually filling the ocean until the exposed crust of the earth parched and crinkled. Massive petrified sand dunes are remnants of a desert created when mountains appeared and the region was cut off from coastal moisture. Some of these deposits, approximately 195,000,000 years old, are called the Shinarump and Chinle formations.

After another thirty-five million years, the mountains eroded to the valley floor. The climate became tropical and jungles thickened with bamboo, palms and ferns. Dinosaurs crawled through swamps and fed at the shores of lakes and lagoons. And then, mysteriously, the giant reptiles began to disappear. Sluggish rivers filled with fallen trees and bones of the vanishing monsters. The detritus petrified into red, brown, green and slate-colored mud. This prehistoric cemetery composed the Salt Wash member of the Morrison formation where most of the Colorado Plateau's uranium had been found.

But this was not where Charlie targeted his search. In fact, he concentrated on an area that the government geologists had investigated and labeled worthless. Instead of scrutinizing the surface of canyon rims and looking for signs of prehistoric Morrison river channels, he used his battered drill to probe a bit further. He insisted that uranium could collect deep underground like reservoirs of oil and then leech upward into the Morrison or Shinarump formations. He looked for downward sloping, or anticlinal, structures that were behind existing claims where small amounts of uranium had been found. He concentrated on the geologically older Shinarump conglomerate.

Bad luck dogged him. The drill pump froze and burst twice. Bits broke. The jeep and trailer were near collapse. M. L. and the kids struggled through a hard winter.

"Hardships never seem to cease for us," M. L. wrote. "We've been out of butane for four days now. Sent our bottle to Grand Junction and they didn't get back with it. If the wind isn't blowing like West Texas we cook on an open fire in the yard. If it is, we go to a log cabin five miles away, deserted, dirt floor and walls, but has a cook stove and table in it. That is if Charles is home with the jeep. Otherwise, we eat Post Toasties, peanut butter, etc. I'm starved for a real meal right now and a cup of coffee without ashes and sand."[11]

When Charlie wasn't tramping in the backcountry, he was looking for someone to finance his search. "I spent more time hunting a grubstake than I did hunting ore," he later said.

Finally, in the spring of 1951, it looked as if an "angel" had arrived. A mining engineer named John Shoemaker offered to stake Charlie in return for a controlling percentage of anything he found.

"Charles finally found himself a rich man who was willing to gamble ... $300 a month grubstake plus a larger drill, water tank, and other equipment necessary to carry on in this business," M. L. wrote her friend. "They've staked much land and things are looking up. Of course, the $300 per month living allowance sounds like three million and will go plenty far out here in the toolies like we are. The only thing I hate about it is leaving here, which I dearly love ... the people, the cliffs, rocks, sunsets and everything. I guess we'll be in Uravan or Naturita, Colorado, within the next few weeks."

The move never materialized. When Shoemaker saw that the Steens had been living on much less than he offered, he changed his monthly stipend to $200. Charlie told him to "go straight to Hell."

It was only weeks later that Charlie came upon a prospector named Dan Hayes working some claims in the Big Indian Wash area on the western slope of the Lisbon anticline. Charlie knew that some small deposits of uranium had been discovered in the area in 1916, but attempts at mining them were unsuccessful. Then in May 1948, Hayes, along with Jim Bentley and W. Y. Brewer, rechecked the region and staked fourteen claims that they named Big Buck. They found some fairly good bodies of ore, but it was riddled with lime. Because of the marginal percentage of uranium in their loads, they were only able to make small shipments to the Manhattan Project processing mill at Monticello.

When the AEC started its diamond drilling project to prove ore in promising formations, the three partners of the Big Buck petitioned that their property be included in the program. The government refused on the basis of previous surveys by their geologists that claimed the district's potential didn't justify the expense.

Still Charlie banked on his intuition. He figured if low-grade ore outcropped on the southwest side of the Lisbon anticline, high-grade could be found down-dip by drilling. Receiving Hayes's permission to check the terrain behind the Big Buck outcroppings, he gathered twelve rocks for cornerstones and some stakes to mark boundaries, and staked out a dozen 600 x 1,500-feet rectangles. He filled out location notices describing the property and claiming it under his name and tucked the papers into tin cans for protection from the elements. He gave the claims Spanish names reminiscent of his happy days in South America—Mujer Sin Verguenza, Mi Corazon, Besame Mucho, Pisco, Fundadoro, Te Quiero, Linda Mujer, Mi Amorcita, Ann, Bacardi, Mi Alma, and Mi Vida ("My Life"). All that remained for him to do was pay $12 to file the claims with the county registrar.

He returned to Yellow Cat to tell M. L. the good news only to find that his mother had suffered a severe heart attack. He immediately caught the bus for Houston, taking the three older boys with him. M. L. stayed behind with the baby and a loaded 30-30 to guard the drill and their meagre belongings.

During the two weeks that Charlie and the boys were gone, M. L. took the opportunity to use the jeep. Her closest neighbor lived seventeen miles away so she would visit friends every time she went to Cisco, the nearest town, to pick up mail or fill the five-gallon water cans at the storage tank.

One Sunday, the Jeep vapor-locked somewhere out in the middle of the desert. There was a twelve-inch hole in the radiator. She was stranded with nothing but a light blanket and one bottle of milk for the baby, and not another car in sight. A cold April sandstorm was blowing so hard that she could scarcely see, but she knew she could not spend the night there. Bundling up the baby, she started to walk, stumbling along the road until a lone truck finally rescued her.

By the time Charlie returned, the jeep was repaired and ready to strike out again. However, the claims at Big Indian had to wait. Rose had recovered from the attack and was convalescing well, but her medical bills were too high to warrant further grubstaking for uranium. Charlie couldn't scrape enough together to make it without her help. Reluctantly, he hitched up the trailer and moved the family to Tucson, Arizona, where he could get carpentry work.

With his family settled once again and a little money coming in, Charlie resumed his efforts at finding a grubstake. He couldn't find a backer for his uranium claims, but he did manage to do some contract prospecting on weekends in the Silver Bell Mining District.

He even had a couple of tempting job proposals. An oil company overlooked his former blacklisting and offered him a position as a field geologist, and he had a chance to work for the Atomic Energy Commission at Hanford, Washington. But he turned them down. He was on a single track. All the while he worked with wood, his mind whistled with the winds of Big Indian Wash.

2 THE EUROPEAN EXPERIENCE

It wasn't long after the train pulled out of "the Mile High City" of Denver, that Duncan Holaday understood why the Denver & Rio Grande Western route to Salt Lake City, Utah, was known as "The Scenic Line of the World." Half an hour's ride from the metropolis, the miles and miles of flat plains seemed to roll into infinity as the Panoramic climbed 2,000 feet onto the eastern shoulder of the Rocky Mountains. A few minutes more, and the view closed in upon them. The long train wound deep into canyons studded with frosted evergreens, drifted snow and mountains that looked like fluted peaks of meringue.

Then suddenly, as if cut off from the world, there was nothing but darkness. The locomotive roared through the Moffat Tunnel, four thousand feet beneath the Continental Divide from which North America's eastern and western river systems made their separate ways. The sensation of speed was heightened as blackness whirled outside the window. Then again, before Holaday's eyes could readjust to the glare and sparkling snow, the engine burst from the western portal of the tunnel to start its long descent into the valley of the Upper Colorado River.

Holaday saw the Colorado River first as a crystalline brook that tumbled from the snowline of the Rockies and gouged its narrow channel through the reddish walls of Glenwood Canyon. Once the waters reached the broadening valley, he watched them widen into a placid, murky stream that nourished the thirsty prairieland where cattle grazed in the shadow of the mountains. By the time it reached Grand Junction, Colorado, nestled between the yellowed, squaw-skirt pleats of the Book Cliffs and the high russet ridges of the Colorado National Monument, it broadened into a mighty river.

It was there that Duncan Holaday left the train on the last day of March 1949. Jack Torrey, of the Industrial Hygiene Section of the Colorado State Department of Public Health, was with him. They were on their way to see Ralph Batie, chief of Health and Safety for the Colorado Raw Materials Division of the AEC.

Batie was concerned about potential radiation problems in the fledgling uranium mining industry. He had asked Torrey's associate, Paul Jacoe, for advice. Jacoe was stumped. He had not dealt with radioactivity. He phoned Henry Doyle, the regional representative for the Public Health Service. Doyle was equally ignorant on the subject and knew no methods of evaluating radiation. The only expert he knew was Duncan Holaday. Doyle contacted Holaday in Washington, D.C., and asked him to come out west and take a look.

Holaday was a rather quiet, tall man in his early forties, with a long, narrow face, slightly hooked nose and slender, tapering fingers. Six years earlier he had been just another staffer with the United States Public Health Service. During the war, he had inspected some eighty different government-owned, contractor-operated arsenals. His group, in charge of occupational health, checked the plants for toxic gases, dusts and other hazards. Later, he worked on the department's survey of the Cemented Tungsten Carbide industry.

Then, in 1946, his chief volunteered him for Operation Crossroads, following the first peacetime atomic bomb test on the Pacific island of Bikini. He completed a six-week navy course on radiological safety and was sent as one of the specialists for the cleanup procedures of ships that had been anchored in range of the nuclear explosion. It was his job to become an expert on radiation exposures. He performed radiography with x-rays, checked for radium and cobalt and sought new electron sources.

The expertise he developed on the Bikini job established his reputation in radiation-related industrial hygiene. He had received many requests for consultation, but this was the first one involving uranium mining. He was interested to hear what Batie had to say.

A driver met them at the Grand Junction railroad station and they drove up Pitkin Avenue past the snow-powdered trees of Whitman Park. Beyond the park Holaday saw the pinkish native rock and weathered wood of turn-of-the-century buildings in the central business district. Along the river were utilitarian sheds of a small industrial section.

Duncan Holaday was one of the first to warn about the hazards of radiation. *Duncan Holaday.*

They crossed the bridge over the Colorado near its confluence with the Gunnison and passed a worn assortment of service stations and truck stops before turning off the main highway towards cemetery hill. Behind the cold iron fence, Holaday could glimpse rows of stark, grey gravestones hunched under bowers of winter-dead branches. Then they descended into an isolated gulch intersected by some railroad tracks and stopped before a small pine-log cabin. The Grand Junction AEC compound. By no means impressive.

All Holaday could see was a log cabin and a crude picket fence that encircled a muddy 55-acre plot. A prefabricated addition was tacked onto the back of the building and a corrugated metal processing mill towered to the south. Discarded Civilian Conservation Corps barracks served as warehouses for hundreds of 30-gallon drums packed with uranium and vanadium concentrates. The Gunnison River flowing on the western perimeter of the narrow canyon was barely visible.

Holaday and Torrey climbed the steps at the front of the cabin. In the security area they signed the visitors' registration sheet, clipped on their clearance badges and waited for Batie.

Presently Batie arrived, welcoming them with apologies. They would have to go down to the basement, he told them. His laboratory was there. He wanted it that way. Alone, where people couldn't monkey around with his equipment.

Every inch of the crowded building was in use. There were no frills. No filing cases. No bookshelves. No carpets. Maps and reports were jammed in cardboard boxes piled on the linoleum floor. The living room of the previous owner served as finance and personnel offices. One secretary worked at a desk in front of an unused red brick fireplace. The manager of the compound, Phil Leahy, had his quarters in the former bedroom in the northeast corner. A few U.S. Geological Survey geologists shared some of the rooms in the frame addition at the rear.

Batie invited his visitors to pull up chairs around the desk in the middle of the institutional-green room. There wasn't much furniture. Long tables along three walls held a hand-cranked pump to collect dust in the mines, some one-liter glass flasks to hold samples of radon gas, a stack of membrane filter papers and mathematical scales. Several bottles of rendered grain alcohol stood in a corner.

Batie got right down to business.

"I think we might have a problem on our hands," he said. "I'm worried that we're headed for another 'European Experience.'"[1]

Holaday and Torrey were both familiar with medical literature dealing with this classic example of the hazards of radiation. Reports dating as far

The AEC compound at Grand Junction, Colorado. *Museum of Western Colorado.*

back as the early 1500s chronicled unprecedented deaths of miners in the Erz Mountains of Schneeburg, Germany, and Joachimsthal, Czechoslovakia, where mines bore rich veins of silver. These mines were also exploited for cobalt, nickel, bismuth and arsenic. Prior to the turn of the century, their pitchblende was mined for uranium to be used as dyes. Later, they became famous for supplying radium to Pierre and Madame Marie Curie.

But the toll in human lives was high. Hundreds of miners in the prime of their lives were stricken suddenly by a wasting illness. They declined rapidly, and died within a year. The annual death rate comprised one percent of the total mining work force. Seventy percent of those who died succumbed to lung cancer.

For centuries the illness that claimed middle-aged miners was known as "Bergkrankheit," or mountain disease. It wasn't until the early twentieth century that doctors suspected that the unusually high mortality rate resulted from exposure to radioactive materials in the mines, especially radon. Numerous medical investigations followed. The resultant epidemiological records of the Erz Mountain mines were reported so thoroughly that they represented the best documentation of an occupational disease to date. Ralph Batie had read the literature. That was why he was worried.

Batie's fears stemmed from a long line of frustrations. From the time of his transfer from the Department of Agriculture to the Corps of Engineers' Manhattan Project in 1942, he had waged a line of one-man battles. His office was in the Grand Junction compound, but his responsibility for health and safety extended to all of the vanadium mills of Colorado and Utah that secretly processed uranium for the atomic bomb. The problem was that every time he sought specialized information or attempted to regulate safety precautions, he met with bureaucratic buck-passing or a lack of cooperation. Washington's hunger for uranium "at any cost" took precedence over anything that might slow down production. The Vanadium Corporation of America, and U.S. Vanadium Corporation, Climax Uranium Company and other mining companies were reluctant to spend thousands of dollars for safety measures in their mines and mills unless *demanded* by the federal government, their only customer. Even the mill workers rebelled against unfamiliar sanitation procedures that Batie attempted to enforce.

At that time the uranium business was in its infancy, and no one could anticipate the possible consequences of radiation exposure. Not much was known about radiation. Geiger counters hadn't even come into existence at that time. Mine and mill workers were told they were going to be handling U_3O_8 (uranium oxide) and V_2O_8 (vanadium), but had no idea what it was for.

"When we first started the mills, the army began to get a little concerned, so they had a radiologist stationed in research in Rochester, New York come out," Batie said. "We were mainly interested in the dust. There was a separation point in the middle of the process where vanadium came out one way and uranium the other. The vanadium came out as a slag and uranium came from the leach through presses as a canary-yellow cake. From there it went into the dryer and, when that came out, it was just like flour. Flying everywhere.

"We sent cylinders with radiation samples to Rochester for analysis but we never found out what it was. They would just tell us it looked like everything was all right. It was the old army game. Communications were not good. The information went to the office of General Groves, director of the Manhattan Project in Washington, D.C., and I don't think even they knew what the radiation tolerance level was. I think it was just ignorance. They were still feeling their way.

"We decided there was a hazard so we followed the radiologist's suggestion and got a doctor to take urine samples. They tested for albumin but didn't know anything about uranium. So everything was pronounced good. We knew there was a problem with radiation but we didn't know how to find it.

"We decided to check with doctors in the surrounding area about people who worked in vanadium mines prior to us. We looked at the vital statistics and found a high percentage of deaths due to acute inflammation of the kidneys. Doctors were calling it Bright's disease. They had no knowledge about what uranium would do to the body.

"After that we ran urinalyses on the mill workers. We had them wear masks. We made them change their clothes after work. But they didn't like change rooms. Most of them wore long underwear or didn't have an extra pair of overalls. They didn't want to bother changing. They would rather go home and shower."[2]

Batie's aggravations continued after the war. When the uranium mills closed, he spent the next few years working at nuclear installations at Oak Ridge, Tennessee, and the Brookhaven laboratory on Long Island. In 1948 he returned to Grand Junction with the Atomic Energy Commission.

By that time, there were more sophisticated instruments and Geiger counters were available. With dust and air samples, they could check radiation levels in the mills. Batie found some readings to be ten to twenty times higher than the suggested tolerance. The floor of one deserted mill was still "hot" after three years of inactivity.

Duncan asked Batie if he had tested any mines. "The mining was really just getting started," Batie explained. "U.S. Vanadium and Vanadium Corporation of America were going at it pretty heavy. They set up diamond drilling outfits and were doing an awful lot of exploration. But most of the mining was being done by lessors, young single fellows, one or two people to a mine, and they were scattered out quite a ways. Most of my radon tests were done up in what they call the Slick Rock area. This was just about on the Utah-Colorado border. That was after Eisenbud and Wolf came out here."[3]

Batie's concern over the radiation threat had prompted him to contact Merril Eisenbud and Bernie Wolf, with whom he worked during his years at Brookhaven. Dr. Eisenbud organized and headed the AEC's Health and Safety lab in New York and later managed the New York operations office. Dr. Wolf, a prominent radiologist, headed the medical division.

Eisenbud and Wolf had already made their mark by cleaning up the beryllium industry. This lightest of all metals traditionally combined with copper as an alloy for electrical contacts and nonsparking tools or as an ingredient in fluorescent lamps. The atomic age introduced new uses. Mixed with radium, polonium or plutonium, it became a neutron generator. It could be used as a moderator and reflector in nuclear reactors and a structural component for aerospace craft.

Unfortunately, it was also highly toxic. Several workers were felled by debilitating diseases. Some developed a chemical pneumonitis. Others suffered a chronic lung disease similar to tuberculosis or silicosis. Worse yet, these chronic cases spread out of the production plants to residents of neighboring areas.

The AEC was the major customer for beryllium. When the health hazards of beryllium became evident, they directed the Office for Health and Safety to find out what was causing the illnesses and come up with a solution. Eisenbud and Wolf did so within six months.

They conducted an exhaustive research program. First, they called a meeting of all of the scientists who knew anything about the subject. The next day, they initiated field studies. They arranged for physical examinations of victims. They undertook medical experiments on animals. They tested air samples in the plants and neighborhoods to determine what elements the people were exposed to.

Their conclusions were simple. The disease was carried in the dust. The solution was basic, as well. Workers were forbidden to wear their work clothes home. In-plant laundries and change rooms assured that there were clean outfits every day. Within the workshops, industrial exhaust ventilation units became mandatory and an ambient air quality standard, the first in the United States, was established. Then, to force industry to maintain that standard, the specified air quality measurement was written into all AEC contracts for the purchase of beryllium. If a company didn't comply, the government wouldn't buy.

Batie hoped that Eisenbud and Wolf might effect a similar program on the Colorado Plateau. He took them on a brief tour of the mills under construction and into several mines. The scientists were appalled at the complete lack of data on radon concentrations in the mines, and at the slipshod practices in many of the mills.

"I took Eisenbud and Wolf about half a mile down the road from the Naturita mill, and the roadbed itself was way over tolerance," Batie said. "The rocks when they hit the surface and air turn canary yellow. Apparently, when the fumes came out and went along the side of this mountain, the rocks there were stained with yellow. And you could put a Geiger counter anywhere you wanted and you could get high readings. I thought Dr. Wolf would go crazy. He and Eisenbud, they just blew their tops sky high."[4]

Eisenbud and Wolf advised Batie to insist on change rooms and better ventilation in the mills. But even more vital, they felt, was an evaluation of the Colorado Plateau mines against the known background of the "European Experience." A firm data base was the first step. Eisenbud and Wolf promised to send Batie measuring equipment and have the results of his tests analyzed at the Rochester lab.

Soon some long, black boxes, each containing four liter-sized glass flasks, arrived. The spherical bottles had inlet tubes that extended almost to the bottom of the containers. Two stop-cocks, coated with silicone high vacuum grease, were at the ends. Batie took the bottles into three or four mines near Slick Rock and sampled air at the face of the ore, the mine entrances, and at the ore stockpiles. According to Eisenbud's instructions, he uncorked the bottles for a few seconds to suck enough mine air into the vacuum to replace that originally present. He attached a glass-wool or membrane filter to the inlet to prevent contamination by radium-bearing dust. After the tubes had been flushed by at least ten times the volume of air they contained, he recorked them. He also took dust samples in the mines and mills by drawing air through membrane filter papers with a hand-cranked vacuum pump. He shipped all of the tests to New York.

When Holaday asked about the findings, Batie sighed deeply. "At some of the more dusty operations in the mills, the concentration of air-born alpha emitters [charged particles emitted from the nucleus of some atoms] was several thousand times as high as those that would be permitted in other AEC installations. In one of the extensive mines, the concentrations of radon gas was over ten times the maximum permissible level which is used in other industries. The readings indicated the probability that severe internal radiation hazards existed in many operations."[5]

Duncan knew that the only standards for radiation tolerance at that time were those established in 1940 by the National Council on Radiation Protection.

Prior to World War II, a number of illnesses and deaths occurred among workers in New Jersey plants where luminous dials for military planes, submarines and ships were painted with radium. The painters worked in small, crowded rooms stocked with hundreds of dials. Ventilation was poor. Even worse, the employees were unaware of any danger and had the habit of forming their brush tips with their tongues. Consequently, they would swallow some of the radium.

The first alarm came from local dentists. The factory workers started losing teeth. Their jaws would crack and crumble and not heal. Coroners' reports indicated an unusual number of deaths among luminous-dial painters due to leukemia. It was suspected that they were absorbing radium from the paint.

Dr. Robley Evans, a physics professor at the Massachusetts Institute of Technology, who was conducting research on methods of measuring radioactive materials, started to investigate the problem at the New Jersey deluminizing plants. At the same time, the U.S. military was gearing up for war; thousands of luminous dials were ordered for ships and planes. Learning of the factory workers' plight, the army and navy wondered about possible danger to pilots and sailors working close to the dials in enclosed quarters. They asked Evans to determine a safe level of radiation exposure, "one that you wouldn't mind having your wives and children experience."[6]

Evans came up with a proposed level of radiation not exceeding 10 picocuries per liter of air. (A curie is a unit of radioactivity undergoing 3.7 x 10^{10} disintegrations per second. A picocurie is 10^{-12} or one-trillionth, curies.) This standard was subsequently adopted by the AEC for all of their uranium refining plants and laboratories. Their health and safety record at these installations was excellent, but privately-owned and operated mills and mines were not under the direct supervision of the AEC. The commission had not published any maximum permissible levels for air-borne radioactive dust or gases. Thus, there were no established standards to which the companies could be held.

In fact, the official AEC posture was conveyed in a letter written on March 15, 1949, by L. Joe Deal of the Applied Biophysics Branch of the AEC Division of Biology and Medicine in response to an inquiry about handling uranium ore. Quoting the medical director, he wrote:

". . . We have replied that prospecting for radioactive ores involves no hazards which are not ordinarily present in prospecting work. We do not believe it is necessary to take precautions against radioactivity in prospecting operations."

". . . I think we can say without any reservations that there will be no radiation hazard in the mining operations," Deal added. "There may be a potential hazard from radon. However, this is rather unlikely since it is my understanding that the radium content of the ores found in this country is quite low in comparison to those found in other countries. Unfortunately we do not know of any printed material that is available on this subject."

But after Eisenbud and Wolf processed Batie's samples, the AEC Health and Safety Division was sufficiently alarmed to notify AEC headquarters in Washington about their findings. They also advised Batie on necessary procedures.

"Eisenbud recommended that we hire an industrial hygienist to collect data and guide the miners and mine and mill operators as to what kind of measurements to take, how frequently, and how to analyze them," Batie explained to Holaday and Jacoe. "He assumed that the AEC would write their established radiation standards into uranium procurement contracts, just as he had successfully convinced them to do with the beryllium industry. But whoever in the AEC was responsible for the health and safety in the mines shot back the word that these questions were the concern of the states. According to the Atomic Energy Act of 1946, until the source material was removed from its place of origin, safety and health matters were not under the jurisdiction of the AEC."[7]

Duncan couldn't understand why the material wasn't removed from its place in nature when it was blasted from the mine.

"They didn't interpret the law that way," Batie said. "They claimed no responsibility until the ore landed in the mills."

Duncan asked Torrey what Colorado intended to do.

Torrey shrugged. The state's industrial hygienists had no expertise on mine radiation. They had no measuring instruments. If they did, they didn't know what data needed collecting or how to collect it. If they knew that, they had no way of testing and evaluating it. Most of all, they had no authority to impose any control measures on the privately-owned mills and mines without a legislative mandate.

"Something will have to be done," Duncan said. He couldn't understand why the Washington AEC office refused to cooperate with Batie. They had a good safety record in all of their other installations. The laboratories. The nuclear plants. This uranium mining end appeared to be a neglected stepchild.

On the train ride back to Denver, Holaday wracked his brain for ways to get action. It was the same old story; nobody was going to do anything about something that they weren't even sure existed. No uranium miners had died yet. There was no proof that underground radiation posed a health threat. In fact, some people insisted that the illnesses and deaths of European miners were due to poor living conditions, chronic respiratory diseases among the population, hereditary susceptibility, even inbreeding. These circumstances, it was maintained, did not exist in America.

When Holaday arrived in Denver, he met with Dr. Roy L. Cleere, executive secretary of the Colorado Health Department, and told him that he thought Colorado had a problem on its hands. After returning to his office in Washington, on April 11, 1949, he wrote a memo to the regional representative Henry Doyle, who had requested his visit. Reporting on Batie's findings, he said:

"The Grand Junction AEC group has endeavored to have this situation corrected but so far have [sic] been unable to get any assistance from the Washington office.

"This situation is one that requires immediate investigation, in order to determine the exact extent of the hazards to which the workers are subjected and to institute corrective measures. At present some 700 workers are involved but the number may shortly rise to several thousand. It is my opinion that every possible assistance should be given to the Colorado Industrial Hygiene Unit in surveying the uranium processing industry and in rectifying the hazards that have been permitted to develop."

Holaday added that arrangements had been made for Jack Torrey and Paul Jacoe to return to Grand Junction on the 17th and spend a week making a preliminary study to determine the extent of the problem. Batie would assist them.

But trouble surfaced at the onset. For some unexplained reason, Jesse Johnson, director of the AEC Raw Materials Division in Washington, sent a man named George Gallagher to Grand Junction. One day, unannounced, Gallagher walked into manager Phil Leahy's office and told him that Johnson had sent him to "take over." Leahy immediately phoned Johnson and asked him what was going on.[8]

"I don't want him to take over," Johnson assured him. "I just want him to get acquainted with what you're doing."

But in Leahy's estimation, Gallagher did start "taking over." He asked questions. He questioned decisions. He even started giving orders. In exasperation, Leahy scouted for other employment in the nuclear field.

During Leahy's job-hunting absences, Gallagher took charge. It was during this period that Batie requested permission to assist the Colorado industrial hygienists in their investigation.

Gallagher wasn't concerned. He didn't go along with the radiation hazard theory. He considered silicosis the only medical threat to miners. He suffered from that disease himself. He knew. So he arbitrarily told Batie that his cooperation with Torrey and Jacoe would be limited. He could lend them measuring equipment and demonstrate its use, but not participate in any actual samplings.

"My job was to teach them how to inspect the mills and what to look for," Batie later remembered. "I got orders from Gallagher and Jesse Johnson to keep myself out of it. I could make one trip through the mills and show them what and where the radiation hazard was. Then I was to give them instruments, the dust sampling equipment, and come back to the office and sit. They did this because they just thought I was a thorn in their side, I guess. They weren't convinced of the hazard. Gallagher still maintained the only problem we had was silicosis."[9]

Batie spent a few days with Torrey and Jacoe. He taught them how to collect dust samples, use a Geiger counter and alpha counter, and then set them on their own with his equipment.

The next thing Batie knew, orders came from Washington cutting off his travel funds and confining him to his desk. Apparently Torrey and Jacoe had reported excessively high dust and radiation readings in the VCA mill at Naturita. To make matters worse, someone in the Colorado Health Department leaked this information to the *Denver Post,* and the story had nationwide coverage. Officials at VCA were furious that the article was published before they had been informed of the findings. They lambasted the AEC. They berated the Colorado Health Department. Somehow, the blame fell on Ralph Batie. He had not even seen the report. He started looking for another job.

The Colorado Health Department was left juggling hot coals. The specter of possible radiation poisoning had been revealed, but further study to assess the problem was impossible without permission to enter the mines and mills. That had to come from the disgruntled mining companies. Finally in August, Executive Secretary Cleere prevailed on state and federal health officials, AEC representatives and managers of the operating companies to meet at the La Court Hotel in Grand Junction to see if they could come to some agreement.

The meeting started bitterly. In fact, it was reported that Denny Viles, manager of VCA, stormed into the room, went up to Paul Jacoe who submitted the damaging report, and said, "Mr. Jacoe, if you ever put your foot on my property, I'll personally pick you up and throw you into a vat of sulphuric acid!"[10] Fortunately, the discussion mellowed after that. The group voted to petition the United States surgeon general to instruct the Public Health Service to conduct a thorough study of the uranium mining industry.

Doyle was particularly alarmed over the situation. Shortly after he had asked Duncan Holaday to come west to confer with Ralph Batie, Doyle had been transferred from Denver to Salt Lake City to set up the Industrial Hygiene Field Station. (Later renamed the Occupational Health Field Station.) The mission of this new office was to assist the health departments of the eleven western states with any problems of exposure to toxic gases or other industrial health hazards. The proposed uranium study would be administered from Dr. Lewis Cralley's office at the Industrial Hygiene Field Headquarters in Cincinnati, Ohio, but its actual operation would be under Doyle's jurisdiction.

In view of the friction between the AEC, the health officials and the mining companies, Doyle decided to conduct some limited research in uranium mines on the Navajo Indian Reservations near Shiprock, New Mexico, and Monument Valley, Arizona. These mines were controlled by the United States Bureau of Mines. There would be no problem getting permission for entry from private owners.

Doyle had heard rumors that there were adverse working conditions at the mines, which were worked primarily by Navajos.

Private and government agencies had approached the Navajo Nation as early as 1941 to encourage them to develop potential mineral resources on their land. The Native Americans agreed, with two conditions: as many Navajos as possible would be employed, and grazing lands would be disturbed as little as possible. A year later, VCA won the bid on a twenty-acre parcel south of Monument Valley that became the Monument #1 mine.

Soon afterwards, Luke Yazzie took an unusual yellow rock to the well-known trader, Harry Goulding. Goulding showed the sample to Denny Viles. It was some of the richest vanadium-uranium ore ever found. VCA got the lease and opened Monument #2, with the understanding that Navajo workers would share part of the profits. But that did not happen. The mining company placed a low value on the mined ore and claimed much of it came from outside the reservation.[11] The Navajos were not even partially compensated for the uranium content until much later.

But Doyle sensed that there was more than financial considerations at stake. He wrote his friend Duncan Holaday again and asked if he would send some air sampling flasks and instructions on how to use them. His inspector would collect samples in four of the Navajo mines and then he would return the flasks to Holaday for analysis at the National Bureau of Standards.

Duncan readily complied and Doyle started his sampling in November. He inspected Monument #2, King Tut #1 and #2 and the Nelson Tunnel. None of the workings had change rooms, toilets, showers or drinking water. The Navajos were not given pre-employment examinations nor a medical program. The nearest health facility was forty miles away on roads little more than trails. There was a first aid kit but no qualified attendant. As for concentrations of radon measured, his samples indicated "4,750 times the accepted maximum allowable concentration on 10^{-8} curies per cubic meter."[12]

When Duncan received the bureau's report some weeks later, he was horrified.

"While I rather anticipated that the samples would show high radon concentrations, the final results were beyond all expectations," he wrote Doyle in February 1950. "The presently accepted maximum allowable concentration for radon gas in dial printing shops or radium refineries is 1 x 10^{-11} curie per liter of air. As far as I can recall, this figure was adopted both because it was quite possible to keep the radon in workrooms at this low a concentration with simple control measures and to allow a generous safety factor.

"Some of the mines in the Schneeburg and Joachimsthal regions have concentrations of radon as high as 10^{-9} curie per liter and in such amounts definite injuries and deaths occur. Even allowing a factor of safety from this level of only 10, all but one of these samples show concentrations higher than 10^{-10} curie per liter and one shows an appalling high figure of 750 times this amount.

"All together these samples disclose a rather serious picture and it is my opinion that a control program must be instituted as soon as possible in order to prevent injury to the workers. The situation is complicated by the fact that we have no assurance that the areas that showed concentrations in the order of 10^{-10} curie per liter may not be considerably higher at those times when a richer ore is being worked. Also these samples do not take into account the radioactive dust that is in the mines, or the external radiation to which workers are exposed.

"I do not think that we can recommend practical control measures without an inspection and survey of each mine, and I also believe that some AEC members should assist in devising control measures, as they have a primary interest in these mines."[13]

But Duncan was to find that the AEC's official posture concerning any controls on radon concentrations in the mines was one of non-cooperation. Any action on the uranium mining situation would be up to the Public Health Service and state health departments. It was they who would have to prove the threat of an impending medical emergency in order to convince the bureaucracy of the necessity for regulating safe standards of radiation.

Fortunately, the surgeon general understood the concern of his colleagues in the West. He granted permission for the field station to cooperate with the Colorado Health Department on a comprehensive study of the uranium mining industry. Dr. James G. Townsend, chief of the Public Health Service Occupational Health Division, knew that Duncan Holaday was the one industrial health specialist with the expertise required for the environmental phase of the research. He asked Duncan if he would accept a transfer to the Salt Lake City field station to do the job. Duncan jumped at the chance.

3 THE DAWN'S EARLY LIGHT

FIRST CAME A BLAST OF HEAT and then the light, blinding, like a battery of flashbulbs exploding in the pre-dawn darkness. Shock waves followed in quick succession with a force that witnesses in a control building 8.9 miles to the south likened to firing a 16-inch coast-defense gun. As the immense fireball ascended it took on a rosy hue and was transformed into a long-stemmed mushroom tinged with luminous purple. Like an opening parachute, the plume climbed up to 43,000 feet while its slate-colored cloud hovered over the 123-square-mile dry lake bed of Frenchman Flat, reducing visibility to a mere 100 feet.

The 22-kiloton Fox, detonated on February 6, 1951, was the grande finale of the Ranger Series of atomic blasts at the Nevada Test Site, 75 miles northwest of Las Vegas, Nevada. The massive flash was seen as far away as Los Angeles. In Indian Springs, 25 miles from ground zero, doors were blown out of their casements, windows were broken and roofs damaged. In one home, bathroom plumbing fixtures were knocked loose and left hanging from their pipes.

Las Vegas heralded this fifth nuclear explosion as a tourist attraction, a pyrotechnic extravaganza that attracted locals, as well as out-of-state visitors and the media. Any gnawing apprehension about fallout or other deleterious effects was dwarfed by a race to snag the best viewing spot along Highway 95 or experience titillating shock waves as far away as the west coast.

News of the once secret nuclear testing in the continental United States spread quickly in the ten days after Shot Able initiated the series on January 27.

"I remember the day nuclear testing started in Nevada—the first blast came without any warning," Gloria Gregerson, of Bunkerville, Nevada, remembered, "No one was informed that it was going to happen. The flash was so bright, it awakened us out of a sound sleep. We lived in an old two-story home, and when the blast hit, it not only broke out several windows, it made two large cracks in the full length of the house."[1]

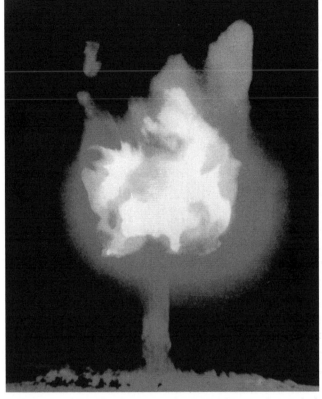

Operation Ranger test shot. *U.S. Department of Energy photograph.*

The Nevada Ranger tests brought the atomic age home to Americans. It was the first time since Trinity, the initial test shot of a nuclear weapon, was detonated at Alamogordo, New Mexico, that a bomb was dropped on native soil. A top secret operation of the Manhattan Engineering Project, which was headed by General Leslie R. Groves, Trinity was fired on July 16, 1945. The plutonium bomb, with an explosive force of 21 kilotons, proved the efficacy of the most violent man-made instrument of destruction in history. Less than three weeks later, a similar, untested, 20-kiloton uranium bomb devastated Hiroshima, Japan. In another four days, a 21-kiloton plutonium device was dropped on Nagasaki to bring World War II to an end.

The specter of atomic warfare became all too real. Mixed with the relief of defeating a formidable enemy was horror of the reality of massive devastation and deep-seated fear of future attacks by other nations that might develop the lethal technology. It was the beginning of a moral debate that was to endure into the next century. Realization of the insidious threat of radiation was yet to come.

"It was apparent that warfare, perhaps civilization itself, had been brought to a turning point by this revolutionary weapon," said Vice Admiral W. H. P. Blandy, head of a task force appointed in 1946 to conduct a new test series of the bombs.[2]

Still military officials knew little about the potential effects of the bomb upon American forces. What if our warships were hit by atom bombs? Had the nuclear age made our navy obsolete? And what about ground troops? What was the bomb's capacity for destruction on land? The consensus was that more experimentation was necessary, but Alamogordo was not the preferred site. The Trinity shot had spread a thirty-mile wide band of hazardous fallout almost ninety miles over the area and the location was considered too small for future detonations.

It was decided that Blandy would conduct future tests outside of the continental United States. That summer he initiated Operation Crossroads to test three 21-kiloton explosions on Bikini, a remote coral atoll in the Marshall Islands. An expendable fleet of damaged Japanese warships and U.S. ships headed for the junk pile was mustered. Shot Able, a plutonium bomb similar to Trinity and the shot that hit Nagasaki, was unloaded from a B-29 bomber on July 1. The device performed adequately but missed part of its target and only sank three ships.

Shot Baker, detonated on July 25 ninety feet under water, was more successful. The explosion wiped out seventy-four ships but the test deposited large amounts of active products from deadly radioactive mists and the vessels "became radioactive stoves and would have burned all living things aboard them with invisible and painless but deadly radiation."[3] After several weeks, when exposure of a decontamination crew became dangerously high, the cleanup project was scrapped and a third test canceled.

President Harry S. Truman had declared that the atom had brought "a new era in the history of civilization." But this powerful technology proved to be a mixed bag. Prospects of unlimited energy for peaceful purposes jousted with the fearsome threat of nuclear war. With this in mind, the United States attempted to implement regulations over nuclear energy. Truman established the civilian-run Atomic Energy Commission (AEC) to assume authority over domestic atomic energy and its development and replace the army-controlled Manhattan Project.

U.S. attempts to regulate the international scene were less successful. When Bernard Baruch proposed that the United Nations establish an atomic development authority to control, inspect and license potentially dangerous nuclear projects worldwide, and then abolish all existing nuclear bombs, the Soviet Union shot down the idea. Among the points of disagreement was Baruch's stipulation denying veto power for permanent members of the Security Council who might attempt to avoid penalties for violations of the regulations. The Soviet Union, lacking nuclear weapons itself and wanting the U.S. to destroy its atomic arsenal before establishment of an international authority, insisted on the veto power and the proposal died.

Soon the Soviets began to spread "an iron curtain" over Eastern Europe and a new global struggle resulted. With the specter of a Cold War came a renewed urgency for a stockpile of nuclear weapons.

By 1947, the United States possessed thirteen nuclear devices, none of which was assembled. Worse, a supply of fissionable material was scarce. Scientists at Los Alamos, New Mexico, suggested that they might be able to use the available resources more efficiently in order to manufacture new bombs. But they would require three more full-scale tests for performance and design validation.

Settling on a site for the Sandstone Series of tests was difficult. There had been problems associated with relocating the natives on Bikini, and the island was too small to accommodate the necessary equipment. Return to Alamogordo was still out of the question. It was decided to select Enewetak, another Pacific atoll that had a larger land mass, less rain and only 142 natives to be moved and could possibly become a permanent testing ground. The secret three-shot project, performed from April 15 to May 15, 1948, succeeded in detonating a thirty-seven kiloton, eighteen kiloton, and a massive forty-nine kiloton bomb. Fallout was mostly contained locally, and the nuclear stockpile was increased to fifty units.

Despite the success, Enewetak was not deemed a logical site for a permanent atomic weapons proving ground. It was too far to transport supplies and personnel. Too expensive. Too cloudy. Too hot and humid. Unpredictable winds. To say nothing of safety and security factors. Once again military experts looked to a continental test site.

That fall, chairman David E. Lilienthal told a Military Liaison Committee that the AEC had one reservation regarding a domestic test site. Both "policy and psychological considerations are strongly against the possibility of holding future tests of atomic weapons inside the United States," he said.[4] But he later admitted that accessibility, financial savings and flexibility in preparing and conducting the project were undisputed advantages. He ordered a site study and classified it "Secret." Codename: Project Nutmeg.

In a fifty-seven page study, navy Captain Howard B. Hutchinson, a meteorologist who had been to Enewetak, reported that continental testing would "result in no harm to population, economy or industry," if conducted on a properly engineered and prepared surface. Devices should be detonated from a tower high enough to prevent formation of a crater or addition of sand, soil and water into the plume of hot gases. Under these conditions, he concluded that most of the remaining radioactivity would be contained in the hot gases and ascend to high levels of the atmosphere and be diffused and dispersed over vast areas. He claimed it did not seem probable that harmful concentrations of soluble radio isotopes would result. [5]

Once having decided testing was practicable, Hutchinson narrowed potential locations to the desert southwest or humid southeast. He favored Nevada, Arizona and New Mexico but allowed that there was a "negligible possibility" of fallout on inhabited areas eastward due to prevailing westerly winds.

Then the Cold War intensified. Berlin was blockaded. China and Czechoslovakia fell to Communism. Atomic spies surfaced. The Soviets exploded a nuclear bomb. The Korean War erupted. As tension grew, the impetus to enlarge the nation's nuclear arsenal escalated.

By 1950, Gordon E. Dean had succeeded Lilienthal as head of the AEC. Little had been done since Project Nutmeg regarding a test site, but with international unrest and a push from Los Alamos to develop a hydrogen bomb, pressure for resumption of testing escalated. A four-shot Greenhouse Series to be conducted at Enewetak in the spring of 1951 was proposed. The experiments would acquire new data on blast and radiological effects of thermonuclear weapons and ascertain their economical and technical viability.

That June North Korean troops blasted across the thirty-eighth parallel to rout the South Koreans, and American forces were committed to the far-off war. The demands of combat overshadowed the importance of Greenhouse and the series was tabled. Project Nutmeg resurfaced. The AEC joined with the Armed Forces Special Weapons Project in a top secret search for a suitable testing location in the continental United States.

In quick order, sites in Canada and Alaska were rejected for inaccessibility, adverse weather conditions, geographical features, etc. The North Carolina coast and Gulf of Mexico were eliminated for lack of federally-owned land and proximity to the Los Alamos laboratory. Utah's Dugway Proving Ground–Wendover Bombing Range was deemed too close to heavily populated Salt Lake City.

The 5,470-square-mile Las Vegas Bombing and Gunnery Range was selected. A virtual desert, it was close to technical supplies, accessible by air and road and already government property, with an existing air base at Indian Springs featuring 6,600-foot runways and housing for 300 to 500 persons. Commission members estimated a "sector of safety" in which a fallout cloud could move eastward 125 miles, where the population count was only 4,100 (not including Las Vegas and Tonopah, Nevada, and St. George, Utah.). Norris E. Bradbury, director of Los Alamos, figured that radiological safety to the public would be much greater than that at Alamogordo.

At a meeting on August 1, twenty-four top scientists met at the Los Alamos Scientific Laboratory to discuss radiological hazards associated with a continental test site for atomic bombs. Alvin C. Graves, head of the Los Alamos Test Division, opened the session by emphasizing that discussions would center on setting specifications for weather condition, yield of bombs, etc., "from a radiological point of view only, omitting insofar as possible the psychological and political implications."[6]

"The discussion is to be based on the assumption that, in the event of a shot in Nevada, meteorologists would pick the actual shot days," Graves went on, ". . . one in which the pre-set specifications as to wind direction and velocity, probability of precipitation, etc. are satisfied; moreover, such favorable conditions must prevail for the length of time specified."

The morning continued with a discourse concerning maximum acceptable radiation dosage conducted by Colonel James P. Cooney, an army doctor. He concluded that scientific studies indicated that a rapidly-administered dose of 25 roentgens whole-body radiation was an acceptable emergency level for one time only. Citizens would be adequately protected with no more than 3.9 rads of external radiation during each series of tests.

(Radioactivity spontaneously emits alpha, and beta particles and gamma rays as atom nuclei disintegrate. Internal harm is caused by alpha particles, beta emissions cause burns and hair loss and can be ingested through the food chain, and gamma radiation can change cell structure as it passes through the body.

A roentgen is an exposure dose of external gamma radiation about the equivalent of a rad. A rad is a unit of absorbed dosage of radiation. Alvin C. Graves testified (Bulloch v. United States) that 25 roentgens would cause some changes in the blood; nausea could occur with 75–100 roentgens, at 100–175 roentgens hair would be lost and half of those exposed to 450 roentgens would die.)

Joseph Kennedy concurred with Cooney's conclusion, remarking that there "is not a probability that anyone will be killed, or even hurt—but it does contain the probability that people will receive perhaps a little more radiation than medical authorities say is absolutely safe."

Benjamin Holzman and George Taylor outlined meteorological considerations and stressed that situations to be avoided were variable—milling winds, precipitation downwind, and cases where the lower portion of the cloud could go over a nearby town.

"The point to be made here is the tremendous dilution which takes place in the cloud," Taylor said. ". . . it could be all over the U.S. in two days."

Holzman added that above 20,000 feet wind velocity could be as high as 100–150 knots. "In that case the dispersion of that part of the cloud (containing, under this assumption, the greater part of the activity) would be very great—it could go over Europe in 48 hours with a 100-knot wind."

But, assuming tests would be made under favorable conditions, the meteorologists agreed that "the only places to worry about are those within a radius of 150 miles."

Renowned scientists Enrico Fermi and Edward Teller figured that a 25 kiloton bomb could be detonated from a tower without exceeding dosage tolerance (3.9 rad) outside a 100 mile radius. Key factors were lack of rain and careful attention to wind direction on shot days by meteorologists.

Fermi reminded his associates that they had reached their conclusions with extreme uncertainty of the elements involved, and suggested that people who might be affected by fallout be warned to stay indoors, shower and take other precautions during testing periods.

The nation's top scientists had put their blessing on the Nevada Test Site. Then, the Chinese attacked U.S. forces in North Korea on November 25, and President Truman seriously considered an atomic response. The AEC went into high gear.

On December 12, James McCormack, Division of Military Application director, agreed that the Nevada site "most nearly satisfies all of the established criteria." But, he added, "Not only must high safety factors be established in fact, but the acceptance of these factors by the general public must be insured by judicious handling of the public information program."[7]

A special committee of the National Security Council endorsed the AEC decision on November 15. President Truman validated the selection three days later.

After Gordon Dean reassured the president that every precaution would be taken, Truman suggested "that it might be well to do it without fanfare, and very quietly to advise key officials in the area of the plans we had for the testing area."[8] The majority of American citizens was kept in the dark.

The day after Truman's approval, officials of the Departments of State and Defense and the AEC drafted a public relations program for the testing. The question of how to break the news to the public became a major issue. "Two aspects came to the fore. The American people needed to be convinced that 1) nuclear weapons testing was a routine activity and nothing out of the ordinary, and 2) radiological safety was under control and nothing to worry about.... The public should be told that 'it has been done before and we can do it again.'... The 'most important angle to get across [was the] idea of making the public feel at home with neutrons trotting around.'"[9]

In the meantime, contractors were hired to survey the property. Of two designated areas, the south side of the bombing range was chosen as the testing site. There, the valleys of Frenchman Flat and Yucca Flat were close to Indian Springs and source materials and were screened by natural barriers to enhance security. The only problem: Frenchman Flat was only 75 miles from downtown Las Vegas, well within the 125-mile limit. This could be a severe problem during winter when prevailing winds blew from the northwest toward the city.

But Los Alamos was in a hurry. Operation Greenhouse had been revived in September despite the fact that there were possible design flaws in the implosion devices. By mid-December, the laboratory had designed a special test series to be conducted at Site Mercury, the Nevada testing range, then named after the old Mercury Mine south of Frenchman Flat. With Greenhouse scheduled for April, time was of the essence. Anything after early February would be too late. Even then, there was insufficient time to prepare for tower shots like those at Project Sandstone. The devices would have to be dropped from a plane. They would have to depend upon the meteorologists to predict safe wind conditions and avoid sending harm in the way of Las Vegas.

Preparations went ahead with breakneck speed the day after President Truman approved use of the gunnery range as a continental test site, despite the fact that neither the president nor the Congressional Joint Committee on Atomic Energy knew that the Ranger Series was on deck. The American public was not informed that a domestic site was even approved.

With the Korean War at full blast, the air force was using the eastern portion of the range for gunnery practice for fighter pilots and the Strategic Air Command conducted bombing runs in the western and northern portions. The AEC, acquiring a sixteen- by forty-mile tract in the southern sector, used the joint occupancy as cover for their secret operation. The agency gave the commanding officer at Nellis Air Force Base approximately $10,000 to make minor repairs of facilities at Indian Springs and Frenchman Flat and suggested he issue a local press release to explain the sudden activity.

But 'Top Secret" was a difficult condition to maintain in the midst of unusual activity around a normally deserted area. Two prominent building contractors from Los Alamos were working at the site, with one of them headquartered in downtown Las Vegas.

On January 2, 1951, the front page of the *Las Vegas Review-Journal* headlined a "Big Indian Springs Plant" being built by the company that constructed "the Los Alamos 'A' Plant in New Mexico." The paper reported, "for the past two or three weeks, plane loads of Federal officials have been arriving almost daily, and with each plane came a Security Officer from Washington." The article hinted that the project "classified as Top Secret" might even skyrocket into creation of three new town sites.

With leaks such as this, the field operations office and Los Alamos laboratory realized a position of semi-secrecy was impossible in view of the site's location close to Las Vegas and "the public fear of atomic weapons." They planned to manage the news by emphasizing national defense and safety and downplaying mention of any potential dangers to the populace. They would make "every effort to educate the local people and also to satisfy their normal curiosity," but also to play on their "local pride in being in the limelight."[10]

AEC headquarters felt differently. Gordon Dean drafted a carefully-worded two-page press release stressing radiological safety measures that would be enforced and the extensive monitoring that would accompany the five-shot tests. He noted that each successive shot would be based on observed results of the previous shot, and the final detonation, significantly higher in magnitude, would be fired only after receiving favorable radioactive data. Besides noting the approval of the president and Department of Defense, the article added that Fermi, Teller and other noted scientists had concurred at a radiological hazards meeting in Los Alamos. No mention was made of when the tests would commence or what they would involve.

By January 8, the beleaguered press release was on the table again. Fermi and another expert removed their names from the release and the new secretary of defense, George C. Marshall, agreed with the Joint Chiefs of Staff that reference to radioactive danger should be eliminated. Rather, the importance of speeding up the nation's weapons program should be stressed, they said. They also objected to the large fifth shot.

Dean, reluctantly went back to the table and cut the list of names and radiological safety information from the release, which he later admitted was "somewhat misleading." He was able to convince the hierarchy that the "big bang" fifth shot was essential, however.

On January 11, the document was released to the media. The AEC rushed to tip off a few hours before the announcement congressmen and state officials that were involved in the test area. Special notice was given to authorities of highly-populated California to allay fears of fallout or contamination of drinking water supplied by the Colorado River. They were assured that tests would only be conducted when the wind was from the southwest, away from the west coast.

That same day, the AEC posted signs at the site and issued handbills alerting the public about the testing range. Under a bold headline "WARNING," the notice announced, "Test activities will include experimental nuclear detonations for the development of atomic bombs—so-called 'A-bombs'—carried out under controlled conditions. Tests will be conducted on a routine basis for an indefinite period. NO PUBLIC ANNOUNCEMENT OF THE TIME OF ANY TEST WILL BE MADE."

The flyer concluded with reassurance that "Health and safety authorities have determined that no danger from or as a result of AEC test activities may be expected outside the limits of the Las Vegas Bombing and Gunnery Range. All necessary precautions, including radiological surveys and patrolling of the surrounding territory, will be undertaken to insure that safety conditions are maintained."

There was no admonition for people to stay indoors, shower, wash their cars or take any other precautions during the testing periods.

On January 12, the *Las Vegas Review-Journal* ran the reassuring headline: "Don't Worry, Folks, Bomb Blast Won't Bother You." The accompanying story noted that Dr. Alvin C. Graves, chief of the Los Alamos test division, "a friendly, earnest man," claimed that "every precaution has been taken to protect residents from any effects of the test explosions." Graves indicated that Las Vegas and its citizens would be protected from "effects of the terrific blast" by interceding mountains.

But Dr. Shields Warren, who had headed a team of navy doctors that studied radiation effects in Hiroshima and Nagasaki, had serious reservations. Warren, who had been instrumental in the Japanese-American epidemiological studies conducted by the Atomic Bomb Casualty Commission since 1947, knew nothing about the Nevada testing program until he read it in the newspaper. He felt that the AEC's Division of

WARNING

January 11, 1951

From this day forward the U. S. Atomic Energy Commission has been authorized to use part of the Las Vegas Bombing and Gunnery Range for test work necessary to the atomic weapons development program.

Test activities will include experimental nuclear detonations for the development of atomic bombs – so-called "A-Bombs" – carried out under controlled conditions.

Tests will be conducted on a routine basis for an indefinite period.

NO PUBLIC ANNOUNCEMENT OF THE TIME OF ANY
TEST WILL BE MADE

Unauthorized persons who pass inside the limits of the Las Vegas Bombing and Gunnery Range may be subject to injury from or as a result of the AEC test activities.

Health and safety authorities have determined that no danger from or as a result of AEC test activities may be expected outside the limits of the Las Vegas Bombing and Gunnery Range. All necessary precautions, including radiological surveys and patrolling of the surrounding territory, will be undertaken to insure that safety conditions are maintained.

Full security restrictions of the Atomic Energy Act will apply to the work in this area.

RALPH P. JOHNSON, Project Manager
Las Vegas Project Office
U. S. Atomic Energy Commission

The AEC posted a warning about up-coming tests of atomic weapons. *U.S. Department of Energy photograph.*

Biology and Medicine, which he then directed, should have been involved in studying fallout effects and the decision to set the 3.9-rad exposure guide for off-site civilians.

When he was later asked why there was no discussion of internal exposure dangers at the meeting of Fermi, Teller and other experts on August 1, 1950, Warren "admitted it was a serious mistake for the Los Alamos scientists to assume that the only risks to the civilians involved external exposures from irradiated dusts."

Warren's early studies demonstrated that fission products, such as iodine, strontium, cesium, and plutonium, were released into the atmosphere upon detonation of an atomic bomb. There had been fallout contamination of fish after the Bikini blasts, and he was concerned about food chain effects on families who lived "off the land" downwind of the Nevada Test Site. Incidences of leukemia rose dramatically three or four years after the bombs hit Japan, particularly among infants and pregnant women. What of the large families living in the paths of Nevada winds?[11]

At the test site itself, things were hectic. It had been a crash program since the inception. Everything had to be shipped in. An abandoned frame building from Los Alamos was rebuilt for a control center, first aid station and decontamination post. A target area was prepared with a blast-proof alpha-recording blockhouse underneath. Shelters for photographers were constructed two miles northeast and southeast of ground zero. A station for scientific experiments was built to the north and west, and two miles south of zero was a wooden shack containing two diesel generators. Cables and electric lines had to be buried. Special lighting for pre-dawn shots had to be installed. All of this in a closet of secrecy.

Finally, on January 25, test officials "tickled the tail of the dragon" with a dry run of the first drop. The special Ranger Series air force crew flew a B-50D bomber to the range and unloaded a dummy bomb on the target to double-check that all operations were "go." They weren't. The non-nuclear high explosive wreaked more damage than expected and the surface radiation monitoring staff was "shown to be unsatisfactory." Officials reported that "complete confusion was the order of the day" at the control center.[12] Still, the AEC scheduled the real Able shot two days later. Shot Baker would follow the next dawn.

Damage control and final preparations proceeded the following day at a feverish pace. The Able device was assembled and prepared for loading on the plane at the Sandia Base near Kirtland Air Force Base at Albuquerque, New Mexico. Two planes for photography and emergency escort were readied for action. At Nellis, teams finalized procedures for sixteen scientific experiments, including field fortification for protection from radiation, thermal effects, thermal and ionizing radiation measurements, analysis of fireball growth, gamma radiation exposure, cloud sampling and tracking, measuring wind and atmospheric conditions. Meanwhile, a formal briefing for test officials was held at 1 P.M. to check Nevada weather predictions for the twenty-seventh. When the report was positive, a final forecast was scheduled for 8 P.M. Conditions remained favorable, so the exercise was ordered to proceed on schedule.

At 1:15 on the morning of January 27, a B-50 drop aircraft with an eleven-man crew from the 4925th Special Weapons Group, took off from Kirtland Air Force Base. Flying at an altitude of 14,000 feet, the bomber, accompanied by a B-50 photography plane and a C-47 emergency aircraft, headed for the Indian Springs Base. At 3:50 A.M., the plane descended to 10,000 feet and a crew started to load a one kiloton nuclear bomb. When the job was completed at 4:34 A.M., the plane climbed to its bombing altitude of

19,700 feet. Two practice runs were made at 5:07 and 5:20 A.M. Reports from the surface indicated calm winds and temperature at minus 2 degrees Celsius. So at 5:27 A.M. the bomb-bay doors opened and the bombing run began.

The second atomic bomb detonated in the continental United States was released at 5:44:05 A.M. It was a rather disappointing show. The device was much smaller than previous bombs and an observer commented it was "less spectacular than those reported for previous detonations, with shorter duration of luminosity of the fireball, slower rise, faster cooling, no real thermal column formed, no mushroom head, and the fission-product cloud rising only to a fairly low altitude."[13]

But according to the Las Vegas Review-Journal, January 28, 1951, Jack Palsgrove, a diesel truck driver, who was near the top of Baker grade, near Baker, California, at the time of the blast, had a different impression. "A brilliant white glare rose high in the air and was topped a few instants later by a red glow, which rose to great heights," he reported. "The bright flash blinded me for a few seconds and gave me quite a scare. I have seen Northern Lights often, but this makes them look silly."

Scores of people saw the flash, felt the jolts, and suffered a few broken windows and triggered burglar alarms, but test officials were quick to put everyone at ease about radiation dangers. Approximately an hour and a half after the explosion, survey crews arrived at ground zero and found relatively minor levels of radiation. According to Dick Elliott, the information officer, "as anticipated, no levels of radiation were found anywhere that could conceivably produce damage to humans, animals or water supplies."[14]

Delighted with the results of radiological tests, the officials relaxed stringent criteria that shots could not be fired unless winds were blowing from "a point somewhat to the south of due west."[15]

Popular newspaper columnist Hank Greenspan assured Las Vegans on January 30, "All tests are conducted under controlled conditions with all adequate safety precautions taken. Ground studies and stratosphere wind drifts are closely checked. The whole structure is thoroughly examined and if there is any possibility of drift over inhabited areas, there is no shooting."[16]

But on February 3, a news item from Rochester, New York, reported that the previous day, "Children played in 'atomic snow' in this city today. Surprised residents accepted without too much concern news that radioactive materials from the Nevada atomic tests had been discovered in the nine-inch fall that blankets Rochester and the surrounding area." The article added that similar conditions existed in Cleveland and Cincinnati, Ohio; Fredericton, New Brunswick; and Chicago, Illinois.[17]

4 DEADLY DAUGHTERS

IT WAS A TYPICAL FEBRUARY DAY in Salt Lake City. Grey. Cold. Intermittent blizzards raged over dirty corn snow speckled with dead leaves exposed by a false spring thaw. Duncan Holaday, sitting in his office in the two-story wooden building that used to be a World War II army barracks, couldn't even see the city in the valley below. Showers of snowflakes blocked out the granite Mormon temple spires topped by the golden statue of Moroni blowing his trumpet. The Walker Bank, Utah's sixteen-story "skyscraper," was invisible. The Great Salt Lake, pewter-colored, disappeared into the whitened west.

The Public Health Service Occupational Health Field Station stood in the foothills of the snowclad Wasatch Mountains, on the east bench of the valley. It was a small island tucked away from the real world. The difficulty Duncan Holaday was having getting people to heed his warnings made him feel even more isolated.

He had been on the phone all morning. Calls to Blair Burwell of Climax Uranium Company, J. H. Hill of U.S. Vanadium, Howard Balsley, secretary of the Independent Vanadium and Uranium Producers Association, and Denny Viles of Vanadium Corporation of America. When his requests for meetings were met with polite acquiescence, he set up a series of appointments at various offices in Grand Junction, Moab, Cortez and Durango, Colorado, and Salt Lake City for mid-March. State and federal health and mining officials would join the discussions. Holaday hoped the meetings wouldn't be a replay of old scenarios . . . his reports on radon counts in the mines and mills . . . his suggestions for ways the companies could reduce the radiation hazard through increased ventilation, use of respirators, and other safety practices . . . their show of concern and promises of change that resulted in little action.

It had been that way since he had started the uranium mine study in the spring of 1950. Now, almost a year later, it was the same. Benign cooperation. A few changes for the better, but no staying power. No fire. The job wasn't getting done.

The move hadn't been easy for Duncan, uprooting his family from Washington, D.C., and bringing them west. A few weeks before Christmas, they had moved into a tree-sheltered home on Third Avenue in Salt Lake's exclusive Federal Heights district. His wife, Pauline, seven-year-old Duncan, Lindsey, four, and their baby, Evelyn, had to adjust to their new surroundings while he spent long hours at the field station.

Getting the study started was a real hassle. The division was under-equipped, under-staffed, and under-financed. The field station had two survey meters for measuring external radiation—X-rays, radium and other gamma ray exposures—but they had no equipment for the alpha measurements of radon, uranium, or radium that were a primary purpose of the project. The staff was small: the director, Henry Doyle, Paul Woolrich, a sanitary engineer, one secretary, two chemists and Holaday himself.

The only funds for the research program were a one-year grant of $25,000 awarded by the National Cancer Institute. Because the surgeon general would not permit one Public Health Service division giving money to another, the grant went to the Colorado State Health Department, rather than the field station. The money was then channeled into the study for travel expenses and purchase of equipment and supplies.

Duncan prevailed on former Manhattan Project colleagues for favors to help stretch his limited budget. He made arrangements with the National Bureau of Standards to do radon analyses. He even talked the Bureau out of their old sampling flasks in exchange for a few new ones. Then he went to the Industrial Medicine Division of the Los Alamos Scientific Laboratory and made a deal for them to analyze urine samples for their uranium content. He persuaded the Navy Radiological Defense Laboratory in San Francisco to do readings on atmospheric samples for uranium and radium. The Instruments Division of the AEC offered, on semi-permanent loan, some laboratory measuring equipment to measure alpha particles on filter paper. The field station checked the samples for vanadium and the Colorado Health Department furnished manpower. The AEC Division of Biology and Medicine and the U.S. Bureau of Mines stood by as consultants.

But, once Holaday was equipped to take samples, he had to find a way into the mills and mines, since the Public Health Service had neither right of access nor any power to enforce air standards. They could gain entry only by accompanying state mine or health department inspectors who made periodic rounds of the facilities. Yet, even then, special permission to conduct environmental studies was needed from the mining companies. And this permission had certain strings attached. After Duncan or his representatives assured a company that they were only collecting information for a Public Health Service study, they had to agree not to publish their findings. Above all, they were cautioned not to alarm the miners.

"The mining companies had given us the right to go in with the implied understanding that we wouldn't walk through there leading a trail of men fleeing the mines behind us," Duncan later said. "This was pretty much standard procedure. You had to make some agreement that you wouldn't wreak havoc in the place. You always wrote up a report that was put out in general distribution, but you were careful not to identify a particular facility. You used A,B,C, etc. Otherwise we'd have never gotten in."[1]

Another problem was finding where the independent mines were located.

"There's a lot of country in the Colorado Plateau and there were a number of mines—nobody knew how many—and it was something of a chore to find them," Duncan said. "You'd go to the ore-buying stations and get the names of the mines, but you'd really have to talk to the ore haulers to find out how to get to them. They didn't bother to put up signs on the roads, of which there was a multiplicity. There were just too damn many roads on top of those mesas, and names such as Calamity Mesa, Outlaw Mesa and Wildhorse Mesa give you some idea of what things were like."[2]

Perhaps the greatest obstacle was the general indifference to the potential hazard. Many old-timers swore that the curative powers of uranium outnumbered the dangers. In the early 1900s, flamboyant medicine show operators filled rattlesnake skins with yellow carnotite pounded into powder and sold the belts as a cure for rheumatism and other "miseries." Some people even drank "healing" potions that were laced with uranium dust.

"It's ridiculous to me," said Mark Shipman, an independent mine operator a few years later, "that right today we can go to other parts of the country and we can pay to go to a health mine and get a one-hour exposure, where that mine operator can show by his statistics that he has gathered over the years that one person out of four does benefit regarding arthritis, sinusitis, those types of related diseases. And yet I come back to our own mine, which contains exactly the same type of radon and radiation situations, and we find ourselves with expenses increasing month by month in order to reduce the radon count."[3]

As for the miners, they were traditionally inured to dangers underground. Hazardous conditions were a part of the mining game. Radiation was just a new name for it. They were making good money. The base pay was $3 an hour, but a contract miner, who was paid according to the amount and quality of ore he mined, could get $10 an hour if he worked with high-grade ore. They weren't going to worry if the high-grade emanated more radon. They were willing to take the risks.

G. A. Franz, Sr., a former Colorado State mine inspector remembered, "Most of the miners thought they were all right. The radon gas was something you couldn't see, taste or smell. They thought as long as it was sand and not silica, they were safe. It was hard to teach these people that something was wrong. We'd tell them they needed more ventilation and they'd say, 'I catch pneumonia when I come into the mines now. There's a draft coming through here all the time.' You'd inspect a mine and find it high in radon and order them to keep out of the mine until such time that better ventilation was provided. You'd turn your back and they'd go back in and work."[4]

Duncan Holaday found a ventilation duct seventy-five feet away from the ore face in one mine. He pointed at it and said to a miner, "That's supposed to be within twenty feet of the face. You're not getting any benefit from that. You have a good possibility of getting lung cancer."

The miner said, "Oh, that won't happen for fifteen years, or so."

"They wouldn't even turn on the auxiliary fan to put air from the main air course up to their working place, because they wouldn't take time to string up the duct," Duncan remembered. "They wanted to make money."

As for wearing respirators, the miners would have none of that. The headgear was uncomfortable. It was hard to breathe in them.

Duncan understood the miners' reluctance, yet he knew that use of respirators would substantially reduce the hazards of radiation.

"I used to wear one myself," he said, "for, say, fifteen minutes, in a mine along with the lessor. And then I would come out and would take the filter paper and wrap it around the probe of the Geiger counter and let him watch the scale and watch the needle go up to the top of the scale, just to try to convince him that we couldn't see what was there."[5]

The demonstration had little effect on miners who often amused themselves when coming off shift by breathing on a radiation counter and betting on whose breath could peg the needle higher.

When Duncan warned that the "European Experience" that had felled so many miners with lung cancer might be repeated on the Colorado Plateau, he was confronted with the long-held argument that American mines couldn't be compared with those in Germany and Czechoslovakia. They were different. The Erz Mountains miners lived in poverty. They were subject to lung disease because they had to walk home long distances in wet clothes and inclement weather. The mines contained silica dust and

The rugged country of the Colorado Plateau made travel difficult and hazardous.

arsenic. Besides, the health of European miners was weakened by centuries of in-breeding among the isolated villagers.

Duncan knew he had to prove the doubters wrong before he could ever get them to heed his warnings. He tried logic.

"That radiation is capable of causing malignant change is known from both animal and human experience," he said. "Some examples in humans are skin cancer in pioneer radiologists, osteogenic sarcoma in radium dial painters, leukemia in radiologists and some of the survivors of the atomic blasts in Japan. It seems reasonable to study our own uranium miners to determine whether or not they are being injured by prolonged exposure to radioactivity."[6]

This sounded plausible to the mining companies. After all, they were proud of their safety record. And so in April 1950, Denny Viles of VCA and J. H. Hill of USV extended full cooperation to the Public Health Service. Holaday and his staff could have ready access to the mines and mills to do their research on the environment. Dr. Louis Cralley, overall administrator of the uranium study, Dr. Harry Heimann, director of the medical phase of the project, and Dr. Wilfred D. David, who would be in charge of the team of physicians, could join him to make a preliminary survey. Viles would even have his personnel relations man, John Maxwell, chauffeur them and facilitate arrangements. Of course, Paul Jacoe or Jack Torrey of the

Colorado Health Department would have to be with them when they inspected the workings. And there was to be no publicity.

Duncan knew he needed strong ammunition to support the premise that there was a relationship between the Plateau mines and those of Europe. The only proof would be in the numbers: the concentrations of radon gas in a liter of air; the external gamma radiation rates; the amount of silica dust, or the nature and extent of exposure to toxic elements contained in uranium and vanadium; how these readings compared to the lethal dosages in Europe.

But they were dealing with a menace that was not only invisible, silent and odorless, it was slow and insidious. It might take years before any symptoms of lung cancer showed up. The disease was likely to be full-blown before it revealed its presence. Meantime, it would be hard to convince people to react against the shadow of an epidemic based only on statistics.

Thus an ancillary medical profile of the miners and mill workers was instituted. It was hoped that the medical facts could correlate with data on the nature and concentrations of materials to which the men were exposed in order to identify health hazards and determine necessary control measures. The mine companies authorized doctors Heimann and David to give their mill employees physical check-ups and even set up temporary chest X-ray units, laboratories for clinical studies of blood and urine, and examining rooms in the plants. In addition, they transported as many miners as possible to the facilities from outlying mines.

Holaday had a gut feeling about what their findings would be before they started. Upon completion of the two-week preliminary survey, his fears were realized. Radon concentrations in the mines proved to be exceedingly high, some almost a thousand times the accepted level of safety. To make matters worse, when information from the physical examinations was tabulated, the results lent little support to Duncan's figures. There was no obvious evidence of developing malignancies. The plateau's uranium industry was too new; miners had not been exposed to radiation long enough for significant effects to emerge. The medical team had gained only a baseline guide for future follow-up studies.

Yet certain evidence was there. The radon counts exceeded those that had killed hundreds of miners in Europe. Certainly now a safety standard for radiation would be set.

"We are attempting to get an opinion as to the significance of such concentrations of radon from the AEC and other individuals who have opinions on the subject in order to see how necessary it is to install control methods in the mines," a hopeful Duncan reported to Henry Doyle. "I

Duncan Holaday and Dr. W. F. Bale inspecting a uranium mine. *Duncan Holaday.*

believe that we should keep these results confidential until we have been able to decide what our policy will be in this matter."

Still, the AEC persisted in turning its back on radiation control in the mines and claimed that this problem was one the states must solve. Utah, Arizona and New Mexico were totally unequipped to sample mine air and had no one trained in any part of the radiation-protection field. Colorado had Torrey and Jacoe, whom Ralph Batie had instructed, but they had no suitable instruments for radon measurements. None of the states had legal muscle or funds for an enforcement program.

As for the mining companies, they weren't prepared to make major expenditures for ventilation and other controls until they were forced to do so by federal or state regulations.

So it appeared to be up to the Public Health Service to press the issue by proving conclusively that the hazard was real and inevitable. They started the comprehensive uranium mining health and safety study in July 1950 to do just that.

It began as a three-month-long endurance test for Holaday and his sampling teams. Their jeeps rumbled over rutted mining roads and scaled smooth domes of slickrock to reach thirty-six of the hundreds of mines hidden among stands of scrub juniper and sagebrush in the hot, dusty desert. They donned miners' coveralls, studded boots and helmets fitted with battery-operated lamps for the total darkness underground. They squeezed through narrow mine drifts tunnelled into sandstone walls and reinforced against collapse by powerful ceiling bolts. Climbing ladders from the main corridors up into small caverns (stopes), they wedged themselves between miners who rammed bone-jarring, needle-nosed pneumatic drills into horizontal veins of ore. Sometimes they could hardly stand upright in the one-man "doghole" diggings they inspected.

In each mine, large or small, they pulled their vacuum flasks from protective metal cases and filled them with air. At the entrance portal. At the ore face. Inside the network of tunnels. They also cranked hand-operated pumps to collect the dust. Enroute between the mines, they gathered similar specimens at sampling and processing mills.

While the environmental engineers made their rounds, Dr. David, Dr. Mitchel Zavon and a laboratory technician carried out the medical portion of the study. They, too, bumped through the Canyonlands in a four-wheel-drive vehicle, but they pulled a housetrailer as well. A truck with an X-ray unit and generator followed behind.

In their mobile examining rooms, the doctors examined approximately seven hundred mine and mill workers. They collected lifetime occupational and medical histories and quizzed workers about personal habits such as smoking, drinking, and sexual performance. They analyzed urine specimens in portable labs. They tested blood for red and white corpuscle counts, hemoglobin estimations and hematocrit values. Two hundred of the miners were checked for atypical blood forms. Finally, the patients climbed into the X-ray truck where pictures were taken of their lungs.

Sometimes the clinics were set up in the mills. More often, the doctors camped with miners in remote backwashes or at crossroads, where contact men fanned out to cajole reluctant "dogholers" to ride in for a free medical exam.

Upon completion of the field surveys in October, Duncan and his colleagues were more alarmed than ever. Most of the mines investigated depended solely on natural ventilation with an air intake stack and an outlet to clear the atmosphere after blasting. If the powder smoke failed to subside in one day, the miners would let it sit until it did. Sometimes it would take up to three days. In the meantime, because of the gaseous air, radon levels increased.

Where mechanical ventilation existed, it was inadequate. In several instances, instead of bringing clean air in from outside, mine ventilators simply circulated air drawn from other underground sources. Some systems forced air to the lowest levels and let it rise to the surface slowly by natural means. Tunnels dead-ended with stagnant air. Dust hung in the atmosphere. Sometimes, where water was sprayed to settle the dust, it came from subterranean sources that were more radioactive than the powder it was supposed to settle.

Consequently, the radon samples were excessively high. The Public Health Service estimated that a maximum allowable concentration of 100 picocuries of radon per liter of air would be safe in a mine. They felt that the standard of ten picocuries per liter that Robley Evans set for industrial plants would not be feasible. You couldn't go into any metal mine in the country without beating that level from the background radiation that is continuously emanated from rock. The European standard of 100 picocuries was more realistic.

One sample taken at the working face of VCA's Prospector No. 1 mine at Marysvale, Utah, showed 26,900 picocuries of radon per liter of air. Another at the foot of the entrance incline registered 14,000. Holaday shuddered when he thought of the consequences of only 1,000 and 1,500 picocuries of radon per liter of air registered in the German and Czechoslovakian mines. Letters were rushed to the mine superintendents as warnings.

"We do not have sufficient data to be able to tell you how long individuals could work in concentrations such as were found in the Prospector No. 1 mine without becoming liable to the development of lung cancer," Duncan wrote Denny Viles in a confidential letter dated October 20, 1950. "Under these circumstances, we feel that it is absolutely essential that the ventilating fan be kept operating during the entire period that men are working in the mine, and that no occupied part of the mine be short-circuited from the air stream. The last time we visited this mine, the main working drift did not appear to have much air moving in it. It would be advisable to introduce fresh air to the working face through a canvas duct extending from the ventilating shaft."

Then followed a game of wait and see. As in other projects undertaken by the field station, the procedure was to prove that a problem existed, devise methods of control, write a report, and then walk away and address the next issue. But this was not a routine case. Something more had to be done.

On the morning of January 25, 1951, Duncan Holaday entered the Federal Security Building South in Washington, D.C., and sought out Room 4058 of the Public Health Service Headquarters. It was his second trip to the nation's capital in a little over a month. Dr. James G. Townsend, chief of the Division of Industrial Hygiene, had called a brain-storming session for experts in occupational health problems related to uranium. Representing the Public Health Service were Joseph E. Flanagan, assistant chief, Dr. Harry Heimann, Dr. Lewis J. Cralley and Duncan Holaday. Also present were Dr. Edwin G. Williams, chief of the Radiological Branch, Dr. Clinton C. Powell, Laboratory of Physical Biology, National Institutes of Health, Dr. Samuel Ingram, Division of Engineering Resources, Dr. William F. Bale, University of Rochester, and Dr. George A. Hardie, AEC Division of Biology and Medicine.

Dr. Townsend opened the meeting promptly at 10 A.M.[7]

"We've made a study of the uranium mines on the Colorado Plateau and found the radon concentrations high enough to probably cause injury to the miners," he said. "We've called this meeting to evoke group thinking on the problem."

Townsend then asked Holaday to report on the progress of the study. Duncan distributed copies of a table showing the various radon concentrations in the mines.

"As you can see," he said, "without exception, all mines showed concentrations that would be considered excessive according to information now available."

Then he stressed that these readings could be lowered to a safe level through mechanical ventilation. He cited specific instances where such procedures had already been put into effect. Analysis of the radon samples taken after the forced ventilation fan had been installed and was operating in the Prospector No. 1 mine at Marysvale, Utah, showed that the atmospheric radon concentrations in the mine had been reduced by a factor of about 500. Both samples were taken at approximately the same location as the one in Duncan's former field survey, which showed a level of 26,900 picocuries of radon per liter. The fan had been turned off for about two hours when the first sample was taken and had been in operation about forty minutes when the second flask was filled. In both cases the atmospheric radon concentration was about 500 picocuries per liter, probably the highest radon level in the occupied levels of the mine because the fresh air was introduced in the opposite end of the mine drift. Most of the pure air went up the incline without passing through the area that was being worked.

"It is encouraging to see how much the radon concentrations in this mine have been reduced by the use of ventilation," Holaday said. "It is my opinion that if fresh air were introduced to the working faces of the drift, the radon levels would be lowered still further."

"Is this easily accomplished?" someone asked.

"Yes. Mine ventilation techniques are fully developed. And it would not be overly expensive.

"One of the first things we thought about was how much it was going to cost. We were not mine ventilation experts, so we talked to the men of the larger companies and obtained a range of estimates that ran from fifty cents to a dollar per pound of ore. We figured if the mining companies wanted to institute additional safety measures, they could add the cost on to the price of uranium."

Holaday explained that the AEC pays the miners fifty cents a pound for development allowance. This money is theoretically used for putting down diamond drill holes or driving exploration drifts. Holaday couldn't see why the AEC couldn't give the mine companies another twenty-five or fifty cents a pound for ventilation allowance. There could be the stipulation that, if they spent the money for putting down churn drill holes and buying fans and gasoline motors to run them, they would get the extra money. Otherwise they wouldn't.

"After all, the AEC pays for all of the health and safety precautions in nuclear plants," Holaday said. "I can't understand why a completely different schedule applies to the miner or mill operators west of the Mississippi, who have to pay their own costs."

Duncan then went on to report his other findings. Dust control in the mines had been fairly good, due to the widespread use of wet drilling and because many of the mines were so-called wet mines. The range of dust was within the legal limits of most states concerned. He didn't consider the silicosis problem in the mines to be acute.

As for external radiation deposited on underground surfaces, Holaday felt that the levels of atmospheric uranium and vanadium in the mine atmosphere did not appear to be sufficiently high to produce chronic poisoning. Measurements of gamma radiation had been made in the mine environment and on the miners' bodies. The results had been erratic. A number of factors in the mines interfered with radiation measurement. Radioactive dust deposits on any surface, and thus concentrates on survey instruments. Radioactive dust is also concentrated on the workers' clothes and thus increases the workers' total apparent radiation dose. Holaday strongly recommended that the workers take daily showers using soap freely, and change work clothes frequently to minimize skin contact with the radioactive dust.

Dr. William Bale then asked for the floor.

"Recent studies [by John Harley at the Health and Safety laboratory] indicate that radon gas is not the true culprit in uranium mine radiation," he said. "We have discovered that it is the decay products of radon—the radon daughters—that are deadly."

Radon's chain of decay was well known to all the health specialists. The uranium molecule is an active, complex unit containing thousands of lively electrons agitating inside of it. Uranium ore contains all of the members of the radioactive family, of which uranium is the parent. One of these members is radium, which is transformed into radon by a process of decay.

Radon is a noble gas like helium, argon and krypton. It is chemically inert and reacts with no other material. Blasting and drilling operations in a mine release the gas into the atmosphere. The radioactive gas has a half-life of about four-days. (Half-life is the time required for a radioactive substance to lose, by decay, fifty percent of its activity.) When it diffuses out from the rock or broken ore into open space, the radon gas then decays into other radioactive elements called daughters. These radon daughters are solid particles. Some of them (the most deadly) are "short-lived" with half-lives ranging from twenty-seven minutes (Radium B) to ten to the minus fourth seconds (Radium C').

Once the decay process is triggered, a chain reaction follows. Radium-226 decays into Radon-222. Radon-222 decays into Radium A (Polonium-218), Radium A changes to Radium B, then Radium C (Bismuth-214), Radium C' (Polonium-214), and so forth. Each element formed by radioactive decay yields into another radioactive element. The electrons always break down in the same order, with the most unstable giving way to the next. Each time the composition of the molecule changes, a new daughter is born. The series, beginning with uranium-238, progresses through thirteen radioactive daughter elements until it finally decays into Radium B, Polonium-210-lead, the deadly sink where the process ends.

Bale went on to explain that radon gas, by itself, need not present a health problem unless it is in very high concentration. Normally, the air in human lungs is in equilibrium with the radon concentration of the air in a room or inside a mine. The radon is inhaled and exhaled without causing damage over a short period. When a person leaves the room or mine, the radon is cleared out of the lungs within a few minutes.

"The vital fact seems to have been almost or entirely neglected," Bale continued, "that the radioactive dosage due to the disintegration products present in the air under most conditions where radon itself is present conceivably and likely will far exceed the radiation dosage due to radon itself and to disintegration products formed while the radon is in the bronchi. This additional dosage . . . is associated with the fact that disintegration products of radon remain suspended in the air for a long time; tend to collect on any suspended dust particles present in the air; and that the human respiratory apparatus probably clears dust from air, and the attached radon disintegration products, with reasonable efficiency. . . . This radiation dose due to Radium C' is some 76 times greater, in terms of alpha ray energy liberated in the lungs, than the dosage that would be calculated on the basis of a simple equilibrium of radon with its daughter products in the lungs."

If the disintegration cycle is completed, Polonium-210-lead with a half-life of twenty-two years, builds up in the body. Some of it is excreted. Much of it collects in the soft tissue and is eventually transmitted to the bone. Thus, the victim is being radiated from within his own body.

A maximum dose of radiation accumulates in the lungs in approximately two hours. It takes a similar amount of time for the radon daughters to be flushed out in clean air. If a wind blows constantly and no radioactive particles remain for more than two hours, there is unlikely to be any damage to the lungs. That's why mechanical ventilation, an artificial "wind," in the mines works. Without adequate ventilation, the short-lived radon daughters enter the body's DNA, the chain of living cells that determine hereditary characteristics, and they knock out a link in the genetic chain. Then another. The body is able to replace the links for a time. But as it is forced to keep repairing the chain over and over again, the DNA molecules undergo such alteration as to become abnormal and develop into a potentially cancerous growth.

"This helps explain various puzzling results in the literature before, where only radon had been measured, with no knowledge of what daughter concentrations were," Bale said.

There had been some very contradictory animal studies made. One group of animals was exposed to radon gas. It killed them very rapidly. Another group, given about the same concentration of radon, lived on happily for quite a few years. Nobody was measuring the radon daughters. The scientists didn't realize that one group would have a much higher concentration of radon than the other, depending upon the rate of air change in the exposure room. People were getting conflicting and unexplainable results from their biology.

When Bale completed his report, Holaday pointed out that these new findings were all the more important because of the unprecedented growth in the uranium industry.

"In the past year, with the working of richer ores, the mining of uranium has changed from a gopher-hole operation to a good-scale, frequently mechanized activity," he said. "As a result, the employment level in these mines may be expected to rise and become stabilized, thus making all the more urgent the need for measures to reduce dangerous radon concentrations."

Holaday pointed out the fact that there were no data available on radon concentrations relevant to domestic operations of this type. The only available literature contained references to European experiences. The type of ore handled and the working conditions in those mines were considerably different from those on the Colorado Plateau. Direct applications of the historic information to the current local situation could not be made.

The health officials, realizing that they were working against time, agreed that something must be done to keep the health and safety levels of the mines on par with increasing production. Radon concentrations should be reduced to as low a level as possible, consistent with good mine ventilation practices. For the time being, the Public Health Service would have to go along with the European recommended standard of 100 picocuries of radon per liter of air. But it would be unrealistic, in the light of what was known to set any *required* level. A mandatory standard set by the AEC, or some regulatory body, could be established only after extensive research and further study.

"This is one problem we can't walk away from," Dr. Cralley said to Duncan after the meeting adjourned. "See if you can't get some controls in there."

A month later, when Holaday sat in his office in Salt Lake City on that snowy February day, he wondered if he could. He had some proof of numbers. Dr. Bale's report was a scientific breakthrough. And he was not alone. Now he had the backing of some of the foremost radiation experts in the country.

5   BONANZA AT BIG INDIAN

A YEAR WAS ENOUGH. A year when every rasp of the saw, each thwack of the hammer drove the dream of uranium deeper. All the time that he worked as a carpenter in Tucson, Charlie thought of nothing but his twelve claims at Big Indian. He tortured himself with newspaper articles and stories in mining magazines.

"More uranium was mined in the Colorado Plateau in 1951 than in any previous year."

"In November 1950, 145 claims were staked; February 1951 tallied 600."

"J. W. Gramlich received the first AEC bonus payment of $9,672 for 2,763 pounds of uranium oxide in .20 percent ore from his Morning Star and Evening Star claims on Lion Creek."[1]

By April 30, 1952, Charlie Steen could postpone his quest no longer. Selling the trailerhouse for $350, he stuffed the family belongings into a two-wheeled cart hitched to the back of the jeep. Excess baggage rode under a tarp on top of the vehicle.

The Steens said goodbye to Tucson and headed back to the Colorado Plateau. Their destination this time was Cisco, Utah, a desolate whistlestop for the D.&R.G. Railroad located a couple of hours west of Grand Junction.

Cisco was a dot in the flat, windy scrubland that stretched from the Book Cliffs on the north to the snowy-peaked La Sal Mountains southward. The town sat like a dry blister on the lip of Highway 50-6. It consisted of a cafe-beer hall, a general store and Buddy Cowger's service station. Little more. The few trailers and lean-to shanties of the local squatters disappeared into the sagebrush and rolling sandhills.

"We arrived looking like 'The Grapes of Wrath,'" Charlie remembered, "but we were in fine spirits and full of hope. We rented a tarpaper shack for $15 a month and stoked the stove with scavengered railroad coal."[2]

But euphoria was short-lived. M. L., exhausted and run down to begin with, developed pneumonia on their first night in Cisco. Worried and deflated, Charlie put her in the hospital at Fruita, Colorado, and faced this

Cisco, Utah in 1991.

latest rout as best he could. His Big Indian claims on the far side of the La Sals seemed farther away than ever.

Somehow he managed to keep the family going while M. L. regained her health. They learned to cope with kerosene lanterns in lieu of electricity. They hauled water in five-gallon cans from the Cisco railroad tank. They dug holes "behind the nearest hill" for a bathroom.

With few close neighbors, they devised their own entertainment. The children escaped from the confines of the one-room shack to a seemingly-endless backyard that invited exploration. There were dinosaur bones and rocks to collect, sequin-sized desert blossoms to admire, snakes, lizards and darting creatures to track.

Charlie kept his dream alive by poring over geological charts and writing letters to friends and acquaintances soliciting their help in financing his search. Meanwhile the wallet emptied, and the grocery bill swelled to over $300.

Then one day much to Charlie's surprise, a check for $100 arrived. Miraculously, another appeared. And another. Even better, his mother came to help. Rose sold her home and furniture in Houston. Now into her fifties and with several marriages behind her, she decided to throw her assets into her son's quest. She gave him the check from her sale and volunteered to join him on the claim to help set up a camp and cook for him. Charlie's spirits rallied. When M. L. had recovered sufficiently to take care of their brood, he and Mama Rose took off for the Big Indian Wash.

Steen's tarpaper shack at Cisco. *Western Mining and Railroad Museum.*

The backcountry had changed from the previous year. Tire tracks scratched red trails over sandy stretches where no signs of life appeared before. Rustic prospectors' camps squatted among the rocks. Lone wanderers haunted the clifftops with Snooper, Lucky Strike, Vic Tic or Babbel radiation counters clicking encouragement.

A few small tailings piles marked a smattering of "doghole" mines. These were nothing but cramped, unventilated, unsophisticated holes in the ground. They were barely large enough to accommodate the one or two owners who crawled inside and clawed out insignificant deposits of low-grade ore. There was no room or money for the mucking machines, slushers, shovel loaders and ore cars of the larger

"Doghole" miner pushes wheelbarrow of uranium ore. *Museum of Western Colorado.*

mines operated by VCA, USV and Climax Uranium. The independent miners sweated over simple jackhammers, compressors, shovels and wheelbarrows. Then they had to truck their diggings to the nearest mill or AEC buying station, sometimes hundreds of miles away.

Charlie's excitement rose when he saw his own claim markers. The forty-eight cornerstones stood just as he had placed them. The tin cans holding location notices remained intact. He walked over the property and showed Rose where he planned to drill. He was eager to get started.

The first task was to bulldoze a four-mile route from the county gravel road up to his claims. He paid miner Wilfred Brunke $500 to do the "cat" work. Rose organized a makeshift cook shack and bunkhouse. Together they hauled food, water and supplies from Moab, over forty miles away.

The next item was a better drilling rig than the little one he used for shallow prospecting. He made a deal with his old friend Bill McCormick to borrow a secondhand diamond drill in exchange for forty-nine percent of the action . . . if any.

He repaired the rickety machine as well as he could. Then he hauled it to the middle of his Mi Vida claim. The locals guffawed.

"The damn fool placed it 1,700 feet back from the rim in a location that didn't have a single indication of surface ore," they said. It was "Steen's Folly" all over again.

But Charlie stuck with his theory. He vowed that uranium deposits existed in anticlinal structures favorable to oil reservoiring. In drilling behind the rim, he violated all proven practices for uranium mining.

He decided to sink his bore hole through the Wingate Formation. The Wingate sandwiched itself several layers below the Morrison Formation and just above the Chinle and Shinarump strata. The AEC had already established that the Morrison was barren in this region. He would bypass this more common uranium-bearing sandstone and target the Shinarump conglomerate. There had been a number of recent discoveries in that layer. By entering through the Wingate he would lessen the drilling distance. There was a better chance to hit carnotite before his patched-up drill collapsed.

Even so, the pump broke. Drill bits wore out. Each breakdown brought more delay, more expense. Merchants at the mining hardware stores cooled when Charlie came in asking for credit.

M. L., stranded in Cisco, kept up a spirited correspondence during Charlie's long absences. "Being an introvert certainly comes in handy in this particular part of the country," she wrote her friend Beverly. "I've learned things about myself that I could have learned in no other place in the world."

She took up painting. She learned to cook the out-of-season venison "5,000 different ways." And, "I've become the darndest rock hound you ever heard of. If we ever have a decent home, I want one room with nothing in it but show cases to display minerals."

M. L.'s one luxury was rolling Bull Durham cigarettes. She compensated for this extravagance by making quilts from the empty tobacco bags. When she couldn't afford to buy tobacco, and that was often; she scrounged for butts along the dirt roads.

"Have found no ore, and have darned near perished, but I have absolutely found it an impossibility to worry," her letters to Beverly stated. "I used to be a chronic worrier, but I'm so happy and as irresponsible as Andy [their third son] when it comes to worrying about what is to become of us. I love it here. I have learned to laugh instead of gripe, and by gosh, enjoy living!"[3]

It was more difficult for Charlie to remain optimistic. There seemed to be one frustration after another. Every time he celebrated some small victory, another misfortune was waiting. It had been so from the beginning.

When he finally got a grubstake and staked his claims on Big Indian, Rose had suffered a heart attack. He had to drop everything and move to Tucson, take up carpentry again. Then, they no sooner saved enough cash to come back to the Plateau when M. L. came down with pneumonia. He had to put the claims on the back burner once more until she got back on her feet. Finally everything started to look better. Some checks arrived from friends to whom he had written. Rose recovered and came to help. McCormick loaned him the drill. But the damn rig broke down again and again. He had to keep shelling out money for repairs and more supplies. Now there were only a few dollars left. Everything hung on this one hole. If it proved to be barren, the tale would be over.

He started drilling on the day before Independence Day, 1952. It was hot, in the nineties. The desert burned. Charlie and Rose stood beside the spindly derrick. It looked like a toy in the vast, empty canyon.

"Well here goes," Charlie said as he fired the motor.

There was a whirring sound. The drill revolved. Grinding, gnawing noises filled the silence as the bit screwed itself into solid rock. The bit, encrusted with diamonds, the hardest material known to man, chewed through the petrified detritus. Inch by inch it probed geologic strata seeking the horizon of ore. Smoke and dust spewed from the deflector as it burrowed.

Charlie sweated over the hole. He began pulling up cores. He laid the foot-long cylinders of sandstone in a row according to their order of occurrence. If he hit something, he would know just how deep it was by its position in the line.

He examined the samples for traces of uranium. There wasn't a sign of the bright yellow ore. He would recognize carnotite if he saw it.

On July 6th, the weather changed. Dark thunderheads gathered to cool the temperatures to just under 80 degrees. By the time the afternoon rains came, the bore reached 72 feet. The cores turned dirty grey. He continued eating into the formation until he had pulled out fourteen feet of the stuff. It was foreign to him. He tossed it on the ground with the other rock lengths and continued drilling. His calculations indicated paydirt at about 200 feet. He still had a way to go.

The summer heat returned. One blistering day seemed to meld into another. The old drill started groaning and smoking, but Charlie persisted. He got down to 197 feet. Then it happened. The stem spun free. There was a loud, whirring noise.

"Jesus Christ!" he yelled, as he dove to switch off the engine.

He knew by the sound that the pipe had broken and most of it was stuck in the hole. He had no way to retrieve it. No more money.

Charlie and Rose didn't talk much as they broke camp. He prepared to drive the 100 miles to Cisco, dropping his mother off at Moab on the way. Absent-mindedly, he tossed one of the grey cores into the back of the jeep. He would at least satisfy his curiosity about the odd material that he did strike.

It was July 18th when the jeep rattled up to Buddy Cowger's service station in Cisco. Charlie was grimy with red dust and his face well baked by the sun. His toes stuck out from the ends of his boots. The smile was gone. He didn't look much like the exuberant dreamer who had left for the Big Indian only a couple of months before.

Cowger was sitting in his wheelchair near the gas pump. An accident with a cyanide gun had left him paralyzed from the waist down, but his handicap didn't keep him from getting in on the uranium fever. At every opportunity he dragged himself in and out of his jeep and crawled around on his hands and knees to scout promising terrain. A brand new Lucky Strike Geiger counter was his pride and joy. Uranium talk was his passion.

"Any luck?" he asked Charlie.

"Lost 160 feet of drill stem."

Cowger shook his head sympathetically.

"Spun off the drill pipe 30 feet from the top."

Buddy Cowger's service station in Cisco, Utah.

Cowger knew that Charlie hoped to hit ore at about 200 feet. He felt sorry for the persistent string bean of a guy who couldn't even afford a Geiger counter.

"I've been testing some cores the kids brought in," he said, trying to cheer his friend. "Don't look too bad."

He put one under the counter and the needle registered slightly.

Charlie reached into the jeep and pulled out his ugly grey rock.

"Hell, I've got better stuff than that," he joked.

"Well, put her under," Cowger laughed.

Charlie jabbed the sample under the counter. The instrument swung clear off the scale.

The men stopped laughing.

"Try it again," Cowger said.

The needle pegged. The counter chattered like static.

"My God!" he shouted. "You've hit it!"

Charlie hadn't recognized the unimpressive grey stone as pitchblende.

M. L. pinned the last diaper on the line. She detested washday. Cold water. Little soap. Four rambunctious boys who thrived on dirt. She hurried back into the house out of the heat.

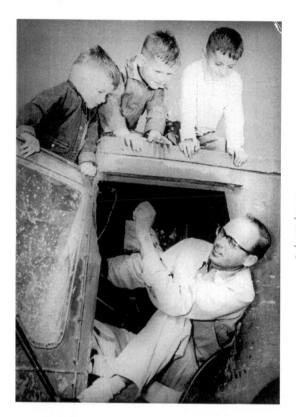

Charlie Steen in his battered prospecting jeep. On top, left to right: Mark, Andy, and John. *Western Mining and Railroad Museum.*

Charlie Steen in the field. *Western Mining and Railroad Museum.*

The sound was faint, at first. Someone yelling. Whooping. She peeked out of the window to see who it was just as Charlie loped across the field, shrieking, "We've found it! It's a million dollar lick!" and ran straight into the clothesline. Her freshly washed diapers hit the dirt with him.

"My wash," she moaned.

"We've hit it, M. L.," Charlie sobbed. "It's a million dollars!"

It took her a minute to believe him. Then all she could do was cry.

The kids didn't quite understand, but their parents' antics convinced them that they should be happy. M. L. insisted on a family picture to com-memorate "the day." Charlie slicked up into some clean suntans and she put on a fresh white blouse, calf-length skirt, ankle socks and "penny loafers." They walked back to the service station and piled the kids on the hood of the jeep for the photo. Then they sped down the rutted road to share their good fortune with the neighbors.

Charlie decided to get control of the area surrounding his strike before releasing the news to the press. He spent the next two months staking almost forty claims along the flank of the Lisbon anticline and adjacent to the Mi Vida. In thanks for their help during the bad times, he invited sev-eral friends to get in on the action. Bob Barrett staked some claims, as did their Cisco neighbors Holly Seeley and Buddy Cowger. Only after firmly establishing "the Steen line" did Charlie announce his find. The story broke in the *Denver Post* on August 30, 1952.

The following week a brief notice appeared in the Moab, Utah, *Times-Independent:*

> Charles A. Steen, field geologist of Cisco, reported last week that he had discovered the mineral pitchblende in the Big Indian mining district, which is nine miles south of La Sal. . . . The pitchblende was found in the first diamond core hole drilled by Steen on the Linda Mujer group of claims which he staked in 1951.[4]

In another story in the same edition, the editor reported:

> By listening to the radio these days it is difficult to determine whether we are enjoying a wave of prosperity or if we are on our last gasp. It's com-forting to look up at those old red rocks and know that there will be very little happening to them in the next four years.

Moab, the dusty little farming community, slept on, unaware of an approaching boom. The Mormon farmers, as industrious and ultra-conser-vative as their pioneer forefathers, paid little attention to the flurry of for-tune-hunting outsiders (they called them "Gentiles") who parked trailers

Moab in the thirties and forties. *M. Irish Collection, Dan O'Laurie Museum.*

and pitched tents along the town's dirt roads. To them, heading out in the "toolies" for a weekend "go" at prospecting was a normal form of innocent recreation. They'd been doing this since the radium days. Therefore, they were unimpressed with the feisty little Texan who claimed a bonanza.

Despite the fact that Charlie's sample cores proved to be primary pitchblende, people didn't believe him. The strike occurred in a much shallower horizon than expected. It was in the Chinle, a formation virtually unknown as a uranium producer. The AEC even said the Big Indian region was barren. Rumors circulated that Charlie had salted his claim.

"Nobody would believe I really had something," he later said. "They accused me of salting the hole with pitchblende so I could sell stock. But it seemed to me the point was, if I salted it, where did I get the pitchblende to salt it with? Canada?"[5]

Popular disbelief in the authenticity of the strike made it difficult to raise money to mine the ore. Charlie insisted on splitting his 51 percent interest in the venture with Rose. After all, her grubstakes kept food on the table those many years. But they were broke. They had nothing to put back into the operation.

To make matters worse, Bill McCormick revealed that he had a silent partner in his 49 percent interest. The partner claimed the discovery was worthless. He didn't want to sink any more money into further drilling and expensive exploration. He wanted out.

Charlie started looking for investors in Salt Lake City, Grand Junction and Denver. To no avail. Then one day in early September he received a letter from William T. Hudson, of Casper, Wyoming. Hudson, his former boss at Chicago Bridge and Iron in Houston, had read the story in the *Denver Post*. He sent his congratulations to the young man whom he had helped through college.

"Well, you surely fooled me when you went off prospecting," he wrote. "Somehow I had always sort of guessed that you would wind up a lawyer, possibly a patent lawyer because of your engineering background."[6]

He asked if there was any way he could help.

Two weeks later Charlie called Hudson. He told him that McCormick's percentage was for sale. He wondered if Hudson might be interested. Hudson said he couldn't invest more than about $5,000. Possibly Dan O'Laurie, another former Chicago Bridge and Iron colleague, would be a prospect. The next day O'Laurie flew down to Moab to talk things over. He liked the proposition. The two men bought out McCormick's partnership for $15,000. And O'Laurie loaned the corporation about $30,000 to start work.

Now Charlie had capital. But not enough, he figured, to sink a number of additional drill holes to prove the field. He decided to put his money into a shaft. He knew it was a gamble, but one he had to take. Once again, everyone thought he was crazy.

The recognized practice in uranium mining was to core a number of samples in a promising area and examine the cuttings for evidence of mineralization. In this way a claim was blocked out and the tonnage of an ore body was estimated. This procedure was important as ore usually occurred in sporadic rolls or pods with unpredictable continuity. It was possible for a deposit to pinch out or disappear entirely.

Moab resident Howard Balsley, who was nicknamed "Mr. Uranium" and was active in the uranium industry from the early radium days, explained the phenomenon in a speech before the American Mining Congress on September 23, 1952:

"Uranium ore is just about the most erratic and inconsistent mineral I know of. Unless a uranium property has been drilled and the ore bodies proved and delimited, one may have a wonderful face of ore today and witness its complete disappearance tomorrow. There is no assurance whatever, so far as I know, of the continuity of an uranium ore pocket or roll."

Phil Leahy, the first manager of the AEC Raw Materials Division at Grand Junction, concurred. Collie Small quoted him in *Collier's*, September 10, 1949:

"Spread a layer of white beans over the bottom of a tub. Toss in one or two blue beans. Cover the layer with a newspaper and repeat the operation until the tub is full. (The layers represent the formations where ore is

Howard Balsley, one of the "grand old men" of the early radium-vanadium era. *Howard Balsley.*

found.) Now take a long hatpin, poke it down through the layers, and see how many blue beans you can hit. That's drilling for uranium."[7]

"It was bad geology, but good economics," Charlie later admitted. If he had drilled more exploratory holes, he would have been out of money. He gambled on sinking the shaft on a single hole and finding enough ore as he went along to pay the costs.[8]

On October 4th, he started a six-by-eight-foot timbered shaft thirty feet southeast of the borehole. Three weeks later, he went to Moab to find a lawyer. He wanted to form a corporation.

Mitch Melich was the only attorney in the town of approximately 1,200 people. He practiced out of a one-room office on the mezzanine of the bank. A handsome forty-year-old, his appearance hinted at his Serbian ancestry—a thick crop of straight, black hair, dark, heavy-browed eyes, olive complexion. He was responsive to the tattered fellow with worn-out boots and patches in his pants who showed up in his office unannounced. He knew poverty firsthand.

Mitch grew up in the copper mining town of Bingham Canyon, Utah. His father, a grocer, died when he was ten. His mother couldn't read or write English. The youngster worked his way through high school. Then, in order to attend the University of Utah, he put in eleven hour shifts during the summers on Kennecott's track gang for $4.25 a day.

As a sophomore in college, Mitch decided to toss his education in favor of a full-time job at the mine. Fortunately, Lou Buchman, his foreman, recognized the boy's potential. He fired him, gave him his check and sent him back to school. Three years later, Mitch had a law degree.

Diploma in hand, it didn't take him long to discover that the legal fraternity wasn't waiting for his talents. There were no openings in Salt Lake City's law firms. A friend talked Mitch into going to Moab.

"What the hell is there in Moab?" he asked. The town then boasted a population of 850.

He decided little was better than nothing. He borrowed $300 from the First Security Bank in Salt Lake and headed south in the fall of 1934. For $100 a year he became city attorney and received a $50 retainer as legal counsel for the bank. His first year's earnings totaled $950.

As the community's only lawyer, Mitch had a stepping-stone into politics. In 1942, he was the first Republican in the district in 25 years to be elected to the state senate. Four years later, his popularity earned him a second term. He ran a third time in 1950, but lost.

"Maybe it was a good thing because that was when the uranium boom started," he later said. "Had I been in the state senate, I might not have become active in the uranium business."[9]

Perhaps it was this recent political defeat that honed his interest in Charlie's proposal.

"I've found a uranium mine and I want to form a corporation," Charlie said. "The mine's in Utah and I'm from Texas, so I think we'll call it Utex—Utex Exploration Company."

Mitch was intrigued by Charlie's self-confidence. He agreed to draw up the papers in exchange for a small amount of stock. Suddenly, the struggling lawyer found himself a partner in an unproven uranium company.

With his corporation formally launched there was nothing for Charlie to do but sweat out the shaft. He oversaw the operation as head geologist. O'Laurie handled finances. Bob Barrett signed on as job foreman in return for 4,800 shares of O'Laurie's stock. They moved a secondhand building to the site and set up a bunkhouse for a small crew. Rose cooked.

It was slow work. Drilling through layers and layers of rock. Blasting the hole with explosives. Mucking out debris with electric shovels. Timbering the sides to keep them from caving in. Foot by foot the gap widened and deepened. Foot by foot the tailings pile grew. Day by day the knot in Charlie's

MiVida headframe and shacks. *Western Mining and Railroad Museum.*

stomach tightened. What if the hole proved barren? What if the rich deposit of his borehole was a fluke? What if the AEC and all the others were right?

October passed. Thanksgiving approached. The mine inched closer to the depth of his original discovery. Still no ore. Charlie knew the miners working for him considered it lunacy to sink a shaft without coring a grid. He prayed that he would prove them wrong.

December first marked his thirty-first birthday, but it was work as usual for Charlie, with no planned celebration. Then the drills reached 68 feet and he got the birthday present of a lifetime. The hole bottomed into heavy, grey ore measuring over fourteen feet of primary pitchblende. A later radiometric assay indicated a staggering 0.34 to 5.0 percent of U_3O_8.

Charlie knew he had the biggest strike in the history of the Colorado Plateau. He had proved his theory beyond a doubt. There was primary ore in the district, and it lay in formations other than the Morrison and Shinarump.

"I can remember that night," he recalled. "My mother and I went down to Moab. It was about 10 o'clock and I had a quart of whiskey. We went to a hotdog stand where the bus station used to be. We were the only ones there. The whole town was locked up. That was the night of the discovery of the shaft and that was the last time that Moab went to sleep at 10 o'clock for a long time."[10]

6 URANIUM FRENZY

So a spindly young Texan with a pretty wife, a gaggle of kids and a nickel in his pocket grabbed for the brass ring and caught it! The AEC couldn't have scripted it better if they tried. They baited the lure with greenbacks, and Charlie Steen, one of the "little guys," got the big one.

What made it even better was that Charlie's story let loose the hoped-for prospecting rush on the Colorado Plateau. He didn't strike carnotite. He discovered pitchblende, lots of it in unexpected places, and made it clear that America's uranium industry was only in its infancy.

Full of quirks and quotes, Charlie became the media's delight. Readers from coast to coast knew the exact moment when he abandoned the tarpaper shack in Cisco and rented a three-room duplex in Moab for $60 a month. They learned that he now had electric lights and a flush toilet, and there was a $5,500 fire-engine red Lincoln parked on the dirt road in front of the house.

Newspapers carried his picture taken while trying on a pair of field boots in Miller's Moab Co-op. The caption told how he gilded his tattered old pair with holes in the toes "to remind me, in case I forget, how 'easy' it is to earn a million bucks."

The press even reported the Steens' first purchases. Charlie got a red jacket with an eiderdown lining, a warm sleeping bag and a portable radio. M. L. opted for a washing machine to replace her scrub board. But Charlie bought her a diamond instead. Since Moab didn't have a laundry or dry cleaner, he had articles of her $1,500 wardrobe, his new suits and the $500 worth of kids' clothes flown to Grand Junction and back for cleaning once a week. As a bonus, they took the new Lincoln on a 7,000-mile tour of Mexico.

Their success stories didn't sit well with their old neighbors, however. M. L. later told *Cosmopolitan* writer Murray Teigh Bloom, "There were dozens of other prospectors living in the vicinity [of Cisco], with their wives and families. All of us were pretty friendly and exchanged information on how our husbands were doing with their claims and how to feed a family on $8 a week, things like that. Then Charlie really hit it, and the craziest thing hap-

The duplex in Moab where the Steens lived after Mi Vida strike.

This log cabin on the outskirts of Moab was Charlie Steen's first home there.

pened. The girls *dropped* me. First there was a general coolness, and then there was a total avoidance. They wouldn't even speak to me on the street. I was terribly upset at first. Then, little by little, I managed to put myself in their place: it is painful to see someone else hit it big if you haven't. In a way, the successful become a sort of living reproach to the unsuccessful."[1]

But Charlie's doings captured the imagination of the rest of the nation. Scores of armchair geologists headed for the Plateau with the hope of emulating his success. It appeared that the government's ploy to trigger a domestic uranium industry was paying off.

The situation brightened even more when the AEC celebrated another headline grabber. A second unknown prospector became an overnight millionaire. This time, besides the "rags to riches" theme, the saga exuded adventure. And intrigue.

Vernon Pick, a balding, middle-aged electrician, turned to uranium prospecting after his motor repair shop in Two Rivers, Minnesota, burned to the ground in 1951. The fire represented more than a business disaster for Pick. It demolished a dream. The soft-spoken, introverted journeyman had set up his machinery in an abandoned flour mill and established a "self-sustenance homestead" on the property. He converted the ancient power-driven mill to a modern hydroelectric plant that furnished heat and electricity.

With his wife, Ruth, and six farm helpers, he produced most of life's necessities on the property. They raised fruits and vegetables in the gardens and orchards, then canned them for the pantry. They caught fish in the millpond. They milked the cows and churned butter and cheese. They collected eggs from the chickens and ate the hens. Lamb roasts graced the dining room table and the sheep wool was spun and woven to make their clothes. They even harvested the grapes for wine. Pick spent his spare time reading the classics, everything from Aristophanes to Zola.

When the blaze ended it all, Pick came away from the tragedy with a $13,000 insurance check. He invested in a black Ford panel truck and a house trailer. After a vacation in Mexico he

Vernon Pick turned to prospecting after his motor repair shop burned down. *Sheldon A. Wimpfen.*

would get a job in a California aircraft factory, he decided. He got as far as Colorado Springs.

There, a seasoned rockhound he met in a trailer park told him about the AEC's incentive program for uranium hunters. He showed Pick some ore samples and told him if he'd go to the AEC compound in Grand Junction, they would tell him how to find some, too. Pick decided this was better than squandering his life's savings in Mexico. He bought a Geiger counter and a book on how to prospect, and drove west.

The AEC knew that the best way to achieve their goal was to help the uranium prospectors in every way possible. Besides the financial incentive program, their construction crews bull-dozed scores of roads into the wild, untouched backcountry. They also established numerous ore-buying stations for the convenience of "doghole" miners scattered throughout the remote region.

In addition, government geologists intensified their search. Huge diamond drilling projects tapped into the desert. A squadron of twelve Piper Super Cub airplanes was mustered for aerial reconnaissance. The planes carried sophisticated radiation detection devices known as scintillometers. The instruments picked up ground radiations up to 1,500 feet away. Daring bush pilots skimmed the ragged canyon rims as close as 50 feet to check potential uranium-bearing formations.

The results of AEC geologic surveys appeared on maps posted regularly on the Grand Junction compound gate. While actual uranium deposits could not be pinpointed, anomalies which indicated the need for further investigation of particular strata were marked. Prospectors who wanted additional information could meet personally with AEC staff members. Pick arranged an interview with Dr. Charles A. "Al" Rasor, chief of the mining division.

Pick admitted that he knew nothing about geology or prospecting. But he said he "had a knack for reading something in a book and then being able to go out and put it into practice."[2]

Rasor was not startled at the naiveté of his questioner. There had been a lot of them. He suggested that Pick read the AEC booklets on prospecting for uranium and then try the Dirty Devil River country of the San Rafael Swell in southeastern Utah. It was remote, virtually inaccessible, and relatively unexplored.

"You can go look for elephants in elephant country and you're going to find a lot of competition because everybody else is going to be hunting there," Rasor advised. "If you've got the stamina and the guts, go hunting where nobody's found an elephant yet."[3]

Pick spent a few hundred dollars on camping and prospecting equipment, some dehydrated food, and a snakebite kit. He left Ruth in a trailer park in Grand Junction and drove to a kickoff point at Hanksville, Utah.

Hanksville was little more than a fleck in the eye of the inhospitable San Rafael Desert. The five peaks of unexploded volcanoes in the Henry Mountains rose to the south. Cathedral Valley lay like a carved sandstone city on the west. Grotesque rock caricatures resembling Snoopy and Disney characters populated Goblin Valley in the shadow of the hogbacks of the San Rafael Reef in the north. Eastward lay Robber's Roost, the hideaway country of deep canyons and lonely buttes where Butch Cassidy and other early outlaws fled the law.

The small Mormon ranching community of Hanksville was not new to Pick. Fifteen years earlier he had spent three days there after his car broke down during a cross-country trip. Andrew Hunt, a local rancher, had befriended him then and fixed his automobile.

When Pick returned in 1951, he looked up the Hunts. They introduced him to a cowpuncher named June Marsing. Marsing had ridden the trails for years and knew the country well. He also had a four-wheel-drive vehicle. Pick knew that he would not get far in the rugged terrain in his Ford, so he asked Marsing if he would like to partner with him and show him around. Marsing agreed. But after about two months, car trouble developed. The two men started squabbling over expenses for automobile repairs and parts. Marsing quit.

Pick then ventured into the badlands alone. He drove his pickup to the roadheads and backpacked into the trackless desert.

"The trouble with most people," he said later, "is that they don't like to walk. They want to go where they can drive a car or ride a horse. To get into the places where I wanted to go . . . the places nobody else had been . . . there was no choice but walking."[4]

For eight months he explored desolate terrain bearing names reminiscent of cowboy lore: Little Wild Horse Mesa, Circle Cliff, The Little Rockies, Poison Springs.

Mishaps plagued him. The truck mired in the sandbar of a dry wash. He came down with arsenic poisoning from polluted water. A scorpion rammed its sting into his back. Cougars tracked him. He saw so many rattlesnakes that he "lost his fear of them." The weather vacillated between blistering heat and treacherous electrical storms.

Meantime his finances dwindled. In the spring of 1952, Pick took a gamble and sank $1,000 into a scintillometer that registered radiation more accurately and at greater depth than his Geiger counter. By June of that year, he had only $300 left.

On June 21, he camped on the banks of the Muddy River that flowed through a desolate area northwest of Hanksville. The water was high from winter runoff and his hike along the turbulent stream exhausted him. Frequent fords across the waist-high torrent were necessary. Stones cut his feet and his wet boots raised blisters. To make matters worse, his drinking water was polluted by the bloated carcass of a cow festering in the weeds.

Sick and discouraged, he began to prepare for the next day's search, and picked up his scintillometer to adjust the dial calibration. The needle was stuck at the high register. It wouldn't return to normal. He wondered if the batteries were weak. But if they were, why would it register on high? With an electrician's curiosity, he started to walk with the instrument.

"It would indicate and then drop off as though the batteries were low," he later recalled. "I walked into a gully and the reading was still higher. I climbed and the higher I went, the higher it read. I was travelling pretty fast now. It was 500 feet to the top of the ledge. I had never seen the counter like that. It was clear off scale. I couldn't tell what was where until it dawned on me that the whole damned ledge I had been walking on, looking for an ore body, was all ore. Nothing but high grade ore!"[5]

He jammed his pick into the rock. It was the Shinarump conglomerate, riddled with canary yellow carnotite. You could see it.

As the adrenaline surged through his body, he forgot the pain in his gut and his bleeding feet. Half-dazed, he scrambled through the brush looking for rocks to build a cairn as a monument to be placed in the center of the claim. It had to be at least three feet high, he remembered. He'd have to fill out papers giving his name and specific details to deposit with the marker. What would he name the claims? He had staked a few others, using the Greek alphabet, Alpha, Beta, Gamma. This would be Delta, he decided. Delta One, Two, Three and Four. He filled out the requisite forms and completed staking the boundaries of the claims with more stones. Then he sank to the ground to sleep the night away.

In the morning, he felt revived, eager to register his strike. He longed to tell Ruth. And Rasor. But he was at least four days away from the pickup and several more hours from civilization, stranded on that damned river.

A raft would be the fastest way back, Pick thought, and so he burned an old driftwood log in two and lashed it in a V-shape with his belt, a bootlace and the scintillometer strap. He piled on his supplies and covered the lot with a tarpaulin. Straddling the awkward craft, he attempted to steer it downstream with a stick. The contraption refused to follow the current and kept crashing against the canyon walls, smashing Pick's feet against boulders and dumping gear. By the time he reached his truck, he was limping and all of the food but dried milk and oatmeal was gone.

Later, Pick claimed that he happened upon Gerald Brooke and Melvin R. "Red" Swanson of the AEC as he drove back to Hanksville. The two government men, concerned about an unidentified pickup truck parked unattended in the wilderness for so long, had instigated a jeep search for the owner. When they saw the battered, weary prospector they offered him coffee and a meal, and he camped with them that night. Next morning, not mentioning anything about his discovery, he continued on to Grand Junction.

Some who heard Pick's story vowed that he met the AEC geologists much sooner. In fact, they accused Pick of fabricating the entire dramatic account that appeared in the *St. Louis Post-Dispatch, Life,* and several other national publications. Pick recounted crossing the waist-high waters of the Muddy River twenty-seven times, attempting unsuccessfully to purify his poisoned drinking water with charcoal, becoming dead-ended in box canyons, hiking for ten days with an untreated scorpion bite, enduring sandstorms and escaping flash floods.

"Although it had become conceivable to Pick that he might die, it was not conceivable to him that he could turn back," wrote Robert Coughlan in the November 1, *1954, Life.* "Part of the reason, perhaps, lay in an obscure and rankling sense of destiny that had always bothered him, even when he was a boy and not doing well in school."

Hanksville residents scoffed at the articles about Pick's hazardous adventures. They tagged his descriptions of fording the surging Muddy River and floating down the raging torrent on a raft as "hogwash." In June, the Muddy is "the usual trickle," they said. As for being tracked by cougars, mountain lions avoid people. They're too shy. And concerning all of those rattlesnakes he claimed to run into, the reptiles hibernate during the months when Pick was prospecting. Besides, during that particularly cold winter, it would be pretty hard to find any "blistering heat."

"Pick didn't discover the Delta Mine himself," they said. "An AEC man tipped him off."

One story had it that Pick stumbled into the camp of an AEC surveying party *before* he hiked the Muddy River. The government geologists had a plane equipped with a scintillometer. A pilot and a radiation observer flew the rims each day to chart the area for possible uranium deposits.

According to this version of the episode, Pick could see that the observer was excited about something. Pick got the man drunk and he told him about high readings he had noted in the Muddy River area. The geologist was not supposed to tell anyone about his findings until after they were posted at the compound. Instead, he pointed out the exact location on a map. Pick then set out on his hazardous trek and staked his claims.

Another rendition amplified the AEC involvement and refuted Pick's description of a harrowing four-day ordeal. It seemed that Pick arrived in Green River, Utah, two days before the supposed discovery day. He came to doctor his sore feet and get someone to accompany him back to the Muddy River area where "he hinted that he was on the trail of something big."[6]

Glen Ekker from Hanksville, agreed to join Pick and take his four-wheel-drive vehicle and a couple of horses. Raymond W. and Samuel W. Taylor published Ekker's account in their book, *Uranium Fever:*

> I discussed the trip to be taken in the morning with Pick, and he said that in the morning we would follow a jeep trail going straight south.
>
> Next morning I got up early and went out alone to locate the jeep trail. I found it and followed it to the slickrock, where I had some trouble following the tracks on patches of sand. At the end of the sandstone area, the jeep tracks stopped, and I followed footprints which continued on. I had experience as a tracker and was able to follow the trail. Then I came upon the first of a series of small stone trail-markers, made by piling stones one upon the other. At this point I went back for Pick, and we followed the trail on horseback.
>
> There were, as I remember, eight or ten markers which guided us to the Muddy River. That last marker ended right at the face of a cliff, at the location which became Pick's Delta mine. . . .

George Morehouse, a former AEC geologist, gave some credence to the latter story. He remembered riding on horseback the day after Pick's discovery. He saw four sets of footprints on the road to the outcrop. These were probably made by the AEC team. On top of the prints were two horsetracks, conceivably made by Pick and Ekker.[7]

Suspicions intensified when Gerald Brooke and Red Swanson, the AEC men who supposedly drove out to rescue the missing prospector, resigned from the commission. Brooke became foreman of the Delta mine. "Swanson was around the operation a good deal in an unofficial capacity," the Taylor brothers reported in their book.

The F.B.I. investigated charges of the pair's complicity in an illegal scam. They found no evidence of wrongdoing. It was confirmed that Pick had worked the area where commission planes flew reconnaissance. Possibly he saw them circling the vicinity that registered high readings and tracked the outcrop on land.

Some twenty years later, Pick stuck to his original story. He maintained that after visiting the AEC compound at Grand Junction and being directed to try the San Rafael Swell, he talked with the Hunts and other former acquaintances in Hanksville. They told him they had lived there all their lives and knew where all of the uranium was in that country. No use looking further.

Refusing to give up, Pick backpacked into the most remote regions he could find. He searched over an ever-broadening area in the vicinity of Hanksville. He found the outcropping of ore within twenty-five miles of town and in plain sight for anyone to see. But it was in an almost inaccessible place where he assumed no one had been before.

He was "shocked" that his "erstwhile friends" circulated so many derogatory remarks about him. He figured that his strike in an area they had judged barren of ore was a blow to them. They considered themselves almost professional uranium prospectors and miners. It was galling that an outsider came in and made the big discovery in their own backyard.[8]

After thorough investigation by the F.B.I. and the AEC security division, the AEC manager at Grand Junction told Pick that he was completely in the clear. A bound volume of typewritten pages over an inch thick covering the inquiry lay on his desk. Pick was not allowed to read it but "would have given a nickel to have had a look."

As for the magazine articles, he found them quite factual in all essential details.

"Do you really think that _Life_ would have maintained their leadership this long if they were in the habit of fabricating stories of this kind?" he asked.

The controversy over Pick's strike intrigued the public for years. A unanimous opinion of "the truth" remained unsettled. But in 1953, with two folk heroes coloring the news, how Pick or anyone else struck it rich didn't matter.

Stories were told and re-told about prospectors of the 1940s whose perceived barren diggings had brought about unexpected bonanzas. Pratt Seegmiller, who lived on the outskirts of the little mining town of Marysvale in southeastern Utah, was one of them. He and his family operated a small souvenir shop at the Big Rock Candy Mountain. Seegmiller was an avid rock-hound and collected an impressive store of mineral specimens to sell to tourists. When he heard that the AEC was buying uranium, he looked at some fifty pounds of canary-yellow rock that he had gathered five years back and wracked his brain trying to remember where he found it.

He stewed all winter. All he could recall was that it was an outcropping at the base of a hill where some old-time prospector had dug around looking for copper. Finally, in April 1948, it came to him. He hiked right to the spot and, in the next few years, staked the Freedom and Prospector claims. He leased the property to the Vanadium Corporation of America. By 1952, the mine proved to be one of Utah's richest deposits. Seegmiller's royalties enabled him "to support my family in a style to which we were not accustomed."

Joe Cooper was a road contractor in Monticello, Utah, a small town south of Moab. In 1946, he and his father-in-law, Fletcher Bronson, paid $500 to buy the Happy Jack copper claim in White Canyon near the Colorado River. But when they started to work the ore, they found that it was contaminated. With uranium. It didn't pay to process the low-grade copper, so they tried to sell the mine. They were about to dump it for unpaid taxes when the AEC announced its uranium buying program. Fletcher and Bronson put the mine into production for uranium and the Happy Jack earned them over $25 million.

"The uranium boom" captured America's imagination. It seemed as if anyone could become a "uraniumaire." You didn't have to be a geologist; you just had to be lucky. People from all walks of life were making the headlines.

A group of high school students staked forty claims and later sold them for fifteen thousand dollars.

Texan Blanton W. Burford and his partners sliced into an eight-foot vein of high-grade uranium ore while they were bulldozing a road into their claims on Rattlesnake Mountain, near Moab. They sold out for four million dollars.

On a Monday morning, after abandoning their equipment to knock off work for the weekend, a county road crew returned to the job and discovered over two hundred claim stakes covering the area. Some crafty prospectors had figured that the machinery had been left there in preparation for drilling, and they wanted to get in on the action.

One day, Paddy Martinez, a Navajo sheepherder, picked up some yellow rocks near Haystack Mountain at Grants, New Mexico. He took the pretty minerals into town and discovered they were the "twentieth century gold" that the white men were talking about. Martinez's find started the mining boom in the famous Laguna and Ambrosia Lakes districts. Paddy didn't make a million from finding the uranium, because he found the ore on property owned by the Santa Fe Railroad. But he was happy with the title of "official uranium scout" and the life-long monthly pension he received courtesy of the land owners.

These and other success stories inspired thousands to give uranium hunting a try. School teachers, insurance brokers, used car salesmen and shoe clerks around the nation converged on the Colorado Plateau in droves. All had high hopes. Few had any knowledge of geology or experience prospecting. They turned to the AEC for help. Mail at the Grand Junction office grew mountainous.

"I was encouraged to read the write up on Mr. Pick in life [*Life* magazine]," wrote one young man. "I founded [sic] it to be an extraordinary little sport story.

"The elder gentleman that incouraged [sic] me to read it, also suggested that if I was an ambitious type of 29 years old young man that I appeared to be, I should try to make some possible arrangements to sacrifice my life for the benefit of humanity, as well as material wealth."

Another letter writer claimed, "I have just purchased a jeep, Geiger counter and pith helmet. Now all that I need is the literature on uranium prospecting."

In a postscript, the correspondent added, "My mother-in-law recently was subjected to intense radiation treatments. She's cured now, but she gets a definite 'ping' from my counter. Now here is what I'd like to know—how much will you offer per pound for her?"[9]

One would-be Steen got no further than a small feature story in *Moab's Times-Independent*.

"A long, lean, lanky boy whose name we were unable to clear, blew into town last weekend and wildly waving a beautiful piece of high grade uranium ore announced he'd made a new strike.

"'It's 35-feet thick,' he said, 'and goes plumb around a mesa, seven miles. And,' he continued, 'it's one of them breedin' reactors you hear about and the stuff is growin' faster than you can mine it.'"[10]

The New Yorker's "Talk of the Town," September 12, 1953, chronicled the experiences of Joe Morris from Ridgewood, New Jersey.

Moabite Jack Turner invited Joe "to hop in the jeep and go along. 'Hop was good!' Joe said. 'I started out from here with 400 pounds of equipment, and I had practically all of it on me when I fell up into that jeep . . . a Geiger counter, an enormous pickax, two hammers, field glasses, a portable fluorescent lamp for testing samples, a mortar and pestle for pulverizing them, two big cameras, a sleeping bag, some canned goods, and half a dozen prospecting manuals. I had so many straps dangling around my neck that one false move and I'd have hanged myself. Turner was smart. He'd only brought a sleeping bag, food, a Geiger counter and a pick not much bigger than a beer-can opener.'"

By the time Charlie Steen moved his family to Moab in early 1953, the town was reeling under the impact of the modern-day "Klondikers." The conservative, in-bred little town was suddenly rocketed into the twentieth century. Everything was changing.

Normally only a handful of pickup trucks and old-model cars rattled over the arthritic planks of the cantilevered bridge spanning the Colorado River north of town. Now the wooden structure creaked and moaned as strings of army surplus jeeps, nine-passenger Chevy station-wagons, Cadillac convertibles and heavy ore trucks (that had to shift into low gear and travel at five miles per hour) rolled into Main Street. Travel through Moab on the highway called "The Navajo Trail" averaged 987 cars within a twenty-four hour period—300 percent over the traffic tally two years before. The city fathers decided it was time for a new bridge.

The 1,200 Latter-day Saints (Mormons) who lived in Moab had always known the face and name of everyone within the radius of 100 miles. They were related to most of them. If someone didn't show up for sacrament meeting at the wardhouse, the "brothers and sisters" started making sick calls. Now the streets were crowded with strangers. People had to take in boarders as the handful of motels couldn't accommodate the crowds. The old schoolhouse, where students from kindergarten through grade twelve recited their lessons, was bursting at the seams.

Just as a housewife notices cobwebs when she is expecting company, the Moab townsfolk started to take a closer look at their community. New neon signs appeared on Main Street. "Paint up, clean-up" campaigns urged residents to give their homes and yards a face-lifting. The county budget was increased to $180,525. After all, Moab was now in the public eye. Universal-International Pictures had chosen the town as the location for filming the movie, "Son of Cochise." And Charlie Steen, now dubbed "The Cisco Kid" by the press, had appeared on NBC-TV's "Welcome Travelers" show. There were even articles about him in *Time, Business Week,* and *True.*

Charlie had no trouble keeping Moab in the news. His colorful one-liners were published nationwide. When his discovery story wore thin, he could deliver a battery of statistics. "We've got twenty-one feet of ore extending from the main shaft in three directions," Charlie told the nation. "So far we haven't had to move a shovelful of muck. It's all ore."[11]

Moab during uranium boom. *E. Shields Collection, Dan O'Laurie Museum.*

When the reporters tired of mining yarns, Charlie rallied with tales of his new-found affluence. "Poverty and I have been friends for a long time," he said, "but I'd just as soon keep other company."[12]

Readers derived vicarious pleasure as Steen's spending sprees escalated. He purchased a Cessna 195 airplane. He bought a couple of pedigreed Dalmatian dogs. He invested in undeveloped acreage atop a redrock cliff north of town. There was no power, no gas, no water and no road to the lofty site, but that was where he was going to build his dream mansion.

"I spent a lot of years down under, looking up," said Charlie. "I always wanted to be up on the hill looking down."

But it was Charlie's nature to share his good fortune. As the first anniversary of the Mi Vida strike approached, he and Bob Barrett decided that it was time for a Utex Exploration Company celebration. It wouldn't do to limit the festivities to stockholders and employees, however. They published an invitation in the *Times-Independent* for a gala open house. The whole town was invited. The party was set for June 6th at the Arches Cafe. There would be an all-you-can-eat buffet. Each woman would receive a corsage. There would be cigars for the men. Ralph Beyer's orchestra would be flown in from Salt Lake City for dancing until the wee hours.

On the night of the big affair, the large room at the rear of the Arches Cafe was transformed from its weekday function as a roller-skating rink to a ballroom. Flower-laden tables surrounded the dance floor. Colored lights played upon a mirrored globe revolving overhead.

Moabites, bedecked in Sunday-go-to-meeting dress, arrived promptly at eight o'clock. The little Turner twins, Freida and Elaine, parted a curtain to reveal Ralph Beyer's society orchestra on the stage. Music blared. Dancers paired. Sparkling tendrils of dust, that had been sprinkled on the floor for skating, filtered up through rainbows of lightbeams to the shining, faceted ball dangling from the ceiling. Children played tag among the churning dancers.

Farmers, miners, merchants and housewives circled the long buffet table to heap their plates with roast beef, ham, chicken, potato salad, tossed greens, fruit, cheese and cakes, and enough soda pop to quench the thirst of the entire community. A steady procession of men craving a nip of the stronger stuff that Utah's laws forbade selling at the Arches, made frequent detours outside to brown-bagged bottles waiting in their cars.

Party-goers got a look and a laugh at the Utex Board of Directors during intermission when the company attorney, Mitch Melich, presented each official an "award of merit." Dan O'Laurie got a yard-long cigar. Charlie received the book *How To Raise a Dog*. When he opened the cover, a paper hotdog flipped up. M. L.'s gift was a toy dog that barked and jumped by remote control. Rose was awarded a cigarette case that sprouted with pop-up pom-poms. Barrett couldn't be found when the time for his presentation arrived so O'Laurie concluded the ceremony by presenting Melich with a box that delivered a hearty shock when a coil burst out as he opened it.[13]

It was all fun and games at the party, but, despite the good-natured public show, all was not sunshine within the Utex Corporation. Charlie was starting to regret his choice of partners.

He hadn't wanted to be president, at first. Past experience had taught him that he didn't have the patience to fuss with details of management and administration. He'd rather be "Chief Geologist." So he named O'Laurie as president, Hudson and Barrett as vice-presidents, himself as secretary-treasurer, with Rose as his assistant, and Melich as company attorney. Barrett was put in as mine superintendent.

Tempers started to fray during the summer of 1953. A wealthy Texan named Dandridge had made an offer of $5 million for the Mi Vida mine. He promised to pay $1 million down and pro-rate the balance out of future ore profits. O'Laurie and Barrett urged the sale. Charlie adamantly refused. His partners suggested that he had earned a rest and should take the cash and learn how to play.

"Hell, I'm having too much fun to play," Charlie said, and he turned down the deal. O'Laurie and Barrett reluctantly folded their wallets.

The Mi Vida kept on producing. A million dollars of ore by August, after only six months of operation. They were shipping two hundred tons of uranium to U.S. Vanadium daily. The ore was so pure that it assayed up to eighty-seven percent uranium.

Charlie and Mitch Melich decided that it was time to build their own processing mill. The few existing operations were having trouble keeping up with the AEC's demands. The mining boom was escalating. There were new discoveries, more shipments. Besides the Mi Vida, there were enough other working mines in the Big Indian District to support an independent mill in Moab. All they had to do was form a group, petition the AEC for permission to construct the plant and finance it. They figured the costs would run in the neighborhood of $10–15 million.

O'Laurie protested. He told Charlie and Mitch that Utex didn't have that kind of money. He didn't like the idea of bringing in other investors. The proposition was too rash.

Charlie was infuriated. He couldn't understand such a lack of imagination, such a conservative attitude in the face of obvious demand for the facility. Their discussions heated. Finally, Charlie flared. He got so mad that he kicked O'Laurie out of the office and changed the locks on the doors. O'Laurie hired a lawyer.

With O'Laurie out of the way, Charlie and Mitch figured it was time to get rid of "the bean farmer," too. Charlie had long disagreed with Barrett's theories of running a mine. He thought the operating costs were exorbitant. Barrett "didn't know beans" about mining. They'd just as well oust both directors and buy out their shares. They went to the mine to tell Barrett.

On the drive from Moab, Mitch convinced Charlie that there might be trouble if two explosive personalities such as his and Barrett's clashed. He persuaded Charlie to go down into the mine and talk to the fellows there. He would gather the rest of the men in the cookshack to break the news.

Barrett was not around when Mitch called the employees together in the mess hall.

"Charlie and I have had a meeting," he told them, "and we have removed Barrett and O'Laurie as officers of Utex Exploration Company. Barrett will no longer be manager of the mine. Charlie will be the new president."

"All of a sudden, I heard this commotion in the kitchen," Melich later said, "and here comes this great big Barrett and grabs me by the seat of my pants and shoves me out of the cookshack. Well, I wasn't going to precipitate a fight because I knew what we were getting into."[14]

"I was stone blind," Barrett remembered, "just waving my arms, and I started in the kitchen on the linoleum floor . . . and linoleum is slick, you

Loading the one-millionth-dollar load of ore at the
MiVida. Left to right: An unidentified worker, Mark,
Charlie Steen and Rose Shoemaker. *Western Mining
and Railroad Museum.*

know. I was like an old Tomcat trying to get around making a noise. . . . At
that time, when I swung at somebody, I didn't miss them. That son of a gun
was like a jumping-jack, down he'd go, and he did it about five times. . . .
Charlie fired me and fired the whole crew."[15]

Shortly after "the Battle of the Cookshack," on December 12, 1953,
Charlie and Rose paid O'Laurie and Barrett $3,272,500 for their shares of
Utex. A mere fourteen months earlier, O'Laurie had invested $15,000 and
Barrett $4,000 in an unproven corporation. Now their stock was worth
$175 a share.

"This settles an irreconcilable conflict of interests," Charlie said when
the sale was completed.[16]

Then, despite warnings that the AEC would never negotiate a buying
contract with him and that major financial avenues would be closed to a
small, independent miner, he gambled against the odds once more. As far as
Charlie Steen was concerned, nothing was going to get in his way of building a state-of-the-art uranium processing mill in Moab.

7 DIRTY HARRY

ANY APPREHENSION DR. ARTHUR BRUHN'S geology students had was relieved by the letter from the Nevada Test Site. For two years the youngsters had felt the earth shake and seen the pinkish clouds drifting over St. George after the nuclear detonations but the idea of getting close enough to actually see an explosion, even from a distance, was disconcerting. It was reassuring to initial the letter as indication that they had read the words affirming that there was "absolutely no danger" and the AEC welcomed Dr. Bruhn and his students "in furthering their education."[1]

Arthur Bruhn's method of teaching "all of the ologies" at Dixie Junior College was far from dull. Affectionately nicknamed "high pockets" for his high-riding trousers, he enlivened his classes with hand-made displays, slide presentations and even poetry. He worked with students on committees, took photographs for the yearbook, produced a school movie covering all of the year's activities, and made a tradition of closing the dances with everyone holding hands and singing "Just For Now," his favorite song written by fellow instructor Mrs. DiFiore.

Bruhn's field trips to explore scenic and interesting sites in the area were always popular. This chance to witness tests of atomic devices was an especially exciting prospect for the students. Their teacher welcomed the rare opportunity to give them lessons in physics, geology, meteorology, history and world affairs that reached far afield from the confining redrock walls of St. George.

The little Utah town was rather isolated in the early 1950s. The excitement of Las Vegas, approximately 150 miles to the west, could just as well have been thousands of miles away. Along Highway 91, a bleak landscape of sage and sand and a small mountain range cut by the Virgin River gorge separated the two communities. St. George boasted nothing like the cluster of towering buildings that looked so out of place in the flat desert of Nevada's gambling mecca. The tallest structure there was the spire of the Mormon temple that rose like a spear of taffy above unpretentious neighborhoods of small homes and businesses.

Ninety years had not done much to alter the little agricultural community since 1861, when Brigham Young, president of the Church of Jesus Christ of Latter-day Saints, sent 309 pioneer families south to "Dixie" to experiment in growing cotton. Cotton production was short-lived, but farming continued and the people remained to lead lives that were simple and focused. The ward (church meeting house) was still the religious and social center. The population grew to approximately 4,000 and boasted a preponderance of children, in accordance with church doctrine advocating propagation. Their tables were generously supplied with poultry and livestock they raised, milk fresh from cows in the backyard, and produce picked from their neat little gardens or "bottled" for storage in the pantry. St. George people were thrifty, friendly, law-abiding and intensely patriotic.

Up until 1953 the residents of St. George and other similar communities downwind of the test site showed little concern about radiation. The AEC had done their public relations well. Persons who had suffered damage to homes or businesses in the wake of a blast were compensated. Safety warnings to keep people away from the firing zone were posted and informational pamphlets were distributed. The public was commended for being partners in this patriotic operation. The word was not to worry, radiation was not a threat. Small doses of radiation over a period of time were no more worrisome than a dental X-ray, and they surely could not produce a cumulative condition. Even when a cloud of dangerously high radioactivity drifted as far as Salt Lake City in 1952, AEC officials assured the public that they could not "emphasize too strongly the absence of dangerous effects."[2]

But the commencement of Operation Upshot-Knothole on March 17, 1953 changed things. In that eleven-test series, thousands of members of the four armed services participated in an exercise named Desert Rock V. Divisions of servicemen witnessed the blasts from trenches at close range, and then attacked make-believe objectives approximately 2,000 meters from ground zero. A few volunteer officers stationed themselves closer to the blast to determine the minimum safe distance for observation. Helicopter units tested effects of over pressure following detonations and hovered in the path of shock waves. Aircraft conducted cloud sampling and tracking experiments and surveyed aerial radiation levels. Rats, mice, rabbits, dogs, sheep, pigs and monkeys were exposed to the blasts with observation of their medical condition afterwards. Buildings, mannequins, automobiles, military equipment and an assortment of fabrics and other materials were outfitted with radiation film badges and placed in display areas for post-shot evaluation.

Military personnel observing one of the bomb tests. *U.S. Department of Energy photograph.*

An Off-Site Monitoring Program, in cooperation with the U.S. Public Health Service, was also initiated. Monitors were billeted to seventeen stationary positions and four mobile units to sample air and land surfaces where fallout was expected to be heaviest. Their methods were painfully simple, however, relying on film badges, portable devices and a modified Electrolux vacuum cleaner. To alleviate growing unrest among those living downwind of the tests, monitoring personnel integrated themselves into the small communities by giving talks to church and civic organizations and visiting schools.

As part of the public relations program, the AEC invited media representatives to join VIPs in viewing the operations. Most visitors were both awed and horrified by the event. A few expressed concerns about safety. Images of Hiroshima and Nagaski were still fresh in their minds. After Shot Annie on March 17, 1953, one observer expressed dismay at "the amount of dust and debris present even two miles distant from ground zero."[3]

A week later residents living downwind of the site were alarmed at the local fallout noticed after Shot Nancy. Salt Lake City's *Deseret News* shared the concern and began publishing articles investigating possible dangers of radioactive clouds. To counter public uneasiness, the AEC distributed a new pamphlet assuring residents that "no community has been exposed to hazardous radiation." They contended that any detectable radiation was well below the accepted minimum standard.

But University of Utah professor Dr. Lyle E. Borst, a former director of the AEC's Brookhaven National Laboratory, had recently completed a study refuting the AEC's contention that cumulative exposures to small doses of radiation were no more harmful than dental X-rays.

Borst commented that when an atom bomb is dropped, "there will be people somewhere on the fringe for whom the difference between life and death will not be the amount of radiation from the bomb, but the amount of radiation they have previously absorbed."

The nuclear expert added, "I don't believe in taking chances. I would no more let my children be exposed to small amounts of radiation unnecessarily than I would let them take small doses of arsenic. The only time a person should expose himself to radiation is when there is a very definite and good reason."[4]

Borst's children were not allowed to play outdoors following a detonation and were required to bathe and wash hands very frequently.

Despite warnings from Borst and a few other concerned experts, the American public continued to be fascinated by the events. A month after Shot Nancy a prestigious assemblage of congressmen, generals, admirals, politicians and CEOs congregated in Las Vegas to witness Shot Simon. The excited participants got up in the middle of the night and were loaded on to buses for the seventy-five mile drive to a viewing area in sight of Yucca Flat.

Shot Simon was not only designed as a weapons test it was to be the largest Desert Rock V troop maneuver to date. Attempting to emulate realistic combat conditions, over 3,000 personnel, including attack troops, radiation monitors, observers in trenches, engineers, airplane and helicopter pilots and 400 persons conducting scientific experiments for the Department of Defense participated.

At 5:30 on the morning of April 25, Simon was detonated from a 300-foot tower just as the wind shifted direction. With a blazing flash the immense fireball rose into a towering mushroom cloud. Instantly a drone plane carrying measuring equipment flew too close to the explosion and disappeared "in a flash and a puff." A whirlwind of flying rocks and choking dust whipped over troops hunkering down in trenches, and sagebrush in the distance caught fire. A huge shock wave vibrated towards the tour buses, striking them with a loud bang, shattering all of the windows and pitching the VIP passengers backwards. Test site officials carrying Geiger counters rushed to check the dazed visitors then whisked them away with reassurances that they were in no danger.

Astounded scientists realized they had underestimated the power of the explosion. Shot Simon produced a nuclear yield of 43 kilotons, three times the size of the bomb on Hiroshima. Due to the sudden wind shift at the time of detonation, radiation at the trench area was much higher than expected. The monstrous fallout cloud produced drifted eastward over Mesquite and Bunkerville, Nevada, heading towards St. George.

Rumors spread quickly in the little Mormon town. Many of those who had gone to the hilltops to watch the flash saw the cloud closing in from the west. It looked bigger than usual. Then came word that for the first time in the history of nuclear testing, the test director had called for roadblocks along U.S. Highways 91 and 93. Cars were being stopped, checked for radiation and then sent to be washed at the closest service station.

Dennis Nelson was nine years old at the time. His life was typical of many small town boys in rural Utah: school, vacant lot baseball games, chores around the house and yard. He paid little heed to happenings at the test site, approximately 125 miles from St. George. But that day he couldn't understand why every-one was suddenly worried about dust and washing their cars. He and his brother, Alma, used to sleep out in the backyard under the mulberry tree on warm spring nights. If there was a test shot, it never occurred to them to wash the fine coating of dust off of the bedding on their old fold-up camping cot before they used it again.

"There was dust stirring all the time, little dust devils," he remembered. "The cars were

Dennis Nelson later became an anti-nuclear activist with his wife Denise.

always covered with dust. I remember they were washing cars in St. George as they came in [from the west]. They would stop the cars at the service station and the AEC had a contract with the station to wash these cars, and the water would run off into the irrigation ditches that went straight into our gardens. We had a watering turn once a week or every two days and it was my job to move the gate to deliver the water. I would put a gunny sack in there and the water would run into the garden. We had a pretty good sized garden and raised a lot of our own food."[5] Dennis couldn't understand why it was necessary to clean off the cars but still all right to let the dirty water drain into the irrigation system.

But Shot Simon had spread even further afield. The AEC later admitted, "Fairly large amounts of radioactive dust fell out over about half the United States after burst 7."[6] That was the half of it. A 7,000 square-mile radioactive rainout was centered on Albany, New York, and it was reported that small amounts of fallout were measured in Canada, Bermuda, Germany and Hiroshima and Nagasaki, Japan.[7]

A little more than three weeks later, on May 19, Dr. Bruhn and thirty students got up in the middle of the night for about an hour's drive to "the Summit," the local name for Utah Mountain. They were approximately half way between St. George and Las Vegas, and would only have to walk a short distance from the highway to get a good view of Shot Harry, the ninth in the Upshot-Knothole Series.

Dr. Arthur F. Bruhn. *Courtesy Kay MacDonald.*

Harry had been ill-fated from the start. The 32-kiloton device was scheduled to be fired from a 300-foot aluminum tower at Yucca Flat on May 2, but there was so much contamination in the wake of Shot Simon that it was postponed for two weeks, then twice more due to unfavorable winds.

Finally, on this dreary, overcast morning, the test manager gave the go ahead. Shot Harry exploded at 5:05 A.M. with such force that the rocks and soil beneath it, the tower supporting it and the cab container and casing holding it were vaporized. A violent updraft sucked up the loose debris and spat it out into the mushrooming cloud.

Winds, emanating from a series of fronts on the west coast, nudged the smoky vapor to 42,000 feet, with a base at 27,000 feet heading east to spread over Lake Mead and north to St. George and Cedar City.[8]

Up on Utah Mountain, the students were stunned. "You saw the flare and then you saw the cloud and you felt the ground shake," Joanne Workman remembered. "What happened was there was a breeze, a good wind blowing, and from it being a clear day, it over a period of time became, not misty, but the air was thick. It had all those dust particles in it. . . . When you're in a sandstorm in St. George you always get that red sand in your hair, but these were gray and black. . . . [I] remember when we used to have clinkers out of the bottom of the stove, those little hard gritty pieces of ash, [and] it was just like that, little black and white things. It was very bad—you know the wind blew it right on you."[9]

Dan Sheahan, and a handful of lead and silver miners, watched as the gray mass headed towards the Groom Mine on the edge of Bald Mountain thirty miles from the test site. "It appeared about as large as the April 25th shot," Sheahan wrote in his diary. "The high and middle part of the cloud went eastward toward Mesquite. The lower tail circled us clockwise from west to north. Fallout here at 6:15 to 6:19 A.M."[10] AEC monitors coming to warn them to stay indoors arrived too late.

At 7:00 A.M., the AEC issued a press release notifying drivers that the cloud was moving north and there might be fallout. Travelers were cautioned to close their car windows and air vents and get their vehicles washed at the nearest station. But downwinders were assured, "If there is fallout, it will not exceed the non-hazardous levels experienced after the April 25 shot [Simon]."

Despite the caveats, at 9:15 A.M. the monitor in St. George got a reading of 7 roentgens in the center of town. (The AEC maximum standard for radiation was 3.9 roentgens over a thirteen week period.)

At 9:30 A.M., Alvin C. Graves, the test director, notified the nearest radio station in Cedar City to warn people to remain indoors. The 10:15 A.M. broadcast was heard by few of the industrious Mormons, who had little time for listening to the wireless.

Over in Bunkerville, Nevada, another Mormon village on the Utah-Nevada border, twelve-year-old Gloria Gregerson was playing under the wide-leafed oleander tree in the back yard.

". . . the fallout was so thick it was like snow," she later remembered. "[We] liked to play under the trees and shake this fallout onto our heads and our bodies, thinking that we were playing in the snow. . . . I remember

Shot Harry, part of Operation Upshot-Knothole, was a 32-kiloton weapons-related device fired from a tower. *U.S. Department of Energy photograph.*

writing my name on the car because the fallout dust was so thick. Then I would go home and eat. If my mother caught me as a young child, I would wash my hands, if not, I would eat with the fallout on my hands."[11]

But more strange things than unusual fallout clouds began to happen on the heels of Shot Harry.

By around 10:00 A.M. when Ms. Workman got home from her field trip with Dr. Bruhn, her face, arms and neck were scarlet and she "felt like my eyes even were burned." When she tried to comb the "clinkers" out of her hair, "the teeth of the comb caught under the skin on my scalp and it lifted my skin right off my skull. . . . I had to lift it up and peel it—and lift it off of my head, in big strips, similar to like a sunburn peels—it just came off in a funny way."[12]

The nineteen-year-old girl continued to lose her hair until she was finally forced to wear a wig for the rest of her life.

Cedar City resident Janet Gordon remembers "Dirty Harry," too. Her brother, Kent, was helping his father and five others herd sheep back from the Nevada winter range to their ranch for lambing when the radioactive cloud descended on them.

"He said it was like a ground fog in the low bushy areas," Janet said. "When he'd ride into it the horses would start to sweat and stumble and could hardly go along. He said he could feel it on his skin; it was almost sticky and it had a kind of taste."

The fallout burned Kent's skin and he suffered terrible headache, nausea and diarrhea. The sheep developed sores around their muzzles and wool fell out in clumps. His horse died a few months later.[13]

The AEC still insisted that "Fallout does not constitute a serious hazard outside the test site."

But complaints continued to surface. On June 12, Dr. Gordon Dunning, head of the Biophysics Branch of the Division of Biology and Medicine, issued the memo, "Data on Alleged Radiation Sickness Attributed to Atomic Tests." The document listed twelve new cases attributed to radiation sickness.

AEC public relations activities were accelerated. A videotape production was quickly produced with technicolor images illustrating how in St. George, "this thriving community went about its business, youngsters on their way to school, housewives starting on the chores of the day, merchants opening up their stores and folks at home listening to the radio program [cautioning residents to stay indoors for about an hour because the wind had shifted directions during detonation of a bomb."]

"There is no danger," the commentator insisted. "This is simply routine Atomic Energy Commission safety procedure."

But later in the tape there is an admission that there "may be an area of real danger that may extend a few miles from Ground Zero. . . . Yes, livestock grazing within a few miles of the site of detonation have in a few instances suffered skin and eye injuries from radiation, but otherwise were in good health. Justified claims by owners have been compensated."

As for St. George, "the amount of radiation deposited . . . was far from hazardous," the film concluded.

It was later established that Shot Harry contributed the most off-site contamination of any nuclear explosion at the Nevada Test Site.

8 THE BURDEN OF PROOF

"THE ONLY WAY WE CAN ever get anything done is to collect bodies and lay them on somebody's doorstep."[1]

Duncan Holaday made the statement in exasperation. It was December 1953. Holaday was speaking at the fifth meeting of the Uranium Study Advisory Committee in Salt Lake City. Another prestigious group of doctors and scientists had gathered. Duncan's colleagues from the Public Health Service were there, along with representatives from the National Cancer Institute, the National Institutes of Health, and the AEC Division of Biology and Medicine.

Three years had passed since the survey on the effects of radon in the mines had been formally launched. The field station and the Colorado Health Department had defined the nature and extent of the environmental problems in the mines and suggested methods to control the radiation hazards. Initially, they had thought this would be enough. That was how they handled most of their projects. But in the case of mine radiation, the usual Public Health Service course of action—"Implore, exhort and attempt to educate"—wasn't working. High radiation counts in the mines remained.

"It appears the condition will continue until evidence of injury to the miners can be developed," Holaday told the group. He had given the matter a lot of thought. He remembered Dr. Louis Cralley's exhortation after the Washington meeting in 1951.

"This is one problem we can't walk away from," Cralley had said. "See if you can't get some controls in there."

Holaday had tried to do just that for three frustrating years. Two months after the 1951 conference, he, Cralley and Paul Jacoe visited the principal mining companies in the Plateau. They apprised the mine managers of the results of the environmental surveys. They explained the findings about radon daughters. They stressed the need for additional ventilation. They pointed out that the uranium industry was expanding rapidly, thus increasing the number of men exposed and the quantity of their total exposure.

Holaday hoped that these face-to-face meetings would clear up any confusion on the part of the mine operators. A year earlier, in 1950, he had been alarmed at VCA manager Denny Viles' misconception about the radiation hazard.

"I went to talk to Denny Viles [at Vanadium Corporation]," Holaday recalled. "He was curious. He said he had heard at a meeting from a presentation by an AEC official [Jesse Johnson, director of Raw Materials Procurement Division] . . . that uranium was so weakly radioactive that it presented no hazard in the uranium mines.

"I told him Mr. Johnson was exactly correct. Uranium was so weakly radioactive that it had no particular hazard. 'Have I ever suggested to you that uranium was a hazard in the mine? The only things I've talked to you about are radium and radon.' He looked at me. I said, 'These gentlemen select their words with great care. They don't expect them to mean anything but what they say.' And I let it go at that."[2]

Holaday telephoned Viles as soon as he returned from the 1951 Washington conference. He told Viles that other experts in the health physics business agreed with him that the situation in the mines was extremely hazardous. He hoped this would move Viles to accelerate his efforts at ventilation. But when he rechecked the Freedom Number 2 mine at Marysvale for radiation a few months later, he was appalled.

". . . they were using a ventilation duct to bring in outside air to various parts of the mine," Holaday said. "It was a small duct about eight inches in diameter. . . . I considered the ventilation to be quite inadequate by any metal mine ventilation standards.

"I took a sample in one area where the ventilation tubing had been stretched in and no one was working," he went on. ". . . about an hour afterwards [I] was unable to get a reading because there was so much activity on the filter paper it had jammed the instrument. It went over the times-100 scale; it went over the top."[3]

During the summer of 1951, the study continued. Two doctors and a technician resumed the physical examinations of miners and mill workers. Holaday's environmental surveys were made. Dr. Bale conducted experiments to determine lung retention of radon daughters. Tests were also made of radon and radon daughter counts in non-uranium mines. These operations had been working for many years without causing an incidence of lung cancer among their workers. Perhaps a determination of their background radiation concentrations could provide some answers.

Duncan Holaday and his teams noted diesel and gasoline-powered equipment underground when studying ventilation problems.

By August, Holaday and his colleague, Howard Kusnetz, had developed a field method of measuring radon daughters based on earlier experiments by Dr. John Harley, head of the chemical laboratory of the New York AEC Health and Safety Division. Radon daughter products were collected on filter paper by drawing mine air through a hand-operated suction pump. The filter paper was removed after each sampling and placed in a glassine envelope for protection. In a period not less than forty minutes or more than ninety minutes from the time of sampling, the papers were taken outside of the mine. There they were measured under a rugged, dry cell battery-powered Juno radiation survey meter that was sensitive to alpha rays. The lightweight unit, which had been calibrated, was put over the filament paper to read the alpha-, beta- and gamma-ray activity of the membrane filters through which known volumes of mine air had been drawn. The system was so simple that the mines could make their own inspections without depending upon the infrequent visits of state or federal mine inspectors. The field station conducted two-day training courses to instruct company employees in the procedure and calibrate their meters.

To make measurements even easier, Holaday devised a new terminology as well. Rather than reading the number of picocuries of radon per liter of air, the process of determining "working levels" of radon daughters was adopted. Measurements were taken on the potential alpha energy, the particles that do damage to tissue, in that liter of air. One working level was the equivalent of 100 picocuries of radon per liter of air. A man's exposure to the radiation was gauged by "working level months," the unit of concentration for a given time. This was the significant factor; how long the miner had worked at that level of radiation.

"We abandoned this picocuries per liter business," Holaday said later, "because that didn't make so much sense, talking to a miner or mine operator and saying, 'Well, you have got so many picocuries per liter in there.' Because your answer would be, 'Does that mean I'm not in good shape?' So we decided to pick the 100 picocuries per liter, the total potential alpha energy that you would get from that [radon disintegration through Radium D, lead], and we would call that 'one' [working level.] Because, then you could tell a miner, you are at five and one is the number you are looking for, or you could tell him you are at .5 and you are in good shape. . . . We just say, for the time being, until we have sufficient data to establish a real solid number, we will use this as a working level. . . ."

By May 1952, Holaday felt there was enough information to suggest remedies for the radiation problem. In cooperation with Dr. Wilfred D. David and Henry Doyle, he wrote an Interim Report for the Uranium Study Advisory Committee. The review described the history of the German and Czechoslovakian mines and the progress of the uranium mine study. It warned of the danger in radiation levels measured on the Colorado Plateau—forty-eight mines showed a median radiation level of 3,100 picocuries of radon gas per liter of air with a maximum reading of 82,800 picocuries. In eighteen mines, the median level shot up to 4,000 picocuries of radon daughters, with a maximum count of 120,000 picocuries. These figures were compared with the 1,500 picocuries that had claimed approximately seventy percent of the European workers from lung cancer.

The report presented methods for correcting any potentially harmful conditions. These included ventilation to maintain one working level (100 picocuries) of Radium A and Radium C' at any work location; wet drilling and other dust suppression methods; use of respirators; better personal hygiene practices among the miners; pre-employment history and periodic medical examinations of the employees. All of the things that Holaday and his colleagues had stressed again and again.

Approximately two thousand copies of the report were disseminated among all of the "people who should be able to do something about it"—government agencies, the U. S. Bureau of Mines, AEC, state bureaus of mines, state health agencies and the mining companies. It was not distributed to the miners. Nor was there follow-up to discuss the findings with the people notified. The field station had no staff or money to call subsequent meetings to review the report. Presumably, the Interim Report gathered dust in the files.

Duncan Holaday's only way to assess the response of the mining companies to the report was by the conditions he found in the mines following its receipt.

". . . our conclusions were that some improvement had been made in some mines," Holaday said later, "that there was no particular push to get the situation in the mines under real control. Some had improved, but . . . more mines were coming on stream all the time, [and] . . . the State Bureau of Mines or State Industrial Commissions, whichever one had jurisdiction, were not about to really use their enforcement authority to require compliance with the recommendation. In other words, that nothing much was going to be done by either federal or state agencies to insist on control of conditions."[4]

As for the AEC, it was cautious. The agency didn't want interim reports of a health study to get in the way of uranium production. Jesse Johnson reacted to the findings of the Interim Report in a memo to the Atomic Energy Commission.

"The Division of Raw Materials is not qualified to pass judgment upon the contents of the report or the conclusions reached therein," he wrote, "but we do feel that the United States Public Health Service is to be commended on its sane and objective treatment of the problem. We believe that the facts should be made public but that speculations or premature and distorted inferences should be avoided. We believe that the health studies should be continued and that the Atomic Energy Commission should lend full support to the United States Public Health Service."

Further expressing his concern over publicity, Johnson wrote, ". . . the report might become the basis for press and magazine stories which could adversely affect uranium production in this country and abroad. Although we understand the report is to have only limited distribution, it is unclassified and it is to be expected that at least some newspapers and magazines will obtain access to it. . . . There is no doubt but that we are faced with a problem which, if not handled properly, could adversely affect our uranium supply."

Johnson concluded by intimating that "the possibility exists that communist propagandists may utilize any sensational statements or news reports to hamper or restrict uranium production in foreign fields, particularly at Shinkolobwe."[5] (United States contracts for uranium imports from the Belgian Congo were still in effect.)

Pairs of advance men gathered research on approximately ninety percent of the mines. *Victor E. Archer.*

Holaday and his colleagues realized that the only thing they could do was continue the study on a long-range basis. The "gag rule" remained in effect. Nothing could be said to warn the miners of their impending peril. If the health specialists called an alarm, the mining companies would close the doors and further research would be prevented. They couldn't alert the press to stir public interest, or the federal bureaucracy would be on their backs. Their only hope was to collect enough solid data to convince mine operators, enforcement officials and legislators of the problem's urgency.

But money to finance the survey was a problem. The field station had been unable to obtain funds to increase its support of the project. State legislatures weren't about to underwrite additional efforts by state agencies. Dr. Cleere, executive secretary of the Colorado State Health Department, decided that the only hope was to request funding from the AEC Division of Biology and Medicine. The division responded with a grant of $25,000.

Holaday and his teams spent the summer of 1953 experimenting with ventilation. Three two-man crews covered approximately ninety percent of the operating mines. The researchers gathered data concerning ownership of the mines, locations, and number of persons employed underground. Then they toured the workings, sketching the mine layouts and ventilation installations.

The mines differed in configuration, the plans being dictated by placement of ore deposits. Flat-bedded deposits were mined through horizontal drifts, usually starting with a short entry tunnelled into the side of a hill or cliff. Some workings had incline shafts which varied in steepness from a few degrees to the vertical. The Temple Mountain mines near Vernon Pick's discovery often originated with a 38-inch diameter vertical Calyx shaft bored through solid rock. The Marysvale District's ore was usually found in vertical fissures. Mining was conducted through inclined and vertical shafts reaching to various horizontal levels beneath the surface. Charlie Steen's Mi Vida used the modified room-and-pillar method of mining.

As the team of engineers walked through the mines, they noted any gasoline or diesel-powered equipment being used underground. They collected dust and gas samples at each working face, in main haulageways and at a few non-working areas or deadend drifts. Ventilation measurements were taken at each sampling station. Carbon dioxide tests were made.

The survey discovered that the majority of mines had inadequate ventilation. Some forced air down eight-inch diameter chum drill holes. The air didn't reach the working face effectively. Most mines had not established any ventilation raises (elevated passages) to provide air courses.

It was also realized that temperature variations between the underground and outside air were a factor. Natural air currents in the cold winter would be greater and reduce the radon somewhat. However, natural ventilation was not sufficient, even in winter, to bring radiation levels to acceptable concentrations. Yet the miners objected to the increased chilling that would result from mechanical ventilators.

"Attempting to drill a face of ore on a cold winter day, the miner finds the jack hammer 'leg' slipping on the ice that now coats the floor of the mine," Gary Shumway wrote in his dissertation, "A History of the Uranium Industry on the Colorado Plateau." "If he for a moment stops the flow of water through the hammer, the water line will freeze solid, making it necessary to stop work to thaw it out. With his clothes splattered with drilling muck, freezing stiff, and his hands numbing, the miner begins to view his alternatives as a choice between freezing to death that day or risking the possibility of lung cancer at some vague time in the future."[6]

Nevertheless, Holaday insisted that mechanical ventilation was a necessity, and should be operated thirty minutes prior to entry of miners and throughout their shifts. Natural ventilation should be fully utilized as an adjunct. He and his associates devised rules and procedures for introducing enough air into the mine headings to lower radiation levels to those recommended by the health service.

The safety engineers stressed that mine layouts, including ventilation raises, should be designed to establish air circulation throughout the underground openings. There should be change rooms and showers available to the workers. Eating and storing food underground should be prohibited.

When the survey was completed, the AEC Division of Biology and Medicine published a report.

"The Biology and Medicine boys thought it was a very excellent report," Duncan said. "So they proceeded to print it up and hand it out to anybody that wanted it, and that is about as far as it got 'til several years later."[7]

The trouble was that only a mine-ventilation engineer could understand it, and most of the mining companies didn't have such an expert.

An exception was Anaconda Mining Company. That corporation, primarily interested in copper, was getting into the uranium business. They developed some mines near Grants, New Mexico, and had the Jackpile Mine on the Indian Reservation at Laguna Pueblo.

Jack Warren, deputy to Anaconda's chief ventilation officer, attended Holaday's ventilation courses at the field station.

". . . when they [Anaconda] found the Jackpile," Holaday said later, "they went in at the edge of the mesa and drove an exploration drift in to determine where the bottom of the ore body was and how far it extended. They had nothing but a developing drift going in and they had a thirty inch vent tube in there blowing air into the place. You couldn't find more than a tenth of a working level any place.

"So, they just carried the vent tube right in along with them. And when we found out the situation [high radon count], they closed up. 'We ain't going to try . . . can't be done,' [they said]. They took off 350 feet of overburden to open pit it."[8] (An open pit mine is exposed to the surface air and, therefore, a radiation problem does not exist.)

John Warren empathized with Holaday's frustration over the lack of cooperation by other mining interests.

"His board of directors were getting tired of seeing most of their competitors get away without spending one nickel for health protection," Duncan said later.

"You know who the bastards are in this business," Warren said to Holaday. "Why don't you get after them?"

"Brother," Holaday replied, "if I had anything but a feather duster, I'd be after them."[9]

Holaday knew "it takes a regulatory agency to pick the boys who are out in left field and make a good example out of them. . . . those who want to fall in line can, without losing their shirts, because they know their competitors are going to have to toe the line, too."

Denny Viles of Vanadium Corporation said as much in a letter to the U.S. Bureau of Mines. "Any ruling they may set up which is possible to live under whereby all operators will operate under the same conditions, we will be only too pleased to do our part."[10]

At the December 1953 Uranium Study Advisory Committee conference, Duncan Holaday reported that the environmental study had gone as far as it could go. The radiation problem had been clearly defined. Ventilation methods to control the hazard had been developed and explained. There was no more that the Public Health Service could do.

But the medical examination program was a different matter. The project as it had been conducted was a hit-and-miss effort that picked up a hundred or so miners here and there. It wasn't producing definitive data. It had been anticipated that the same miners would be re-examined when medical teams visited areas where they had been previously. However, miners were migratory. They moved from job to job. Few repeat examinations were possible. What's more, the very terrain of the Colorado Plateau made it difficult to reach, or find, all of the workers. The fact that the majority of miners had had three years or less of working experience (not enough time for cancer to become evident) also compounded the problem.

The conferees agreed that a major epidemiologic study was essential. An extensive cohort of miners must be examined and their cases followed for an indefinite number of years in order to ascertain their basic state of health. In addition, a survey of inactive and retired miners, who might have quit their jobs due to health reasons, should be included. Essentially, a census of the entire Colorado Plateau uranium mining industry was necessary.

In early 1954, the National Cancer Institute agreed to assume primary responsibility for organizing and conducting the expanded medical study. Pope Lawrence, chief of the Field Investigations and Demonstrations Branch, coordinated the program. Dr. James Egan was transferred to Salt Lake City to be in charge of the epidemiologic segment. The AEC's Grand Junction office, Public Health Service, field station, and Colorado Health Department cooperated.

It was an expanded, modern-day version of the old country doctor. Two pairs of advance men drove jeep station wagons into the backcountry to locate the mines. At each working operation, the organizers got lists from the foreman of all of the present employees and those who had left within the past year. Then they talked to the miners themselves. They told them about the study that would "be used only for statistical purposes" and said that the workers could get free medical examinations if they participated in the program. An agreed-upon time for each physical checkup was set to coincide with the schedule of an examining team, which would follow in several days.

Two separate medical parties operated from self-sustaining field camps, which were established and maintained by the Grand Junction AEC office and located as close to the mines as possible. On the appointed days, the miners arrived at the camps for their examinations. They came in four-wheel drive vehicles, trucks, run-down jalopies and sleek new cars. Some brought their wives and children along for the ride. Others came directly from work.

The first stop was the X-ray truck, furnished by the U.S. Public Health Service Tuberculosis Division. The miners stripped to the waist, pressed chests to the plate, and held their breath for the picture. Then they proceeded to the Public Health Service laboratory trailer. There a clinical technician collected samples of blood and urine. The last stop was the examining trailer, where a physician completed the miner's medical and occupational history and gave a physical exam. The miners were told that the findings would be confidential. However, should any abnormalities be discovered, personal physicians would be notified.

"If the initial census of miners in the field activities of this summer yields significant numbers," said a project memo, "then the group examined will be set up as a 'population at risk' to be followed for ten to twenty years by every means available.

"The success of this project depends upon the acquisition of future annual census data to reveal those individuals who disappear from the original population group in order that they may be traced." This could be accomplished through tracing death certificates, motor vehicle and state tax records, contacting physicians, hospitals, and the Veterans' Administration, and collecting published obituaries. It would be a death watch.[11]

The intensified medical study was completed during the summer of 1954. Approximately 1,400 uranium miners—white, Indian and Mexican—throughout the 125,000 square-mile Colorado Plateau were examined.

But there were unavoidable problems that limited the efficiency of the data. Extremely high environmental temperatures contributed to accelerated sedimentation rates in many of the lab tests. Shipment of blood samples to more sophisticated laboratories was often delayed. Sometimes the power output of the generator was too low for proper operation of the centrifuge, thus indicating falsely high hematocrit readings. The laboratory trailer tilted whenever anyone entered or exited, having an adverse effect upon the erythrocyte (red blood cell) sedimentation rates.

Illiteracy and lack of education of many of the patients also affected the report. Often the miners misspelled their names, forgot social security or Veterans' Administration numbers, or were unable to give permanent addresses or names of next of kin. Many couldn't accurately recall information for their histories. As for notification of personal physicians in case of abnormal findings from the X-rays or examinations, few of the miners regularly went to a medical doctor. Many patronized osteopaths, chiropractors or naturopaths who were not likely to be professionals in treatment of cancer or other organic disease.

But the epidemiologic study finally caught the attention of government leaders. On October 5, 1954, Utah Governor J. Bracken Lee addressed a letter to Lewis L. Strauss, chairman of the AEC.

"The increase in uranium mining activities in this area and the method of operation of these mines have resulted in the exposure of the workers to large amounts of radioactive dust and gas," Lee wrote.

"Medical examinations of the workers have shown that up to now there have been no apparent ill effects on their health as a result of exposure to radiation. The effects of chronic exposure to radiation are long in developing and probably will require many years before they will become apparent. We cannot wait until evidence of damage develops before taking measures to control the situation.

". . . However, we do not wish to promulgate regulations unnecessarily restrictive," Lee went on. "We realize that extensive control measures will be beyond the economic capacity of some mines, and we would not care to see any mines shut down needlessly."

Lee concluded by requesting a meeting of interested parties involved in uranium mining throughout the West. The Seven-State Uranium Mining Conference on Health Hazards would convene to formulate working standards for radioactive dust and gas in the uranium mines and recommend control measures.

On February 22 and 23, 1955, the conference was convened in the marble halls of Hotel Utah. For the first time, the problem was tossed in the ring. There were AEC bigwigs from Washington and New York. Health enforcement officials came from Utah, Arizona, Colorado, New Mexico, Idaho, South Dakota and Wyoming. Operators of large mining companies attended along with single "doghole" miners. Medical educators, attorneys and mining engineers were there. The public was invited to attend the meetings. Proceedings were open to the press.

"We hope that by the end of the second day, we will have agreed on a realistic standard maximum allowable concentration of radon and its degradation products," said Chairman Otto Wiesley, head of the Utah Industrial Commission. ". . . we hope that. . . . we'll be able to appoint a committee of enforcement officers and operators. . . ."

Every effort was made by the professionals to reach their audience. Experts forsook technical language for lay terms. The complexities of radon and its decaying daughters were simplified so that everyone would understand.

"I could lay on my wrist a chunk of an element that was emitting alpha particles and, in general, I wouldn't get hurt because the horny layer of my skin would stop the particles," Dr. John Z. Bowers, dean of the University of Utah Medical School, explained. "However, if an element which is emitting alpha particles gains entry into my body either by inhalation, through a laceration, or by injection into my blood stream, then I get hurt."

Bowers compared the acceptable standards for radiation to taking aspirin.

"For example," he said, "we will take three aspirin tablets to take care of a headache. . . . If we give the same person thirty aspirin tablets, he will get a violently upset stomach. . . . There are amounts that are safe, there are amounts that are harmful. . . . The problems of radiation, radioactive elements, radon, and so forth, are no different from those of any other toxic agent."

Then came the stronger medicine. Health and safety professionals discussed the matter of ventilation—the only method of controlling this radiation. Uranium mines are different from conventional hard-rock mining operations, the audience was told. While the generation of silica dust in many mines is a problem only during the working shift underground, the emanation of radon and its decay daughters is continual. If ventilation is turned off, concentrations may rise to fantastic levels. And if miners are allowed to work under those conditions, they are susceptible to lung cancer.

Attorney Paul Cannon added a further dilemma. Development of lung cancer by a miner was within the occupational disease law, he told the group.

"That law is very strict with regard to proving that the cause was in the mine and so on . . ." he said, "but if the experts can testify that a man picks this up from a uranium mine, we're liable for it now as an occupational disease."

"You will never know when a case might be filed against you as an operator," Otto Wiesley added. "It might be twenty years from now, even more, and you will be liable."

Wiesley explained that a Supreme Court ruling for occupational diseases declared that the statute of limitations begins after the disease is diagnosed and the probable source determined.

"That simply means, gentlemen," Wiesley continued, "that if fifty years from now a miner developed cancer of the lung, he wouldn't have to file an application until he knew that he had the disease, by name, until he knew where he got it . . . or until he knew he was disabled."

As the picture grew greyer, the meeting's goal seemed further away than ever. Everyone agreed that a standard for radiation was desirable, even necessary. The discussion kept coming back to levels of one hundred picocuries of radon and three hundred picocuries of the daughters. But talking about it and achieving it were two different matters. The mining companies were concerned about the cost of mechanical ventilation, as described by Duncan Holaday and other safety engineers. The price could range from a fraction of a cent per ton to nearly one dollar.

"The Bureau of Public Health has been giving examinations for a number of years, and has yet to come up with ill effects," said Virgil Bilyeu, superintendent of Utex Exploration Company. "Therefore, we don't want to set a standard that would be impossible to meet. . . . Yesterday we talked thousands of cfm for ventilation of the mine. Now, we have done some tests at Utex and found that in a 15 x 20 heading with fans blowing three or four thousand cfm, our average radon concentration was around 520 picocuries per liter. I feel that a hundred would be impossible to meet."

Matt Rowe, operator of the small La Salle Uranium Company agreed.

"We have stopes [tunnels] at different levels," he said. "We don't have all this nice big ore like Utex has and like Joe Cooper and some of the rest of you. We have a lot of small places to work and they certainly aren't where you can drive a nice straight drift. You do a lot of work on your hands and knees, a lot of slushing in small areas, a lot of very difficult places to ventilate. We could not possibly change the air every four minutes. We could not possibly get down to one hundred picocuries. It is just impossible. . . . We want to do all that we possibly can. But I want to tell you right now that Anaconda with all its resources couldn't get down to that one hundred picocuries figure that you're quoting here."

Nevertheless, it was decided. The conference voted that minimum standards for a safe working level would be one hundred picocuries of radon and three hundred picocuries of decaying daughters.

"There comes a time when you've got to break off the theory and get down to actual everyday mining," Otto Wiesley said. "We're going to set up the standard that we're going to shoot at even though we can't attain it."[12]

And that's just what they did.

9 "THE FUTURE OF AMERICA"

JAY WALTERS, JR., wasn't the type of man you'd expect to find in Jack Coombs' address book. Coombs, a Sigma Chi at the University of Utah, had graduated in 1950 with a B.S. in Business, after which he began his career with J.A. Hogle and Company, Salt Lake City's oldest and most prestigious brokerage house. One year later, Coombs left Hogle's to join the ultra-conservative Harrison S. Brothers firm as a partner. The young man had an air of honesty and dependability. He moved in Utah's better circles. With his dark wavy hair, blue eyes and athletic build, he was popular with the young-married country club set.

Walters, on the other hand, was sixtyish, had a swarthy complexion, a bay-window and thinning strands of straight, greying hair. An elk's tooth dangling from his watch chain rivaled anything he had in his mouth. He favored seedy double-breasted suits with suspenders and cowboy boots. His conversation was laced with obscenities. His wallet was usually empty. Walters was a product of the old days when brokers formed mining companies on the floor of the stock exchange, raised the money and then went out and broke rock themselves.

The only thing the two men had in common was a genius for salesmanship. They could sell you a stringless yoyo. Whether you needed it or not.

Coombs first met Walters in August 1953. He was wrapping up details for a manual of over-the-counter securities sold in the Intermountain Region of the western United States. The booklet listed issues not traded on the stock exchange: U.S. government securities, municipal bonds, bank and insurance company stocks, utilities, shares in large corporations and small "growth" companies. Since there was no registered over-the-counter market, securities dealers negotiated with one another for the price of these offerings via a battery of direct telephone lines and teletypes that connected their offices. They would ask for current quotes on a particular stock, dig out the scuttlebutt on what their competitors were doing with it, and then decide if they were going to hold firm or raise or lower the ante. Sort of like a poker game.

Jack Coombs needed to get information about a mining company Walters headed named Cedar Talisman. He wanted the date and state of incorporation, the price range of shares issued, properties owned by the company, and names of the directors. When he bumped into Walters on Main Street one day, he arranged a meeting to get the disclosure facts. But soon after they got together, Walters wanted to talk about something else: uranium.

"As the conversation progressed, he brought in the subject of uranium," Coombs said later. "I had heard of the mineral in Chemistry class at the University, but I didn't know anything about its use. Walters said uranium would be a new industry. It was now an infant. Investment circles didn't know about it yet, but it offered fabulous opportunities. Walters envisioned an atomic industry."[1]

"Uranium is the nucleus of tomorrow," Walters had said. "This uranium business represents the industry of the future and is going to change the habits and the way people live throughout the world."[2]

He said he couldn't understand why the local brokerage houses were so passive about this promising investment field right in their own backyard.

Walters had a nose for a good promotion. When he heard about Charlie Steen's strike, he left his home in Arizona and high-tailed it to Moab to buy a few uranium claims. Then he drove up to Salt Lake City, where he booked a room at the Newhouse Hotel on Main Street and Fourth South, half a block away from the Salt Lake Stock Exchange. He started checking the offices of local stock firms to see what they were doing with uranium. They were doing nothing.

"Christ! I can't understand it," he told Coombs. "Uranium is the future of America!"

Walters' eyes flashed as he spoke of the industry's potential. The United States was desperate for a domestic source of uranium. So desperate, in fact, that the federal government was willing to pay for every bit of ore that could be produced. The AEC promised $10,000 bonuses for new lodes of high-grade. They paid up to $50 per ton on .3 percent ore. They pitched in six cents per ton for haulage costs, built roads, constructed mills, offered the services of geologists. And Uncle Sam was the only customer. The miners had a guaranteed market for their product. This was the only deal of its kind in our history.

Granted, there were drawbacks. The majority of mining operations were one or two-man "doghole'" mines, that were little more than holes in the ground, The spotty nature of uranium deposits made it necessary to drill several prospects to locate the ore bodies. Exploratory drilling fees were so high that most ventures needed thousands of dollars to prove their claims. Big mining companies were not interested in backing the little guys and the independents couldn't hack it on their own.

This was where the investors could come in. Why not raise the capital by issuing over-the-counter penny stocks? Penny stocks were within the reach of almost everyone. Three million shares at one cent each added up to $30,000, enough to finance a lot of the Colorado Plateau ventures.

Jack was only too familiar with the ultra conservative investment scene in Utah. The Salt Lake Stock Exchange, spawned primarily from the state's silver, lead, and copper industries, had ninety mining companies listed after seventy-five years of doing business. Utah's capital city had a mere handful of stock brokers. There had only been two or three penny stock offerings. Jack had been involved in those.

"The first issue I ever did was John Morgan, Jr.'s Uinta-Wyoming Gas and Oil," he said later. "I ran into him one day in front of the Walker Bank building and he wanted to raise about $50,000. I said, 'Sure, I'll do it.' So I called up the guys in 25 Associates and Coombs Associates [investment clubs composed of a group of Jack's buddies]. I sold about $20,000 worth of stock and John Morgan about thirty thousand. Then Harry Brothers said maybe it should be registered with the Securities and Exchange Commission."[3]

Utah's securities laws were very liberal. A statute enacted in 1925 adopted a state act by which the Utah Securities Commission theoretically operated on the basis of "we'll look at a deal to see if we think it's any good and then let you sell it. We won't 'approve' it, but you can sell it."[4]

There were three methods of underwriting new corporate enterprises in Utah. An intrastate offering could be made when a group of Utahns needing a relatively small amount of money raised funds entirely through sale of stocks to state residents. The issue was under the surveillance of the Utah Securities Commission. No formal filing with the SEC was required.

A "short form" filing with the SEC under Regulation A had to be made when $300,000, or less, was raised with some sales in other states. A full SEC registration was needed for interstate sales exceeding $300,000.

Jack checked with his lawyer, Bob Cranmer, and found that Uinta-Wyoming was subject to the SEC's Regulation A because some of the shares had been sold outside of Utah.

"So we had to offer everyone their money back and make this offer in recision and get an offering circular printed," Jack said. "We sold the stock for a penny in February. In August, we got the issues cleared and started trading at eight cents. This was the first issue I ever knew of that had been filed with the SEC."[5]

Jack had enjoyed his initial bout with wheeling and dealing. Uinta-Wyoming had been so successful that he followed the same pattern later with Justheim Petroleum and Morgan Oil. Walters' suggestion that he get in on the ground floor with this new uranium industry intrigued him.

Walters told Jack about his claims in Moab. He wanted to form a little company and drill the properties. He had talked to Sam Bernstein, a lawyer, about setting up a corporation called the Uranium Oil and Trading Company. All he needed was $50,000. He asked Jack if he would raise the cash.

Jack was tempted. It wasn't a hell of a lot of money and it would be fun to see if this uranium thing would turn out to be the latter-day "gold rush" that Walters prophesied. But he could visualize Harry Brothers' reaction to such an offbeat proposition. Harry was a good securities analyst. A conservative one. He'd think his partner had lost his senses by tackling a penny uranium stock.

"Talk to Lyman Cromer [president of the Salt Lake Stock Exchange] and Earl Havenor [a reputable broker nearing retirement]," Walters said. "They'll tell you I won't go south with the money."

"I'll give it a go," Jack said. "But I'll have to find someplace to operate away from Brothers' office."

"Great," Walters said as Jack started to leave. "And by the way, could you spare a twenty for my hotel bill?"[6]

Frank Whitney knew that it was not his superior coffee that attracted the swarms of people clutching money in their hands. When he arrived at the coffee shop on the fourth floor of the Continental Bank building that Monday morning in August 1953, he had to plough through the bodies to unlock the door. At six A.M. they filled the hallway all the way to the elevator.

The previous Saturday afternoon, Frank and his brother, Dick, had met their friend Jack Coombs for a drink. Jack was all excited about his conversation with Jay Walters, Jr. He repeated Walters' dramatic predictions about a coming "uranium boom," and told them that he had agreed to sell penny shares for one of Walters' promotions. The only problem was that his partner, Harry Brothers, refused to be associated with a character like Walters. Jack had to find someplace else to peddle his stocks.

"Use our place," Frank said.

Jack thought it over for a few minutes. Maybe it would work. He could spread the word and handle the paperwork at the office. All Frank had to do was write out orders and collect checks in the coffee shop. It was as simple as that.

"It will be kind of a lark," Jack said, warming to the idea. "No one will get hurt. Nobody would lose more than twenty or thirty dollars. It will be something to talk and laugh about."

So it was decided. Jack would let Frank sell the Uranium Oil and Trading stock *literally* over-the-counter.

Coombs spent the weekend on the telephone. He called all of the 25 Associates and Coombs Associates he could reach. He told them that for a few bucks, they could get in on a new bonanza with penny uranium stocks. All they had to do was show up at Whitney's Coffee Shop. For every one thousand shares of Uranium Oil and Trading Company stock they bought for ten dollars, they could get a cup of coffee for two dollars. (Whitney's commission.)

By the time Jack arrived at eight on Monday morning to check the action, there were fifty buyers waiting at the coffee shop to give him their money. The tiny restaurant was jammed. The ten red vinyl stools at the counter were occupied. Men were packed, three deep, around two of the formica tables. Frank was frantically writing receipts for checks at the third table, dragged up by the door. Frank's brother, Dick, had taken a long break from his job upstairs in the bank to help pour coffee.

The Whitney brothers, a hash slinger and a $300 a month C.P.A., sold $50,000 worth of stock, at a penny a share, in the first week. After two months, they were having so much fun that Dick left his calculator and Frank sold the coffee shop. They established the Whitney and Company brokerage house with a typewriter and a telephone in a corner of Clare Reese's insurance office, down the hall from the coffee shop in the bank.[7]

Jack Coombs was amazed at the initial run on Uranium Oil and Trading. But he was disappointed that the stock hadn't climbed to three cents within sixty days, as Walters had anticipated. So he called his brother, Bill, who worked for the Hot Shoppes in Washington, D.C.

"I told him that there was a little uranium company and to pick up some stock," Jack remembered. "Willow [Bill] was always a very enthusiastic salesman. Before we knew what had happened, the stock was three cents bid. Within a week, Willow, with his Washington friends, had the stock at five cents, and all the stock traders along the wire had shares.

"This is what makes the markets. A customer visits a broker. The broker buys for the customer. The order clerks see the trade and the order is executed by clerks at the other house. As a few more orders come in, the

The Whitney brothers and their wives enjoy an evening in Las Vegas. Left to right: Frank, Marge, Dick and Abbie. *Abbie Whitney.*

clerks and customers' men start inquiring. Then they start talking to friends. The stock keeps moving up and they are all buyers."

Jay Walters, Jr., headed for New York to drive up prices even more. He talked his friend, Plato Christopulos, into joining him.

"Walters came in one day," Plato remembered, "and he said, 'Plato, you and I are going to take a trip to New York. I'm going to show you how to make money.'

"So we go to New York and check into the hotel. He's on floor fourteen and I'm on nine. We immediately go to work. We saw this one broker, Julius Meier. So he introduces me and, right away I'm like a virgin going to a party because I've never been with this kind of company. I've been with brokers, but I never knew this kind of action.

"So the first day we stood around and Walters said, 'Plato, when I tell you to nod your head, nod your head!' In other words, he was using me for a guinea pig, you know, a patsy. So they're sparring back and forth and all I'm doing is listening.

"Julius Meier starts to line up his customers for special interviews . . . and I'm telling you, he's got some pretty prominent boys.

"The first interview was this lady. When she walked in, I didn't think she'd have enough money to buy coffee.

"Walters says to her, 'The reason I want you to buy this stock is because uranium is something that's going to change the whole universe, something that's coming into the future.'

"She started to ask questions and she happened to ask, 'Well, how do you get to this stuff?'

"'Don't ask me how we get there,' he says. 'Christ, we drive up with the jeep to these places. In fact, the other day when we were up that way, we lost fourteen men. The God-damned truck dropped over!' "The lady says, 'Really?'

"And he says, 'Well, that's how tough it is. That's why the price is so expensive. Because of terrain. Isn't that right Plato?' I says, 'You betcha!'

"He says, 'By the way, how do you feel today?' She says, 'My rheumatism's bothering me a little.' 'Rheumatism! Wait a minute! This is pitchblende. Try this. Every time I have a little rheumatism, I use it myself. Put that in your hand and squeeze it. Squeeze it! Harder! Harder!'

"So, for about fifteen minutes she's squeezing that pitchblende. Now I'm starting to laugh and I'm biting my tongue because this is so hilarious.

"He says, 'Plato, why don't you go down and get a couple of cigars.'

"So I come back and she's still got this God-damned thing.

"Finally she released it, and he says, 'How do you feel?'

"'I feel great,' she says. 'No pain. No pain at all.'

"He sold her about $50,000 worth of stock."[8]

Walters was on a roll. People started calling him Colonel Walters and stories about his trading talents became legion.

"He could maneuver those securities so fast that it would make everybody furious," said Bob Cranmer, "because he was in, make a buck and get out when they were about coming in."[9]

Cramner had been the attorney for Uranium Petroleum Corporation of America. When he had a falling out with the management of the company, he was left with a block of about one hundred thousand shares, worth twenty cents a share. Cranmer couldn't sell the stock because someone in the company had put a stop order against his certificate with the transfer agent.

"So I went to Jay Walters," said Cranmer, "and said, 'Jay, I've got this stock and I can't sell a share of the damned stuff.'

"'Give it to me,' Walters replied. 'And how do you want to split the profits?'"

Cranmer figured he wasn't getting anything on the stock as it was. He offered Walters thirty percent.

"So he went to work on the transfer agent in New Jersey," Cranmer said, "and got it broken out of the block transfer down into street certificates in my name. Then he conveyed the impression to brokers in New York (minor houses . . . in those days what you'd call 'bucket shops') that he was shorting the market. And boy, the minute these people in New York thought they had Jay Walters short in the market, they ran the hell out of the market so that he would have to come back, theoretically, in five days and buy at a higher price. When the five days came, he walked into New Jersey and threw down and covered the bets. . . . The market before he started this maneuver was twenty cents, and he ran it to sixty cents."

It was said that whenever Walters wanted to sell someone some stock, he'd wake up the person with a phone call at about two in the morning.

"That's the time to call," he'd say. "Midnight! Never call anybody in the daytime. They won't buy anything in the daytime."

Al Bain, trader at J. A. Hogle and Company, used to quip, "Be careful of Jay Walters. He wouldn't give you the soles out of his spats."

Others remembered Walters as a good promoter, a good salesman, and a real friend. "I don't think he ever took advantage of anyone. He felt if you invested in any of his companies, you'd make nothing but money," said broker Max Guss.

With Uranium Oil and Trading in his pocket, Walters launched Alladin Uranium Company. Coombs and the Whitneys got busy again. They promptly sold $43,000 worth of stock. In a matter of weeks, both of the uranium issues had closed. The market prices were up around five hundred percent above the opening listings. The fever had started.

One day that fall, Hal Cameron and Dewey Anderson walked into the Whitney's office. Hal owned a business selling batteries and automotive electrical parts. Dewey bought oil and gas leases for the Three States Natural Gas Company of Dallas. He collaborated with the corporation's geologist, James D. Edson, in checking for promising properties. Once a purchase was recommended, he had a working agreement that he could sell any minerals the firm didn't want to a Salt Lake syndicate. The syndicate was formed by Dick Muir and Ed Dumke, who sold stock for J. A. Hogle and Company, along with Ed's brother Zeke and a few other young entrepreneurs.

Dewey was very methodical about purchasing mineral rights. First he would check land titles in the county courthouse to be sure the information given by the avowed owners of properties was legitimate and they had the right to deed the minerals. The law stated that, if a person bought a piece of land that had been deeded or patented from the federal government, the

Dewey Anderson takes a golf break from his uranium promoting. *Ellen Anderson.*

parcel was considered one entity until it was split. If the person wanted to deed the mineral rights to his property to someone else, the parcel would become two entities. One person could own the surface rights. Another could own the minerals underground. Once the ownership was split, it could never be put back together.

Dewey told the Whitneys that he had stumbled on a deal that could make them all some money. He had been in Huntington, a tiny village in southeastern Utah, about twenty miles south of the coal mining town of Price. He was negotiating with a couple of Swedish farmers named Shorty Larsen and Bud Nielson, and their wives, for oil and gas rights on some of their land.

"I have a unique way of buying minerals," he said later. "I take hundred dollar bills down and just keep laying them on the table until the farmer and his wife decide they want to sell. I have a top figure, but sometimes we get it for a lot less. Let those hundred dollar bills flutter on the table and it gives them quite a psychological charge.

"I was down one night buying an eighty-acre tract on Clearcreek," he said. "When I'd finally purchased the mineral from these people, one of the women in her calico dress could see that she was going to get some new dresses.

"'If you can get that much out of an old piece of taxed coal ground,' she said, 'what could you get out of some good uranium claims?'"

The Larsens and Nielsons went on to tell Dewey Anderson that they had given Charlie Steen a couple of hundred dollars when he was trying to get enough money to sink his borehole. When Charlie made his strike, he showed his gratitude by directing them to some open land along the Lisbon Fault. The two farmers staked a couple of claims.

"I didn't know what uranium claims were worth," Dewey said later. "But the Mi Vida had come in, and just northwest of the Fault, another mine called Cal Uranium had quite a good show of ore. So, being in the oil business, I figured if you were in between two strikes, you had a pretty fair chance. I didn't know what to offer them, but asked if they'd give me an option and name a price. So she suggested $20,000 for the claims with a twelve-and-a-half percent overriding royalty."

Dewey agreed on the terms and asked the couples for a thirty-day option. He was authorized by the Three States Natural Gas Company to spend up to $20,000 on a project if it was approved by Edson. Knowing that the geologist was in Price, he left the farm and rushed up to see him.

When Edson saw the maps and data, he became very excited. He urged Dewey to write a check immediately. But Dewey was obligated to notify the Three States Natural Gas Company first. He tried to reach Pappy English, one of the vice presidents in Dallas. He couldn't be found. So Dewey returned to Salt Lake and got the syndicate together.

"I told them what I had in mind and they about threw me out the window," he said. "They didn't want any part of it. Didn't like uranium. Didn't think it was worth anything. A flash in the pan."

Next he approached Texas millionaire Clint Murchison and several major companies. All of them refused to get involved. The price was too high.

"So I decided to go it alone," he said. "I went back to the Swedes and told them if it was worth $20,000 for a thirty-day option, if they'd give me a sixty-day option, I'd give them $40,000."

They agreed.

Once Dewey had the option, he had to raise the money to finance it. He talked his friend Hal Cameron into investing in a partnership. But they were soon low on funds. Then they learned about the burgeoning penny stock market and approached the Whitneys.

"We want to start a uranium company," Dewey told the brothers. "We'd like you to sell the stock."

He told them about these uranium claims right by Charlie Steen's Mi Vida! It was a natural. They couldn't miss. In minutes, the four young men who scarcely knew what a corporation was, much less how to organize one, were discussing potential board officers and methods of raising at least $60,000 to pick up the option and pay organizational expenses.

In a few days, the newly-formed board of directors met. Geologist Paul Walton was named vice president. He agreed to guarantee $80,000 in return for a percentage interest. Prominent attorney Hal Moffat handled the legal matters. Bob McGee, a certified public accountant, oversaw the books. The Whitneys said they would put shares of the company on the over-the-counter market.

All that was left was deciding what to call themselves. Frank Whitney suggested Federal Reserve Uranium Company. The name sounded substantial, important. But someone else wondered if "Federal Reserve" smacked too much of government involvement.

"Well then, why not just Federal Uranium?" Frank said.

Since it was to be an intrastate offering, it wasn't necessary to register with the SEC. Federal Uranium was listed with the Utah State Securities Commission.

While the organizers were putting Federal Uranium together, Max Cohen, a vice president and attorney for Three States Natural Gas Company, had plans of his own for the Nielson-Larsen claims. He had seen the memo Dewey sent to Pappy English when he tried to interest the company in the uranium deal. Without telling Dewey, or even his partners, he offered the Huntington farmers $200,000 for the claims.

"They [Nielson and Larson] did everything in the world but kill me to get that option back," Dewey said later.

"One of the Swedes had said we could have an extension on the option. We figured one partner can speak for another, unless the other speaks up. So we thought, 'Good, we have an extension.'

"Then, at six o'clock one night, Hal and I came back to the office and there was a telegram sitting there that said, 'You have your $40,000 in our home before twelve, midnight, or your option's up.' "

Dewey slammed the message down on the desk. How were they to get $40,000 just like that? Business hours were over. The banks were closed. Worse yet, midnight was only six hours away! It took at least three hours to drive to Huntington. That was under usual conditions. Tonight there was a raging blizzard outside. Roads were so slippery that people were bumping fenders all over town. So what would the steep canyon highway east of

Provo be like? The 7,477-foot-high Soldier's Summit? There was no way they could meet these demands! But he was damned if he'd let Cohen and the Swedes put one over on him.

"I called Paul [Walton]," Dewey said, "and he put the money in the bank. We jumped in the car and drove from here to Huntington with the check.

"We got into Huntington about 11:30 P.M. (Huntington's about as big as your hat. I think both the entering and leaving signs are on the same post!) The Swedes weren't home. They were on a big drunk, thinking they had their claims back. But there was a beer joint open. We got the people in the beer joint to verify that we were there with the money, then sat and waited for the two Swedes to come in. They came in about three-thirty, drunker than billygoats. They saw us and couldn't believe it. They said they wouldn't accept the check because it was after twelve. We proved that we were there before twelve."[10]

Federal Uranium was on its way.

With the option in hand, the Whitneys got to work. On December 30, 1953, they announced the offer of 7.2 million one-cent shares of Federal Uranium. On New Year's Day 1954, they moved into a one room office on the eleventh floor of the Continental Bank. They hired a secretary and hung a sign on their door that read: "Whitney and Company—Since 1953."

The weather was so mild that the members were still playing golf at the Salt Lake Country Club in early November 1953. Jack Coombs was on his way to the course when someone behind him started honking and waving. When Jack pulled over to the curb, the man drew up alongside and rolled down his window. Jack had never seen him before.

"Hi," the stranger said. "I'm Ray Bowman. I want to get into the uranium business."

Bowman was selling real estate with his brothers, Ralph and Jack, for their father, Ned J. Bowman, who had an office in the Walker Bank building. The boys had never had any experience with the stock market. Their first taste of it came when they joined the throngs of novice investors who crowded into Whitney's for "two dollar cups of coffee." The penny shares of Uranium Oil and Trading Company they bought there skyrocketed to three cents, five, and eight. It didn't take them long to decide that the uranium business looked a lot more exciting than real estate. They wanted to meet with Jack and find out how to get involved.

"Find Jay Walters, Jr.," Jack told Ray. "Get some claims from him and form a company."

The Bowman brothers jumped right into the ring. They found Walters and bought some claims. Then they filed an intrastate registration for a uranium company. They christened their new enterprise Uranium Corporation of America. Obviously, anything with "Uranium" and "America" in the title would be hot, they thought. It was. Jack Coombs quickly sold $30,000 worth of the stock at two-cents a share.[11]

By this time, others were jumping into the parade behind Jack. He'd have to lengthen his stride to keep the lead. But his partner, Harry Brothers, was too conservative. Jack knew that he would have to leave Harry and form his own company. Near the end of 1953, he opened doors of Coombs and Company in a vacant room of the KALL radio station above the Utah Theatre. Noland Schneider, an attorney fresh out of law school, scrapped his developing practice and started selling penny stocks for Coombs, instead. Jack's brother, Willow, left Washington, D.C., and joined the staff.

Then one day, the Bowman brothers came in to talk to Jack about a new proposition. It seems their friend, Bill Lang, had a lot of connections in New York. Bill had an "in" with some of the principals at the large brokerage firm of Reynolds and Company. He told the Bowmans if they could get some really good uranium claims, the Reynolds people might come in on the deal. Coombs and Ray Bowman decided to drive down to Moab and take a look around. It was on this trip that they first met Mitch Melich and Charlie Steen.

Mitch and Charlie were in the process of easing Dan O'Laurie and Bob Barrett out of Utex Exploration Company. They were also buying up claims along the Lisbon Fault in the vicinity of the Mi Vida mine.

Jack and Ray heard that Utex had recently purchased the Little Beaver property. The claims looked good. There were indications of ore and they were right next to Steen's big strike. Bowman offered Utex about $32,000 for an option to buy the property. Then the pair returned home to invite the Reynolds bigwig to return with them and pick up the option.

But when the threesome arrived back in Moab, Steen and Melich were entertaining the president of Homestake Mining Company and a principal of Lehman Brothers.

"We think they had already taken an option from Homestake," Coombs said later, "because when we went down there . . . they stuck me and Ray in a room and we sat and played cards for two days."

The Coombs-Bowman deal never materialized and the Little Beaver claims ended up with Homestake.

It wasn't long before Mitch and Charlie invited Ray Bowman's brother Ralph and Jack Coombs to return and discuss another proposition,

Jack Coombs and Ray Bowman on a fishing jaunt in Mexico during the boom days. *Jack Coombs.*

however. On December 1, 1953, Charlie and Bill McCormick had paid Dan Hayes, Joe and Don Adams and Edward Saul $50,000 for an option to purchase the Big Buck claims for $2,000,000. (This was the mine that Charlie decided to drill behind when he started work on the Mi Vida. "I could have bought the Big Buck for $10,000 then," Charlie later said.)

The Big Buck claims, recorded in August 1948, were the earliest uranium workings in the Big Indian district. Known as the Adams group, the fifteen unpatented mining claims had proven reserves. Within five years, the mine had shipped approximately 3,700 tons of ore that averaged a uranium content of .36 percent. Because of the incredible ore deposits in the neighboring Mi Vida, Charlie and Bill hoped that further development of the Big Buck would uncover similar lodes.

But Utex, a privately-held corporation, was recycling much of its profit back into production. Charlie had invested heavily in opening his own drilling company. Neither he nor Bill had ready cash to launch another project; they needed outside capital. So it appeared obvious that, since they had the Big Buck and Bowmans had the financial connections, collaboration would be a natural. They offered to assign the option to Bowman and Coombs.

The two promoters headed back to Salt Lake City to figure out how they would raise two million dollars. Ralph found his answer in the pages of the *Salt Lake Tribune*.

The newspaper reported that Joseph W. Frazer, co-founder of Kaiser-Frazer Corporation, was in town. Ralph knew that the well-known automobile executive had resigned from his firm nine months before, after forty years in the automotive industry. He had time, money, and he had a name. Frazer would be a hot prospect.

Ralph got on the phone with Frazer and made an appointment to see him in his room at the Hotel Utah. In a matter of hours, he had the New York industrialist excited about this new uranium industry that was going to "change the world." They would form a company, Ralph told him. Frazer would be president. It mattered not that he knew nothing about mining. He could invest a portion of the two million and let them use his good name to help raise the balance as a stock issue. He would be a figurehead.

Frazer accepted the offer. He didn't have a lot of money to speculate on the deal, but he could introduce Ralph to some New York stockbrokers who could underwrite the project. He would also help enlist a few nationally-prominent persons to serve on the board. Ralph said, "Fine," and they flew to New York.

The first promoter Frazer introduced him to was Fred Gearhart, of Gearhart and Otis, Inc., with an office on Wall Street next to the American Stock Exchange. For 300,000 one-cent shares, Gearhart became the brains of the company's stock dealings. He later transferred 100,000 of his shares to Crerie and Company, of Houston, Texas, who helped with the under-writing.

Next they retained New York attorneys Aaron Holman and Newton Brozan as general counsel to the company. They promised the law partners 50,000 shares. Holman was named vice president and Brozan became secretary-treasurer. Texan oilman Lucien Cullen and John A. Roosevelt

(Franklin D. Roosevelt's son) joined the two lawyers, Frazer, Steen and McCormick to complete the board of directors.

On January 25, 1954, Charlie Steen and Bill McCormick formally assigned the option for the Big Buck claims to Bowmans' new enterprise called Standard Uranium Corporation. Besides the $50,000 cash for the option, Steen and McCormick received 750,000 shares of common stock at the price of one cent per share. The company promised to repurchase 200,000 of these shares for one dollar each, should Steen and McCormick wish to exercise that option six months after the date of issuance and for one year thereafter. Finally, Standard Uranium committed to a $100,000 drilling contract with Steen's Moab Drilling Company.

In May, Gearhart opened the public offering. Standard became the first uranium company to be fully registered with the SEC. The 1,430,000 shares of the company's common stock with a par value of one cent sold for $1.25 per share. On the first day, all 1,430,000 shares were sold out and the stock started trading at two dollars. Stock purchased at a penny before the public offering catapulted up 12,500 percent with the opening price.

Ralph Bowman, with a $125,000 finder's fee and 159,000 shares bought at twelve-and-a-half cents per share, was a millionaire.[12]

Zeke Dumke, a young Salt Lake insurance broker, could see that Charlie Steen was terribly uneasy. He had been brought in late to a cocktail party that was being held in a suite at the Hotel Utah. It was the fall of 1953. Zeke's father-in-law, Rulon White, had invited a few friends for drinks prior to the annual Utah Manufacturers' Association banquet. By the time Charlie arrived, the guests were well into their second martinis. The manufacturers' wives, rustling in taffeta and silk, pulled away from conversations with friends and worked their way close to the overnight celebrity. Steen had been rocketed into public recognition via *Collier's* and *Cosmopolitan*. The ladies homed in to glean some first-hand gossip.

"Is it true your wife and children were going without food or milk?" they asked.

"How big is the diamond you gave M. L.?"

Zeke could see that Charlie was miserable. He elbowed through the crowd to see if he could get the poor man a drink.

"Yeah, Scotch," Charlie said. "Bring it to me in the bathroom."

Zeke was interested in talking to Charlie. Dumke was a member of the syndicate that worked with Dewey Anderson on oil and gas leases. When Dewey had offered the syndicate the Nielson-Larsen uranium claims that later became Federal Uranium, Zeke and his partners had felt that it was too big an investment with too little time to check it out. Now Zeke wanted to know what Charlie thought about the Federal deal.

When Charlie answered Zeke's knock on the bathroom door, he was sitting on the toilet lid. He motioned for Zeke to sit on the edge of the bathtub. While they sipped their drinks, Charlie made a few offhand remarks about being rescued from cocktail parties and then Zeke asked him what he knew about the Federal claims.

"I know of ten mining claims that are better," he said. "And they're for sale."

Charlie went on to tell Zeke that he had given the interest to these claims to people who had helped him out during the bad times. He had retained the royalties. He was anxious to have the properties drilled to find out if there was any ore on them. That was the only way his friends would realize any value on the land.

Charlie had come to Salt Lake to look for financing. He was anxious that the money for this project be raised in a public offering. The Mi Vida was completely private. People had been after him for a long time to give them a place to invest in uranium. Since these claims adjoined his original find, he felt that he could tell potential investors that this was going to be an honest run. A complete risk, but well located.

Zeke was interested. His brother, Ed, had recently returned to Salt Lake after getting his degree in business from Columbia. He had started with J. A. Hogle and Company, but was now affiliated with A. Payne Kibbe, a local broker. Perhaps they could figure out a deal.

Charlie knew Kibbe. He had financed some exploration work for Utex. He agreed to talk with them.

By December, an agreement was reached. Kibbe, Ed Dumke, and their attorney Max Lewis, met with Steen at his room in the Hotel Utah. They organized Lisbon Uranium Company. The underwriters purchased Steen's ten claims for $65,000. They gave him an overriding royalty of twelve-and-a-half percent and guaranteed a $100,000 contract for Moab Drilling Company. In addition to Steen's claims, they had optioned eight adjoining properties from "the Swedes," Nielson and Larsen, and a number of claims from Bob Barrett. On January 15, 1954, Lisbon Uranium Company offered 1,079,000 shares of common stock to the public at twenty cents per share.[13]

It seemed that the uranium companies couldn't spring up fast enough. Once Uranium Oil and Trading Company unscrewed the lid, mining operations spilled out all over the marketplace. Alladin. Uranium

Corporation of America. Federal. Standard. Lisbon. Timco. Jolly Jack. Consolidated, Apache, and scores of lesser corporations.

New stock brokerage houses materialized like fruitflies. Where there had been twenty traders in 1953, by the next spring there were one hundred twelve. The offices ranged from closet-sized "bucket shops" with folding chairs and a bare incandescent lightbulb to Muir, Dumke and Light's establishment beneath crystal chandeliers on the mezzanine of the once-elegant Newhouse Hotel.

Even Art Davis's "greasy spoon" on Main Street saw market action. The Grabeteria was a popular spot where harried businessmen stood at narrow shelves and bolted down roast beef sandwiches and Morrison meat pies while they scanned girlie magazines hanging from nails along the wall. The infant Doxey-Merkely brokerage house started trading uranium shares among steaming pots and dirty dishes in a cubbyhole at the rear of the cafe.

In April 1954, Dick Muir, a graduate of the Harvard Business School, quit his job as a customer's man at Hogle's with the intention of joining his father and brother in the produce shipping business. He asked his family one favor before starting, however. He wanted three weeks to delve into the uranium happenings in southern Utah. He had borrowed $1,300 to stake his adventure and he had a unique idea.

"The pattern of property development in the district created a situation that offered a remarkable opportunity," Muir wrote in an article for the *Harvard Business School Bulletin*. "The original claim-stakers gradually sold off their property to outside bidders. Prices for raw, wildcat claims soared, some going for as much as $10,000. These original owners, though primarily interested in cash, generally retained a royalty.

"Such royalties were of great potential worth, representing an interest in the gross value of any ore produced, free from any development costs or responsibilities. In some cases, if the holder of the working interest failed to perform satisfactorily, the entire claim reverted to the royalty holder. Then, too, the cost structure of a mining operation was frequently such that the royalty interest represented the only profit likely to come from it."[14]

Muir discovered that, in the scramble to buy claims, people were forgetting about the royalties. He knew that his $1,300 wouldn't go very far in buying actual claims, so he decided to go after the royalties.

Dealing with the farmers was a unique experience for Muir. One of the ranchers was Jim McPherson, who "had a face just like it had come from that red rock, and he looked like he could wander out by the corral and a horse would run under him." Muir learned that you had to go in the back door when negotiating with a man like McPherson, "a quiet, able guy who had contended with his own environment, which was pretty rugged, strong and silent."

"You'd ask a question, and there was no immediate response," Muir said, "because he was sitting there and he was whittling. Then you'd look out of the window and wait awhile and kind of shuffle your feet around. Time would elapse that was really embarrassing. Of course, it was only embarrassing to me because Jim and Buck [Kirk] didn't talk much.

"Finally, Jim would say, 'Well, what do you think, Buck?'

"And this was maybe seven or eight minutes later. Then, finally, they'd negotiate the agreement. But, I tell you, if Jim wasn't speaking, he was thinking plenty, because I think he got a great deal more than I ever did."

Muir's procedure in approaching the original claim-stakers was to say, "Look, you've sold off most of your claim, and you've got a fair amount of money. You've retained a working interest (maybe ten percent). If they've [the buyers] really got something, five percent royalty on the claim will do you a lot of good and it's [selling the other five percent] an opportunity to get more money for your claim."

He would offer to buy half of the royalty interest on the claim. And, more particularly, to take an option, because then he could make his money really go further.

"And so, in general, I took mine right off the top," he said. "The royalty holder got his first out of the gross income. What was left, went to cover the expenses and then, if anything was left over, that became the profit to the working interest."[15]

Muir succeeded in "blanketing the major ore trend." He spent $1,000 of his $1,300 taw on assembling royalties, ranging from one to four percent, on over four hundred claims. In the process, he ran into his former Hogle colleague, Ed Dumke, who was bidding against him on some property in Moab. They decided to join forces and formed the brokerage house Muir, Dumke and Company. Muir didn't return to the family business.

But Dick's royalty caper was not finished.

One Friday evening, a friend brought a man into Muir, Dumke's office. The man needed $15,000 by the following Monday to exercise an option on a block of uranium claims known as the Radon, Hot Rocks, Cord and Daniel Ruddock groups. All of the claims were located in the Big Indian district. Muir arranged to get a one-year loan of $14,000 from another stockbroker, W. D. Nebeker. Nebeker, who had just refinanced the old Kentucky-Utah Mining Company, agreed to make the advance on

Dick Muir, "Skip" Light and Ed Dumke were the youngest members of the New York Stock Exchange. *Dick Muir.*

behalf of that company. At the same time, Kentucky-Utah put up $750 for three-fourths interest and Muir, Dumke $250 for a quarter interest on a ninety-day option on the entire package of the Radon properties. Terms of the option were the payment of $280,000, with one-fourth of the amount in cash and the remainder in stock.

Two weeks later the Radon claims were in ore.

"Unfortunately, we didn't begin to realize the ultimate value of the royalties which our option secured," Muir said later. "Caught in a tight spot on another deal, we traded our option for Kentucky-Utah stock, a corporation which eventually merged with Federal. Our $250 investment showed a market value of $88,000. . . . But what was really the killer on the thing was that each percentage of royalty on the Radon, for a period of about nine years, paid between $40,000 and $50,000 per year." If Muir had

had the cash to exercise his option, his interests would have been worth a quarter of a million dollars per year for the nine years.

Still, his profits from the Radon transaction stood him well. A year later, twenty-eight-year-old Muir, and Dumke, twenty-nine, joined by another partner, Given A. "Skip" Light, twenty-seven, became the youngest persons ever admitted to membership in the New York Stock Exchange.

"Now that we have accomplished our aim so soon," Dumke told *Time* magazine, "we have the longest potential experience of any partnership in the country."[16]

The rash of over-the-counter penny offerings spread throughout the west. Towns like Moab and Monticello, Utah, where the word "stock" had never meant anything but "cattle," were peppered with tipsters. Denver, Colorado, caught the speculation fever. Indians on reservations in New Mexico and Arizona wanted their share of profits from the "yellow dirt."

It was a madness. Grocers gave uranium stock tips along with loaves of bread. Schoolteachers organized investment pools. Society matrons refused to go to bridge club without a new over-the-counter issue to tout. Young parents chattered more about uranium stock than Dr. Spock.

Sometimes uranium shares were even used for barter. Dewey Anderson traded automobile dealer, Wayne Stead, a 1953 Pontiac and 25,000 penny shares of Federal Uranium for a 1954 convertible. They figured that made the new $2,500 car worth about $18,000.

Another time, Anderson gave Al Holman $100 worth of Federal for five pairs of new shoes.

"Those shoes turned out to be worth about $300 a pair," Dewey said.

The brokers scrambled from six in the morning until eleven at night. Sometimes they even found checks and buy orders wedged under their doors at home. Customers bought anything. They would wait for hours to get to the trader and then ask, "What's new today?"

The broker would answer, "XYZ Company."

"What's the price?"

"Two cents."

"Who's in it?"

"We don't know."

"Where are the properties?"

"We don't know."

And the customer would say, "That sounds great! Buy me $1,000 worth."

Business Week commented, "All you have to do is put together a company with 'uranium' in the title and you've got a going thing."[17]

New stock notices were being advertised on the radio and television. Newspapers ran special listings of uranium stocks. Jack Coombs started the column, "Uranium Topics" in the *Salt Lake Tribune*. He wrote:

> Boom! . . . Boom! . . . Boom!
>
> Everything's booming on the Salt Lake uranium market. Stocks that were selling for 20 cents a share five months ago are now offered at $2 or more. . . . Penny stocks have zoomed to 30 and 40 cents.
>
> It's a big lottery. . . . It's a giant slot machine with uranium shares instead of bells.
>
> Naturally, in the wake of the boom have come some interesting stories.
>
> There's the story of a man who came in a local brokerage house to sell some stock, and before he could complete the transaction, he'd made several hundred extra dollars.
>
> And how about the fellow who got on the end of a long line in a brokerage house last week, and signed up for 10,000 shares of stock in a new firm.
>
> Uranium stocks are selling so fast the ink isn't dry on the certificates before they're sold over the counter.
>
> Will it last?
>
> Nobody knows.
>
> Is it healthy for business?
>
> Who can say?
>
> It's like eating a turkey dinner with all the trimmings. . . . Let's enjoy it while we may.[18]

The pace was frantic. The players were young. The fortunes came fast and easily. A kind of electric energy consumed the dealers and promoters. They couldn't let down. After working twelve hour days, they gathered in their offices to drink, shoot craps and play poker until three or four in the morning.

"We'd fight like dogs to make a dollar, but, after working hours, we were all friendly, just like one big happy family," said Plato Christopulos.

Gambling stakes were high. Stock salesman Bob Bullen bought a new convertible one day and had lost it at cards by the next morning. When Ray Bowman boasted that his brother, Jack, could drive a golf ball farther than anyone, Jack Coombs argued that his dealer, Bob Pearson, could do better. A challenge match at the Salt Lake Country Club was arranged. The loser would pick up the tab for cocktails and dinners for the employees of both firms, plus a cash prize for the winner. Side bets of one thousand dollars were placed. Coombs won.

The new millionaires spent their money almost as fast as they made it. They built lavish homes, bought their wives fur coats, went sportsfishing in Mexico. Expensive cars and private aircraft were commonplace. The Bowman brothers had a single payment left on a new Beechcraft Bonanza airplane when they had to fly to Texas on a promotion. When they got there, they discovered they had to take six people in a four-passenger plane. They promptly traded their airplane for a larger Twin Beech. When they completed their presentation, they turned back the new plane. They had lost the Bonanza, but the Twin Beech "did the job for the day."

"It was a get-rich-quick business," Ray Bowman said. "People were buying penny uranium stocks without looking to see what they were taking. But nobody pushed them into it; it was a madness. I was buying uranium property the same way; without looking. I couldn't afford to look. If I did, somebody else would shove ahead of me and make a million dollars while I was looking. I would hate to tell you the deals I lost out on by not snapping them up fast enough."[19]

On May 27, 1954, uranium stock sales in Salt Lake City hit a daily record with seven million shares. A buying phenomenon known as the fabled "five days in May" followed. Jay Walters, Jr., had hit the nail on the head.

10 THE COLOSSUS OF CASH

IN THE CANYONLANDS of southeastern Utah there is a rock formation they call "Jacob's Chair." The gigantic sandstone throne commands a broad expanse of ruddy desert interrupted by deep, jagged cracks and phallic pinnacles. Some mighty being might rest in this solitary seat of honor to view the wonders he has created. Jacob's Chair was Floyd Odlum's favorite geologic structure in this land he so admired.

"That chair's too big for me," he would say when flying over the broken landscape, "and I'm not sure I'd be worthy of sitting in such a big one. But I hope someday I'll deserve an honorary chair in this interesting and important place."[1]

An honorary chair was not out of the question for Odlum in 1954. Already he was known throughout the world as "Fifty Percent Odlum" or "The Colossus of Cash."

The sixty-two-year-old business magnate had come a long way from his boyhood days in Union City, Michigan. There, as the son of a Methodist minister, he had to work his way through school. He picked fruit and berries. He dug ditches. He stacked lumber in a sawmill. He was even a jockey on an ostrich at the race track.

But by the mid-1950s, Odlum had gained a reputation as a financial wizard. His status was further enhanced by his second wife, the flyer Jacqueline Cochran, who made aviation history. She had established numerous women's air speed records and was the first woman to make a totally blind landing. (Cochran later became the first woman to pass the sonic barrier and exceed the speed of sound.)

The Odlums lived in an elegant New York apartment with a view of the East River and white-gloved servants attending them at table. They also spent much of the year in Indio, California, at their extensive ranch, where such dignitaries as presidents Dwight D. Eisenhower and Lyndon B. Johnson were frequent guests. There, under thick groves of fruit trees, fragrant jacaranda and date-bearing palms, the Odlums and their friends enjoyed a private nine-hole golf course, an Olympic-sized swimming pool, tennis courts, a skeet range and a stable of Arabian horses.

Odlum had earned his millions by buying and selling multimillion dollars enterprises: Consolidated Vultee Aircraft. RKO Studios. Birdseye Foods. Greyhound Bus. Sunray Oil Company. American Trust Company. Paramount Pictures. Stinson Aircraft. Franklin-Simon Department Stores and the Sherry-Netherland Hotel. He once had held a mortgage on the Empire State Building. High-powered trading was his passion and he was a master at it. He prided himself on never making less than forty percent on a major turnover.

Odlum's entrepreneurial career had been launched at an early age. When he was sixteen, he moved to Boulder, Colorado, and talked the registrar at the university into entering him as a freshman. He didn't have any money, he told the registrar, but he would give the school a personal promissory note. Disarmed by this unusual approach, the official accepted the enterprising teenager into the student body.

The young man was as good as his word and paid off his debt promptly, with cash to spare. He earned his tuition by working as an assistant librarian. He operated a laundry. He sold real estate. During the summers, he rented four empty fraternity houses, then operated them as hotels for tourists and students. He took over the school paper and the dramatic society and ran them at a profit. By the time he was a senior, his reputation for business acumen was confirmed. Under his 1912 graduation picture were the words, "Manages to get his hands on everything that makes money."

Odlum went on to complete three years of law school, specializing in mining and irrigation law. After graduation, he moved to Utah, where he earned top grades in the state bar exam. Even so, he couldn't find a job as an attorney in a Salt Lake City firm. Instead, he signed on to help the Utah Power and Light Company organize a legal department. Without pay. The manager promised to consider paying Odlum if and when he proved himself. After ninety days it was clear that he had, and he was put on the payroll, at a salary of fifty dollars a month. When Hortense McQuarrie became his bride, he received a twenty-five dollar raise.

Then in 1917, Odlum's fortunes took a sudden upturn. A man named Sidney Zollicofer Mitchell, of Electric Bond and Share, based in New York City, had been buying up a number of small western utility firms. Utah Power and Light was one of his acquisitions, and during the negotiations in Salt Lake, Mitchell met Odlum. He liked the young lawyer and invited him to move east and apprentice in the law firm of Simpson, Thacher and Bartlett. At twenty-five, Odlum became Mitchell's protege, and soon was representing the financier in complicated financial dealings around the globe.

Odlum's own investment interests started shakily when Mitchell offered to let him buy Electric Bond and Share stock at the underwriter's discount. Odlum got a friend to lend him the collateral in return for a fifty-fifty partnership in the deal. He borrowed $30,000 from the bank. Then—zap!—the market took a drastic plunge, and Odlum was heavily in debt. He learned the grim realities of speculative investing the hard way. But the whirlwind world of high finance left Odlum with an addiction for the stock market—and a nervous stomach.

Things brightened in 1923 when George Howard, a fellow attorney, joined Odlum in an investment pool. Each man, and their wives, contributed $9,900 to form the United States Company. Within a year, the budding investment enterprise was worth $51,000. Two years later, the bottom line was $748,000. As the decade neared its end, assets were in the millions. The company had taken in forty-five new stockholders and changed its name to Atlas Corporation.

It was then that Odlum, with uncanny perception, foresaw the Great Depression, and turned it to his benefit.

"With Hartford of A&P Stores and Wall Street banker Paul Warburg, Floyd Odlum ranks as one of the handful who forecast the coming of the great depression," wrote Martin L. Gross in *True* magazine. "But Odlum alone had the gimmick in the national disaster that not only made it possible for him to survive, but to more than triple his millions in the midst of a depression."[2]

Odlum figured that stock prices had soared as high as they could go. They must come down. Consequently, he started to sell, and he kept his earnings in cash.

I saw a strange situation," he told Gross. "The public's confidence had dropped so much that the investment trusts were priced lower than the value of the stocks they owned. For five dollars you could buy a share of Trust X that owned securities of railroads and factories that were selling for ten dollars a share.

"It was a wonderful spot for someone with nerve and cash. But the only way to take advantage of it was to buy control of these investment trusts and then sell the companies they owned. I decided to try it."

Odlum's methods were well considered. He bought voting stock, always careful to avoid forcing the prices upward. Sometimes he paid with his personal Atlas stock, other times in cash. After gaining control of an investment trust, he would rapidly sell its companies. Then the process was repeated. Again and again.

It was after his profitable sale of Consolidated Vultee Aircraft in 1953 that Odlum turned his attention to the uranium industry. That transaction, along with a few smaller deals, had left Atlas with twenty million dollars in its pockets. Itching for a new investment opportunity, Odlum homed in on the burgeoning action on the Colorado Plateau.

Western mining was not new to Odlum. The United States Company had been involved in the vanadium boom during the 1920s. Odlum's Atlas corporation controlled more than one hundred and fifty claims in Utah and Colorado. It also operated a mill at Rifle, Colorado. When the Manhattan Project started buying uranium from the old vanadium dumps in the 1940s, Odlum's dormant Waterfall Mine, which was about five miles southeast of Steen's Mi Vida, was one of the suppliers.

"So partly because of that," Odlum said later, "and partly because of my natural inclination, when uranium came in as a metal the government wanted real badly, I had sort of a desire, one way or another, to get into it."

The Atlas board of directors didn't share Odlum's fervor, however. "They didn't want to do any prospecting or anything that smacked too much of an adventure or too much risk," Odlum said. "But, they did say that if we found some already developed and proven properties that they might go into that. Then I asked them if they had any objection to my doing some prospecting on my own. And they had none."[3]

Odlum put out feelers in early 1954. In order to be accessible to the promoters, he went to Salt Lake City and booked a suite at the Hotel Utah. There he found the investment prospects interesting—and unusual. What appeared to be a boom in penny stocks was fomenting. The only problem was that it showed more puff than product.

For example, Federal Uranium Company, which had sold $72,000 worth of penny shares almost overnight, was moving up. The prospectus admitted that the board of directors had paid $60,000 for their claims and retained a scant $12,000 for working capital. Not much to engage in any major mining activity. Especially when half of the working interest for any claims was promised to those who found ore. But the investing public didn't seem to be concerned by this disclosure, if indeed they bothered to read it. The stock climbed quickly from a penny to two-and-a-half cents, and then a nickel.

It was the same story with other issues. Lisbon Uranium, which opened over-the-counter in January at twenty cents, was selling at thirty-three cents in February. Standard was going strong and scores of new stocks were coming on the scene.

Odlum made a tour of the southeastern desert country to see what was happening in the actual mining of uranium. The big producers were Steen's Mi Vida, Cooper and Bronson's Happy Jack and Vernon Pick's Delta. Odlum envisioned purchasing one of these giants and then merging a number of smaller operations into it to form one large conglomerate, financed with eastern capital to explore and develop the known ore bodies. This would give him a major position in the industry. When rumors of his interest spread, promoters and owners of unproven claims began to line up at his doorstep.

Odlum, now aging and suffering from severe arthritis, conducted his business in hotel rooms or at home. He didn't like offices.

"A desk is a storehouse for junk," he would say, "bad for posture, and a setting for cold conversation."

Usually Odlum's meetings were held in various rooms of his New York apartment or at the California ranchhouse. There would be one group of petitioners in the den, another in the lounge. Still others awaited their turn in the living room or various chambers. Sometimes at the ranch, Odlum would leave his dark leather armchair and ease his slight, bent body into the swimming pool and talk over deals while he gently exercised his crippled legs. It is said, when he was dealing with the millionaire Howard Hughes, that he would accommodate his paranoid associate by conferring in the bathroom with the water taps running. Or, sometimes, Hughes would insist on holding their discussions while driving around the block in his car.

One of the first developers to approach Odlum was A. Payne Kibbe, the Salt Lake City broker who issued over a million shares of Lisbon Uranium stock in January 1954. His initial offering had been sold. Exploration work on the ten claims he had purchased from Charlie Steen had commenced. There was some indication of ore, but not enough to warrant sinking a shaft.

In the meantime, three claims adjacent to the Lisbon properties became available. They were owned by Shorty Larsen and Bud Nielson, the Swedes who had sold other uranium leases to Dewey Anderson to form Federal Uranium. The claims looked good. Charlie Steen had sunk a drill hole eight feet from the Swede's property line and hit a highgrade deposit. But the title was in question. Kibbe and his associates didn't dare put the properties into Lisbon and issue stock for them to raise additional capital. But they had a $75,000 option on the land.

Payne Kibbe checks a Geiger counter at an ore face. *Evelyn Kibbe.*

So in April 1954, Kibbe contacted Odlum. He told him that the company intended to develop and mine the claims. He proposed that Odlum buy the optioned properties to make this possible.

The very mention that Floyd Odlum was nosing around the uranium industry immediately affected the market. Lisbon stock skyrocketed from forty cents to ninety-five cents in three days.

Then other factors entered the picture. The public became infatuated with the seductive penny market and started playing it like a crap game. Brisk trading increased stocks six, eight, and sometimes twenty times the opening price. Mines that had been engaged in exploratory drilling began to tap into uranium. Companies that had started on a shoestring found major financing. By May, Salt Lake City's over-the-counter market went crazy.

The whole barrel seemed to explode at once. Standard Uranium, the only corporation with a New York underwriter, made the first registered SEC offering of 1,439,000 shares at $1.25. The price soon zoomed to three dollars. Lisbon stock climbed to $1.75. A brand new company named Timco Uranium announced it had made five strikes of commercial grade ore. Its stock soared to fifty cents, fifty times the original subscription price of a few weeks earlier. Cromer Brokerage Company traded 6,729,000

Floyd B. Odlum often conducted business from his swimming
pool at the California ranch. *Associated Press.*

shares of Apache Uranium at three cents in a flat seven days. Consolidated
Uranium, Yellow Cat, Alladin, Great Basin, and hordes of other companies
joined in the frenzy.

Federal's board of directors decided to capitalize on the excitement.
Dick Whitney gave Jack Coombs a call.

"Federal wants to place some stock out of town," he said. "Have any
ideas?"

Coombs did. A New York broker with J. May and Company had called
him asking for a good uranium stock for some brokers in Philadelphia to
distribute. Whitney gave Coombs 400,000 shares of Federal at a penny
under market price.

Then on Saturday, May 22nd, Coombs got another call. It was a broker
in Houston, Texas. He also wanted some Federal to sell.

"And listen to Walter Winchell tomorrow evening," the caller added.

On Sunday night, the familiar staccato patter of the famous newscaster came on the air.

"Good evening Mr. and Mrs. North and South America and all the ships at sea. Let's go to press!"

Then Winchell announced that Federal Uranium Company had hit ore. The samples assayed at about 0.4 percent uranium oxide. Best of all, pending stockholder approval, the company was reducing its outlay by converting working interest and about half of the twelve-and-a-half percent royalty interest to stock. A good buy, Winchell proclaimed.[4]

On Monday morning, mobs of customers waving checks and cash jammed the Salt Lake brokerage offices. Telephones jangled. Teletype machines chattered. Brokers closed their doors early and took the phones off the hook to catch up on paper work. A number of new underwritings hit the street before noon.

"One of these [stocks], which came out at one cent a share, immediately jumped to six cents a share," wrote Bob Bernick, business editor of *the Salt Lake Tribune*. "It boasted about the shortest prospectus in local underwriting history . . . a legal-sized piece of paper with a map in the center and details of the claims, proposal, et al, in fine print around the map."[5]

In that single day, seven million shares were traded. Lyman Cromer, president of the Salt Lake Stock Exchange, sold a million of them. Jack Coombs traded 750,000 shares. The prices jumped all over the map. Timco went up to fifty-five cents. Lisbon hit $3.25.

Federal Uranium, whose announcement of a strike had blown the lid off, experienced the craziest fluctuations of all. On Friday, the stock had closed at twelve-and-a-half cents. Winchell's Sunday night bombshell blew the price up to fifty-four cents the next morning. But by the end of the day's trading, the shares were down to twenty-four cents. People were buying and selling so frantically that the brokers got burned. Many of their customers, who had bought at the high level, refused to pay when the price dropped more than half by closing time.

"Well, if the current uranium boom hasn't done anything else, it's accomplished one thing," Jack Coombs wrote in his newspaper column. "It's stopped people talking about McCarthy and the Army and started them talking about Apache, Lisbon, Atlas, Redman, Federal, Atomic, Congo, Walters . . . and the other uranium stocks on the Salt Lake over-the-counter market.

"In fact, uranium is the biggest piece of news to hit the Salt Lake area in plenty of months.

"Housewives discuss uranium issues over the backyard clotheslines, bank tellers talk about uranium while counting money . . . no doubt about it, uranium has caught the town and the nation's fancy. It's a magic word."[6]

After the extraordinary "five days in May," the market settled somewhat. Bernick's column, "Up and Down the Street," reflected on the situation:

"During the week, the common question was: 'What had happened? Is this the end?'

"There were as many answers as there were investors. Some of the theories were remarkably naive . . . such as: 'I hear there's a plot by "big interests" to drive the values down so they can buy out these little firms cheaply.'

"Others bemoaned losses: 'I dropped $3,000, $18 of which was my own money!'

"Most brokers here were of the opinion that the area would not again see the 'foolish hysteria' which characterized the uranium market during the last days in May and in early June."[7]

But Floyd Odlum looked at the larger picture. He was not only enthusiastic about the mining of uranium, but he was convinced of the importance of nuclear power worldwide as a future energy source.

"By 1975, I believe more than seventeen million tons of ore will have to be mined for the single purpose of fuel for power plants in production and under construction," he later told a meeting of the Colorado Mining Association.[8]

Odlum figured that since Lisbon had claims adjacent to Federal, and Federal had hit ore, Lisbon looked promising. In late June, he met with Kibbe again.

"Mr. Odlum, being a sophisticated individual, was made acquainted with the title problems and chose to take the risk and move forward," Kibbe said later. "He succeeded in making the acquisition, the titles were cleared up, and then he traded the Swede property, which was just north of the Lisbon claims, into Lisbon. He also purchased two claims from Steen and one, I think, from Barrett, right in that same area. This gave a solid block on which there were no [proven] ore reserves. In fact, at that point, he paid [Steen] half a million dollars for one of the claims . . . the Nixon claim. He traded that package after he purchased it for Atlas, to Lisbon for Lisbon stock, which gave him seventy percent controlling interest in Lisbon."[9]

Thus, Odlum was officially in uranium. But a developed mine, one with proven deposits, was what he really wanted. He deployed some investigators out in the field to see what was for sale.

"My representative from Albuquerque called me in New York and said there were a bunch of crazy men around there making some noise," he said later. "They were in a big fight. They had a mine and they wanted to sell it for twelve million dollars. I could turn some geologists in there and take ninety days to look at it and, if I didn't find over twelve million dollars worth of ore, I didn't have to buy it. Well, I employed some geologist from Salt Lake City to give me a report and he reported that there wasn't anything to it. That was Steen's Mi Vida mine."[10]

[Charlie Steen later claimed that Odlum had invited him to New York and tried to buy the Mi Vida. Steen refused to sell. "I'd sell anything I have at a price," he said, "but not the Mi Vida."][11]

Then Odlum's associates approached Joe Cooper and Fletcher Bronson. The two Monticello, Utah, men had worked for Odlum's United States Company at the Waterfall Mine during the old vanadium days. Bronson was in charge of the mill and Cooper was the mine foreman. When the mine closed, they went out prospecting and ended up buying the Happy Jack claims from Sam and Edgar Hayes. They came close to selling what they thought were worthless copper properties when the uranium content of the ore suddenly produced a bonanza. By the time Odlum negotiated with the pair to buy their workings, they wanted over twenty-five million dollars. Atlas Corporation didn't consider the price worth the risk. (A year later, Barlu Oil Company purchased the Happy Jack for thirty million dollars.)

"Then along came Pick with his Delta mine," Odlum said. "Everybody was trying to get it. I thought it would be a good foundation for a larger enterprise, so I negotiated with him and bought it for Atlas Corporation."

Vernon Pick had had a successful first year. He had blocked out approximately 300,000 tons of ore with an in-place value of forty dollars per ton. The Delta's production averaged 1,500 tons of uranium per month. Rumors buzzed that only an insignificant portion of the mine's potential had been realized.

But Pick had discovered that developing a claim was no easy matter. Especially one like the Delta. There, in the Temple Mountain region that was so remote and wild that a vehicle couldn't penetrate it, the ore was buried under a thick sandstone cap. To make matters worse, the deposits lay in a steep escarpment that towered 4,500 feet above the valley floor. Hauling drilling equipment up to mine the discovery was a monumental and expensive task. Pick had to mortgage his truck and trailer to buy drills. Ore had to be packed out by horseback until he could afford a bulldozer to carve a road out of the tangled rock and brush.

Keeping a cool head with his newly-gained wealth, Pick had paid off his prospecting debts in the first three months. He lived in the old trailer until he had built the road and purchased the heavy machinery. Only after he was clearly in the black did he indulge himself by buying a ranch-style house in Grand Junction with a wet-bar in the basement, a yard full of peach trees, and a Cadillac sedan in the garage. He spent his free time tinkering in a small workshop. He was attempting to invent a radiation counter that would tap deeper ore than the existing scintillometers.

"The one thing that it's [money] bought for me is time," he told Peter Wyden, a staff correspondent for the *St. Louis Post-Dispatch*. "Time to do all the things I want to do. I'm not sure I know what to do with that kind of money. I've never been exactly money-greedy and so I'm just letting it jell, getting used to it."[12]

It wasn't long before a dream of establishing a full-scale laboratory research center took precedence over a mining career. He would move to California and build a complex where state-of-the-art radiation detection devices, water desalination units, and other inventions could be developed. He began to entertain bids from potential buyers of his mine. In the summer of 1954, he turned down six million dollars. A few weeks later, Odlum offered ten million. With Odlum's proposition, Pick saw his dream within reach. He insisted on cash, and an amphibious airplane to facilitate exploratory work in Canada. Odlum complied by drawing a check for nine million dollars on the Manufacturers Trust Bank and converting a used PBY plane to Pick's specifications.

"Pick had had a lot of publicity by this time," said Nels Stalheim, a future president of Atlas Corporation, "and it was supposed to be a mine with unlimited reserves. It was a big splash. Odlum thought if you got into the uranium business with a big splash you'd accrue lots of other properties.

"He had two or three different geologic opinions on it. Ray Wimber [of Salt Lake City] was a little conservative in the reserves. But there was another fellow from Grand Junction, Walt Verner. He estimated a bigger reserve. That was the basis on which Mr. Odlum bought the property."[13]

Odlum had his own version about his decision to purchase the Delta mine. ". . . it was because I heard and followed the direction of strange voices," he told the National Western Mining Conference in Denver. "These voices came from within rocks and little invisible characters talked to me about weird things. There were a lot of these characters, who said they were organized together into what they called uranium 'atoms.' Each atom had a central group commander called 'nucleus.' Nucleus claimed that he and his associates had been waiting for billions of years to go to work as slaves for Man, but had only received the call recently. Nucleus said he and his fellow nuclei had to go through a form of suicide to perform their work, which was accomplished by changing a part of their own mass or bodies into energy. Nucleus said that the work task was started by having himself shot by one of his associates called neutron.

"These characters boasted extravagantly what they could do if I would buy Vernon Pick's mine. They said that within three or four months, they would make it worth more than double my cost, but they would by no means stop there."[14]

Odlum named his new mine the Hidden Splendor. Unfortunately, his "little voices" had steered him wrong. This dive into uranium mining proved to be a glaring error in his long line of brilliant dealings. His geologists and engineers had measured the pillars of ore that had been removed and calculated the amount of uranium that had already been recovered. It was difficult to do much long core drilling to estimate the reserves that might be buried deep within the mine. They simply assumed that the remaining ore would at least equal that which had been mined. They were wrong. The ore pinched out and the Hidden Splendor became known as Odlum's Hidden Blunder. Punsters quipped that the splendor was so well hidden that they never found it.

Gossip swirled around Pick once again. People claimed that the former electrician's mine wasn't "in ore" when he sold it. It was said that he put out long, exploratory holes. Or that when he ran out of ore, or couldn't prove any uranium reserves on the Geiger counter, he inserted plugs in the tunnels. Afterwards, if potential buyers came in with radiation counters, they would get an indication of ore at every face they checked. Pick was also accused of a form of reverse "salting," taking what ore there was out and then leaving a trace and covering his tracks. None of the accusations was proved.

Despite his difficulties, Pick realized his ambition. He bought an 830-acre hilltop ranch near Saratoga, California (by San Jose), and built a large home and a 7,420-square-foot research laboratory. The PBY was hangared in the nearby Salinas Airport, which the uranium millionaire leased.

All the while that Odlum was negotiating on Lisbon and the Delta Mine deals, he eyed the progress of Federal Uranium. "They had nothing but a smell in the ground up near the Steen properties," Odlum later said. "Rather deep, too. But it had some titles around it and they needed some money for development. It was still just a prospect. And I said I would turn in my claims and some cash, but not for their stock, which was so overpriced. It just wouldn't make sense. I told them that I was going down to the Argentine to be gone for several months, and if their stock was selling, maybe in the twenties, I'd sit down and talk to them about it."[15]

When Odlum returned from South America, Federal was down to about fifteen or twenty cents. Odlum was ready to move. With renewed rumors of his interest, Salt Lake's financial community started to crackle.

In late December 1954, Federal's president, Reed W. Brinton, met with W. L. Davidson, a self-proclaimed preacher who headed the Davidson Syndicate, one of Odlum's primary Albuquerque affiliations. The syndicate represented a diverse group of companies with holdings throughout the Colorado Plateau. The association was looking for an umbrella corporation to give themselves a common identity. Federal seemed a likely prospect.

On December 28th, the two entities agreed to merge as Federal of Utah. Stock commenced trading at $3.75 in January. In a single month, the price soared to $5.25. By February, shares sold at $8.12.

Obviously, this new concept of merging uranium companies was profitable. Consequently, other firms made tracks to Odlum's suite at Hotel Utah. "There were fellows that would come in with claims and Odlum would call meetings to have these things approved," Reed Brinton said. "We tried to talk him out of some of his acquisitions. . . . We felt that he was watering the company too much by issuing shares for dubious properties. But he was in control. He did it that way and he was willing to take anything. But ultimately, what we accumulated had enough value that Federal became very profitable."[16]

By late January, Odlum had swallowed up so many companies that it was decided that the conglomerate would have to form a whole new corporation. In April, Federal Uranium of Nevada was incorporated. The merger agreement not only included Federal of Utah and the properties of the Davidson Syndicate, but it brought in Kentucky-Utah Mining, Howell Mining, Utida, Western States Mining, Interstate Mining, Santa Fe Uranium, United Incorporated and U and I Uranium. Federal became one of the largest companies on the Colorado Plateau that was exclusively engaged in mining. It also triggered some of the most fascinating exercises on the penny stock market.

In order for Odlum to accomplish his take over of the local companies, a "standard of value" had to be established. Odlum used the market price of Federal of Utah in this instance as the stock had been trading for a number of years. Walter Winchell's mention of the company on the air had attracted stockholders all over the country, and Odlum's personal involvement added the excitement of "big business" and high finance. In a way, Odlum brought credibility to the local uranium stock play as Charlie Steen's discovery of the MiVida proved the viability of uranium prospecting.

Odlum made a reverse split back to one share for every fifty when he acquired Federal of Utah. This cut the existing ten million shares down to two hundred thousand, thus providing a very thin market, as well as a high price which was easy to maintain.

When Odlum's merger was announced, takeover fervor roared to a high. There was no such thing as a restriction on insider trading. Officers, directors, stockholders, accountants, consultants and stockbrokers were all aware of the companies being acquired and the ratios on the share exchange basis. The prices of all of the stocks started to move up, and the prices were being determined by the price of Federal Uranium of Utah, which was leading the parade of acquisitions. Two shares of Federal of Utah was established at one share of Federal of Nevada. Odlum kept the old Utah corporation at about nine dollars, so the Nevada stock was worth eighteen dollars per share.

The market on the other merged corporations was much lower. Kentucky-Utah traded at ten to one. Howell Mining was twenty-five to one. Utida sold at thirteen-and-a-half to one, and Western States Mining went at a monstrous four hundred to one, and so forth. Thus a substantial price disparity for Federal Uranium of Nevada existed. If a person exchanged shares of Federal Uranium of Utah, the shares in the surviving corporation Federal Uranium of Nevada should have a market price of eighteen dollars. But if one exchanged shares of one of the other takeover companies, the market value would only be ten dollars.

"The arbitrage opportunities looked awesome. You'd sell for eighteen dollars and buy back for ten dollars," Coombs said later. "But," he added, "there were problems. You had to have the shares of Federal Uranium of Utah to sell or borrow them, and no one would loan you the shares because they wanted to take advantage of the arbitrage."[17] Coombs was quick to recognize the possibilities. He started a market in Federal Uranium of Nevada on a "When Issued" basis. (Not legal today, brokers at the time could defer delivering stock certificates until they were issued. If the certificate was never issued or a merger did not go through, the trade would be cancelled.) Coombs sold Federal Uranium of Nevada at seventeen dollars a share and bought back Kentucky-Utah and Howell Mining on the basis of ten-dollar Federal Uranium of Nevada shares.

"For five days I had the market all to myself," Coombs said. "This was probably my biggest trading coup in the market."

The spread closed down to less than one dollar as other brokers became involved. Coombs had made over seventy-five thousand dollars, but there was one problem. He had purchased $225,000 of local stocks and the brokers were delivering the stocks for payment. Coombs didn't have the money to pay them. To make matters worse, he didn't have the "When Issued" stock to deliver to the brokers he had sold it to because the certificates had not been printed yet.

Coombs hurried down the block to the Continental Bank and Trust Company. Rushing into president Walter Cosgriffs office, he said, "Walt, can you loan us $200,000 for a few days until we receive the Nevada stock?"

Cosgriff agreed. But it was sixty days before Coombs could pay off the loan.

11   SUCCESS AND SUBPOENAS

SUCCESS TO CHARLIE STEEN was something like baking in a sauna and then dashing outside to roll in a snowbank: gratifying, exhilarating, and sometimes downright uncomfortable.

It started quickly. Once the Mi Vida mine was in production, Steen's lifestyle cartwheeled. No longer was he forced to shore up a ramshackle tarpaper shack to house his family. He lived on a lofty mountaintop in a $250,000 mansion that had a swimming pool, a greenhouse, and a separate cottage for servants. The days of pounding the pavement to find a grubstake were behind him. He was now an established executive. Utex Exploration Company. Moab Drilling Company. Mi Vida Company and Big Indian Mines, Incorporated. The man who had once camped out with a single helper and his mother as a cook had bought and remodeled the old Starbuck Motel in Moab to accommodate scores of his employees. Later, as their numbers increased, he even developed a 160-acre residential subdivision for them. He named it Steenville.

Steen enjoyed his new riches and notoriety. The clock on the mantel next to his bronzed work boots was always set for the cocktail hour. He entertained prominent people like Elliott Roosevelt, automotive industrialists E. L. Cord and Joseph Frazer, presidential cabinet members Stewart Udall and Orville Freeman, and Utah's Governor J. Bracken Lee. Hollywood celebrities—Henry Fonda, James Stewart and Dorothy Malone—were among the notables the Steens hosted in a continuous round of parties. Lavish dinners catered by famous chefs introduced Moab to unfamiliar delicacies such as Blue Point Maine oysters on the half-shell. Clyde McCoy, Bobby Bears, mariachi bands from Old Mexico, and Lola Montez with her entire ensemble of dancers were flown in to entertain guests.

Steen's employees and the local citizenry were not left out of the festivities. By 1954, the Utex Discovery Party had graduated from the Arches Cafe, where the first anniversary extravaganza took place, to an immense new hangar at the airport. Three thousand guests circulated throughout the

Steen built his hilltop mansion because he always wanted to "be up on the hill looking down."

Steen's bronzed prospecting boots commanded n important place on the mantel. *Western Mining and Railroad Museum.*

evening. Caterers ordered fifty hams and stirred up three-quarters of a ton of potato salad. Pickles arrived by the barrel. Whole heads of cheese and truckloads of bread were delivered for the occasion.

The entertainment topped anything that Moab had ever seen. The internationally-famous Caroliers clanged their bells. A chorus line squirmed and kicked. A magician tricked the eye and a comic "did his stuff on parallel bars." Dancers cavorted until sunrise to popular hits played by a dance band imported from Salt Lake City.

Steen wasn't adverse to "throwing away his money" on parties, but recreational gambling did not usually appeal to him. He figured he got more action in the mining game.

"Some sixty-five million years ago uranium was shot up from the center of the earth," Steen told Maxine Newell, author of *Charlie Steen's Mi Vida*. "Now I come along and tell my drill crews, 'Over there's where you'll find it.'. . . It's a gamble. You can lose $5,000 in five minutes drilling. A diamond bit costs $1,000; five hundred feet of pipe $1,800. The tables of Las Vegas bore the hell out of me."[1]

But a "friendly little poker game" was different. He played frequently with Bill McCormick, Mitch Melich, Bill Hudson, Andy McGill and Dr. I. W. Allen. These were no penny ante sessions. The stakes escalated as the players' uranium interests grew. One hand was so memorable that Steen commissioned Utah artist Dean Fausett to memorialize the event in oil.

McCormick, Melich, Allen and Steen were seated around the felt-topped game table in Steen's amusement room. Hudson and McGill were watching the play. Dr. Allen shuffled the deck and started to deal out five cards to each player.

McCormick opened his hand. He held three tens.

"Fifty dollars," he said. He placed his bet and then leaned back to light a cigar.

Steen was looking at two pairs. "I'll call," he said. He tossed in some chips.

"I call," said Melich, pushing out a stack of counters.

"Call," said Allen, adding his fifty.

"Cards?" Allen asked.

"One," said McCormick, flipping out a discard.

"One," said Steen.

Melich and Allen each took one card. Melich was drawing to a straight. Allen was drawing to a flush. Neither man hit, so they folded and threw in their cards.

Artist Dean Fausett memorialized the famous poker game in oil. Left to right: Mitch Melich, Bill McCormick, Bill Hudson, Charlie Steen, Andy McGill and Dr. I. W. Allen. *Mitch Melich.*

McCormick blew out a plume of grey smoke. "I'll bet a hundred," he said. He moved more chips toward the heap.

"I'll raise you five hundred," said Steen.

The two men raised and counter-raised. The pot doubled. Tripled. Soon it was up into thousands of dollars. Finally, $10,000 was riding on the pot.

McCormick leaned back in his chair. "What have you got?" he asked Steen.

"Two pair."

McCormick grinned. "I've got four tens," he said, and started to wrap his arms around the mound of chips.

Steen reached out and laid a hand on McCormick's arm.

"But my two pairs are Aces," he said. He had been dealt four natural Aces before McCormick caught his fourth ten on the draw.[2]

Steen couldn't seem to lose. It was as if everything he touched turned to thousand dollar bills. And his new style of high-living came easily to him. He chased sportsfish along the coast of Baja California in his sixty-five-foot yacht, the *Minnie Lee.* He cheered the Kentucky Derby runners in the company of Texan oil barons. He reveled in expensive pranks. When *Salt Lake Tribune* humorist, Dan Valentine, asked Santa for a ton of uranium in one of his Christmas columns, Steen complied. He had 2,240 pounds of the ore dumped on Valentine's driveway on Christmas Eve. Stuck in the

Dean Fausett oil painting of Charlie Steen playing with his four sons and Butch, the Dalmatian. *Western Mining and Railroad Museum.*

pile was a sign that read: "Merry Xmas, Dan. I intended to send you a 'Gold-plated' Cadillac, but you said you wanted a ton of uranium ore . . . so here it is. Happy, happy New Year. Charlie Steen."

When Steen took a cruise to Europe, he sailed first class on the *Queen Mary.* He walked his mile around the promenade deck in the company of the rich and famous. But Charlie Steen, always a nonconformist, was not about to be intimidated by the snooty "social set" aboard the luxury liner. Every night he strutted into the elegant dining salon with his fashionably-gowned, bejeweled, and now blonde, wife on his arm. Everyone in the room was dressed formally. Everybody but Charlie Steen, that is. He wore a business suit. Other diners looked askance as they followed his progress through the hall. When Steen later attempted to get into a bridge game, he was rebuffed. The Steens were snubbed as "nouveaux riches" throughout the crossing. Then, on the last night, when all of the other men had packed their tuxedos, Steen appeared for dinner wearing black tie.

"I wanted to show them that I had a dinner jacket," he said.[3]

All in all, Steen preferred traveling on the airlines. For shorter trips, he used his Cessna 195. Business took him all over the United States. Maintaining a killing schedule, it wasn't unusual for him to have trips to

Washington, D.C., Chicago, New York, El Paso and St. Louis almost back
to back. One time he was so heavily booked for speaking engagements that
his attorney wired: "For hellsake get off TV. Suggest you hire John Wayne.
All rigs stuck. Come home quick."

Often his flights were frivolous. He flew to Salt Lake City for weekly
rhumba lessons. When there was a special television program that he wanted
to see, he would have his pilot circle the plane over Moab so that he could
watch the tube where the reception was not hampered by mountains. One
time, after attending a one-hour meeting in Grand Junction, he impulsively
outfitted his crew and took off on a spontaneous thirty-eight day lark.

Steen enjoyed the self-gratification and power that came with wealth.
But he retained his innate spirit of generosity. He recognized the changes
coming over the unsophisticated little cowtown of Moab and felt somewhat responsible for them.

Moab was being touted as "the Uranium Capital of the World."
McCall's magazine claimed it was "The Richest Town in the U.S.A." Writer
Elizabeth Pope noted that there were twenty millionaires for every two
hundred and fifty citizens. This was forty-eight times as many millionaires
per capita as there were in the entire country. The "undersized, unlighted,
uncertified and mountain-surrounded airport" ranked second in the
nation for private planes per capita, Pope reported.[4]

The town was about ready to burst. There were shortages of everything. Facilities that had comfortably serviced Moab's twelve hundred
inhabitants the year before were stretched to accommodate an exploding
population of seven thousand. Water lines were overtaxed. Electricity
within the city limits was generated by a small diesel engine and a hydroelectric plant. Sometimes in the middle of a movie the power at the theater
would go off. The audience, herded outside, would sit patiently at curbside
until the picture could resume.

Facilities at the one hospital in Moab became inadequate. Eleven different religious denominations of the Community Baptist Church took
turns worshipping in the Mormon Chapel. Children at the crowded
schoolhouse were sharing desks and even sitting on the classroom win-
dowsills. Sometimes they took instruction out-of-doors.

There wasn't even an undertaker in the hot desert town. When the
lights went on in the basement of the Mormon church at night, it was the
signal to the neighbors that someone had died. The telephones started
ringing and word spread through town. Everyone was asked to deliver ice
throughout the night to preserve the body.

Housing was critical. People were renting rooms in private homes, crowding into a handful of motels. Mobile trailers filled vacant lots. Tents were pitched along the ditch banks outside the city limits. Some uranium seekers were even sleeping in their cars.

The Midland Telephone Company had a two-position board and five long-distance circuits. Suddenly mobs of uranium promoters were frantically attempting to call Salt Lake City, Denver or New York. In desperation, the phone company used its Western Union printer to Grand Junction to relay messages. From Grand Junction, the dispatches were teletyped to Provo, Utah, for communication to the desired destinations.

Emma Dalton, Grand County recorder for twenty years, resigned. She "couldn't keep up with the job." In sixteen years, she had used a single record book to register mining claims. Now she filled up one of the thick, heavy journals in a week.

Steen did what he could to remedy the situation. He bought and developed buildings and land for offices and housing. He donated $50,000 towards the construction of a new hospital. He gave the schools property to build on and contributed acreage to four denominations to construct churches. He transferred the title to a spring to the city for a fraction of its actual value.

"When I hit Moab in the beginning, there probably wasn't more than a half dozen indoor toilets in town and damned few window shades," Steen mused. "Riley's, for instance, that little fly-specked drug store there, they lived during the summer for the tourists and during the winter they lived on buckskin. I certainly left Moab a much better place to live than when I first got there."

Yet Steen was plagued by small-town resentment. Some locals were aggravated by the fact that a fellow from Texas could come into their country and take from the land what was "rightfully theirs." Petty jealousies surfaced.

"I put my well on the city water system," Steen said, "and got a bill for $1,200 for one month. When I went down to complain, the City Council said I was trying to take advantage of the taxpayers by expecting a special rate on my water."[5]

Steen's family was not spared similar annoyances. The boys lived under the onus of being from "the big house on top of the hill." Schoolmates instigated fistfights. When Johnny Steen got his first car, he was repeatedly challenged to "drag Main Street." Finally, the four young Steens escaped to military school.

M. L. Steen was unprepared to handle the envy generated by her designer clothes and the $132,000 worth of jewels she wore.

"I love the rings and other jewelry Charlie's given me," she told *Cosmopolitan* writer Murray Teigh Bloom. "But they made a lot of people jealous when I wore them to parties. One night, while doing the dishes, I wore a diamond necklace and earrings Charlie had just given me. I couldn't resist turning to our guests and joking mildly, 'The things you have to do to keep a maid nowadays!' Everyone smiled, but I could see they didn't think it was really funny."[6]

Perhaps Steen's biggest aggravation was the tidal wave of "goat pasture" prospectors and boiler-room stock promoters who bilked the public into investing in worthless properties "that were right next to Charlie Steen's."

"They'd take a great big map on a large scale and move their property right next to mine," Steen said. "That was a favorite trick."

It wasn't long before he learned that subpoenas followed in the shadow of success. Almost on cue with the Mi Vida mine getting into production, Steen found himself entangled in an extended series of lawsuits.

One prolonged session of court battles was waged against claimjumpers. Opportunistic prospectors scrutinized Steen's original twelve claims with the intent of discovering flaws in their recording that would enable them to "overstake" the prospects. It was just like the gold rush days.

A man named Charles Yetter professed that he had uncovered a small body of unrecorded territory in the midst of Steen's prospecting line. He staked out a triangle right in the middle of the property which he stated had not been covered by valid mining claims. Hearing this, Steen exploded. On February 4, 1954, he placed a front page ad in the *Times-Independent*.

"To Whom It May Concern:

"In April 1953, Charles Yetter, a consulting geologist, visited the Mi Vida mine of the Utex Exploration Company and stated that he represented New York capital and wished to purchase the Utex property. He was given confidential information concerning Utex and adjoining property."

The ad went on to state that Utex refused to sell. Yetter returned in July with representatives of the Simpson Mining Company and Engineer Associates of Grand Junction. The men jumped Steen's claims and the Big Buck, owned by Donald Adams and Dan Hayes.

"The attempted locations by Yetter and his associates were made secretly, clandestinely, surreptitiously and fraudulently," the ad continued, "and with the attempt to steal and take from those who, in good faith, after considerable hardships, have located their claims according to law.

"I am advised that this 'Claim Jumping' is a common practice of Yetter and his associates on other valid located property on the Colorado Plateau.

"I regret that Mr.Yetter and his associates are bringing back the conditions which existed in many of the old mining camps when committees of vigilantes were formed to protect their properties against 'CLAIM JUMPERS.'"

On March 11th, Yetter and Simpson entered a million dollar libel suit against Steen.

"We sued him in the Federal Court in Salt Lake City with a blue ribbon jury," said the plaintiffs attorney, Charles Traylor of Grand Junction. "The court ruled that it was libelous per se and all we had to do was prove damages. But, it was during the uranium days and all those people's businesses had picked up or tripled, and they couldn't show much damage. I think we only got about $25,000."[7]

Another litigation resulted from a two-year lease in a portion of the MiVida mine. On February 11, 1953, shortly after mining commenced on the property, Dan O'Laurie leased a two hundred-by-six hundred foot section of the mine to G. and G. Mining Company, owned by Archie Garwood, R. C. Gerlach and W. E. Bozman. O'Laurie made the deal while Steen was spending Christmas in Houston. When Steen returned, he had his first quarrel with O'Laurie over the agreement. He considered it a financially ignorant move. But the company was bound by a written lease.

Three years later, the two parties were in court. The fight centered around an ore-bearing sandstone pillar that had been painted green.

The MiVida consisted of a network of "rooms" and pillars that were engineered by running a series of twenty-foot-wide drifts (side tunnels) through the ore body. The blocked-out pillars measured eighty square feet. These blocks were further developed by driving additional twenty-foot drifts, which left alternating twenty-by-eighty-foot rooms and pillars. Thus, a portion of the ore was mined and the pillars were left to support the back of the workings. When the extent of the ore body had been mined in this manner, crews started at the far end of the property and "pulled" (drilled and blasted) the pillars along the strike towards the entrance. A few pillars were left in place as large, and for as long a time, as possible to avoid collapse of the mine. Sometimes, removed pillars were replaced by timber roof supports. Mined ore was then loaded with shovel loaders into a Utex Company diesel truck and a G. and G. Company trailer for transport to a surface stockpile. Both companies shared a haulageway that had been driven into the mountain by G. and G. Mining Company.

All went well until the summer of 1954. Then began the battle of the pillars. G. and G. wanted to "pull" all of the pillars in the common haulage and passageways. Utex objected. Garwood and Gerlach claimed that they had title to the pillars in question because of their lease. Otherwise, Utex should buy them. Steen figured that the lessees had no right to try to force him to purchase pillars that were rightfully his anyway. G. and G. had made enough money on their lease. He was entitled to the rest of it.

Finally, on August 23rd, Garwood, his attorney George Dilts and his mine superintendent Joe Trudgeon squared off with Steen, Mitch Melich and MiVida superintendent Virgil Bilyeu. They gathered in the wood-pannelled office of Utex Exploration Company in Moab. No one was smiling

"We have prepared a map for you and want you to look at it," Steen said, unrolling a map of the MiVida mine in the middle of the floor. "And we have marked areas to be mined and areas not to be mined. The areas painted red can be mined. Those in green must be left."

"Are we going to be paid for them [the green pillars]?" Garwood asked.

"No," said Steen. "Not unless the court says so."

"It seems to me that we have the right to mine the pillars, all of them, under the terms of the lease," objected Superintendent Trudgeon.

"Not unless the court says so," Steen repeated.

Discussions ended.

On October 20th, Garwood and Gerlach sent a letter to Steen. Utex Exploration Company had the right to purchase all, or part, of the pillars along the haulage and passage ways of the MiVida, they stated. Based on the current AEC purchase price and the quality of the ore already removed, the estimated total of 11,326.5 tons of uranium, with an average value of .40 percent uranium oxide, would be worth $215,203.50. If Utex didn't want the whole package, Steen could have the option to purchase pillars along the haulageway only, at $97,689.45. Or he could take those along the passageway, valued at $89,687.60. Should he choose either of these alternatives, he would be required also to buy the pillars where the haulage and passageways joined. The price for that section was $27,826.45. Steen had thirty days to accept or reject the offer. If he refused or if G. and G. had not heard from him in the proscribed time, they would commence pulling all of the pillars.[8]

Steen was on the phone to his attorney before he had reached the signature on the letter. Then he got hold of Superintendent Bilyeu. The offenders were soon off the property.

Around the 18th of November, trouble erupted. Bilyeu walked into the G. and G. workings and found the miners drilling pillar number nineteen in

Jumbo drills in the Mi Vida mine. *Western Mining and Railroad Museum.*

Mucking ore in the Mi Vida. *Western Mining and Railroad Museum.*

Charlie Steen underground with his son Mark and his pure-
bred Dalmatian, Butch, who was always at his side. *Western
Mining and Railroad Museum.*

preparation for inserting dynamite and blasting it. The pillar had diagonal
green stripes painted at the base. It was one of the columns that Steen
wanted left standing. Bilyeu ordered the men to stop. They ignored him.
The superintendent left.

In a few minutes, he returned. This time Bilyeu carried a prospector's
pickaxe. He was followed by a group of about twenty miners, some of
whom were brandishing monkeywrenches. A DC loader cat tractor and a
large Utex dumpster used for hauling ore chugged behind them.

"Stop that drilling," Bilyeu said. "If you don't stop voluntarily, we'll
stop you. And we'll go to any limits."

One of his men picked up a piece of pipe. Another leaned against the
rock wall and slowly beat the ground with a two-by-two board.

The machine-gun rattle of the drills persisted. The muckers and haulers went about their work.

Bilyeu grabbed a piece of cable on the line that carried air to the drills. He tied it to the dumpster. Then he signalled the dumpster driver to back away. The air line came down. The drills fell silent.

Still the G. and G. men kept on hauling and loading ore.

Bilyeu waved to the cat driver. He motioned him to drive the tractor to the bottom of the incline. Then he told the dumpster operator to park at the top. The haulageway was totally blocked.

All the while, the Mi Vida miners watched quietly. They swung their wrenches and makeshift weapons, but made no move to violence.

The G. and G. drillers hooked a large rubber hose where the steel air line had been broken and got the drills working again. Bilyeu nodded to the cat operator. He drove the DC loader up against the air line and mashed it. The drilling stopped once more. Bilyeu and his men left.

The next day, the Utex men were back. This time, they removed the DC loader cat from the bottom of the incline and drove the dumpster about halfway down. Then they turned it crosswise in the tunnel. There was scarcely room for a man to pass, much less any heavy mining equipment. Work on the pillars ceased. The G. and G. miners were idled. They sat out their shifts by playing cards. A Utex guard stood by and watched. The impasse lasted until November 23rd when Garwood and Gerlach moved out their equipment.

A lengthy and bitter lawsuit followed. Finally, on July 10, 1956, United States Federal Judge Willis Ritter awarded the plaintiffs $244,488.58, with eight percent interest per year until the amount was fully paid. In his opinion, Judge Ritter stated:

". . . the lessor owner evicted these lessees from the mine, prevented their having an opportunity to follow through on their thirty-day notice to mine these pillars or have lessor buy them; and the net result of the whole situation is that Mr. Steen has now acquired that mineral body and has not paid a dime for it. He acquired it, it seems to me, by his own wrongs."[9]

As the lawsuits multiplied, Steen's idealism turned sour. It wasn't just claimjumpers and avaricious business associates who were after him. Many who had been close to him and with whom he had shared his good fortune challenged him in court.

"I have experienced the reasons why most prospectors sell and run," he told the Colorado Mining Association.

One courtroom battle was waged against Pete Byrd, Steen's roommate during his freshman year in college. Although Byrd had only seen Steen once in fifteen years, he heard about Steen's doings in the fall of 1952, when his sister sent him a clipping from the *Los Angeles Times*. The brief article had a Grand Junction dateline and said, "A geologist by the name of Charles A. Steen claims to have found a fabulous deposit of uranium near Moab, Utah." Byrd's sister had attached a note, "I wonder if this is the Charlie we used to know." Byrd, in turn, scribbled, "Well, if it is, congratulations," and sent the notes and article to the Grand Junction paper. He asked them to forward it to Steen. There was no reply.

Then, unexpectedly, one day in December, Steen appeared at Byrd's ranch in Texas. He urged his former roommate to move to Utah. He had a plan to homestead some land in a kind of syndicate development. They would take over Bob Barrett's lease in Fisher Valley and expand from there, using uranium drill rigs to bring in water wells. Byrd liked the idea.

"So I loaded up my family and borrowed a thousand dollars," said Byrd. "I bought a two-bale cotton trailer and loaded all our stuff and headed for Utah. We got about thirty miles and the trailer broke down, so I hired a cattle truck and we just loaded the trailer on the truck with a horse and half a dozen cows, and moved to Utah. . . . But, we never got off the ground. I broke my foot. Then the money started coming in and Charlie wanted an airplane. And a party that lasted for fifteen years started."

Steen purchased his Cessna 195 on May 16, 1953, and Byrd became his pilot. Byrd flew for him until January 1954.

"In that time, I flew half as much as I flew ten years in the Marine Corps," Byrd said. "It was a real rough way to make a living."

After seven months, the two men decided to purchase the Moab airport. Steen offered to help buy the run-down facility and put his old college "roomie" into the aviation business. Byrd would resign as the Utex pilot and take over management of the airport operation. Two years later, they were in a lawsuit.

Byrd and Steen had widely divergent views concerning operating procedures. Steen, with his usual candor, made his opinions known. Byrd reacted by placing an ad in the *Times-Independent*.

"Charlie Steen is not connected in any manner—financially or otherwise—with my company," the notice read.

Steen sued.

The battle centered on whether the Utex Aviation Company, later called Byrd Aviation Company, was formed on the basis of a partnership or a loan. Byrd claimed that Steen's investment in the airport was both.

"While I was out of town, prior to getting into this thing," Byrd later said, "he'd signed a contract with a boy named Fred Frazier to buy the airport for $20,000, which was double the amount we'd agreed we'd pay him. When I questioned him about it, he said, 'Don't worry about it, it's all tax money, anyway.' I ended up paying Fred Frazier the $20,000 completely and, in addition, Charlie later claimed that all the money he'd put in was loaned, rather than a partnership contribution. He claimed we owed him $12,000."[10]

"I've only been a plaintiff twice in my life and he was one of them," Steen said, recalling the Byrd incident. "Byrd let it [the airport) go to pot and he made the mistake of telling me he wasn't going to pay, so I took him to court and I won."[11]

Steen contended that the money he invested in the airport was strictly a loan. After Byrd placed his ad in the paper, Steen thought the pilot would return the money. But months passed and the loan was not repaid.

"He went around town, telling people I wouldn't dare sue him for it because it would make me look like the world's biggest cheapskate," Steen told Murray Teigh Bloom. "After a few months of wrestling with myself, I sued him anyway, and got my money back. Maybe it was small of me, but that's what I did."[12]

The judge ruled that when somebody claims a person owes them money and that person makes a payment, the debtor has acknowledged the loan as a valid obligation.

Steen gradually learned to steel himself against the hurt of these confrontations. So many of his gestures of generosity had backfired. One by one, those whom he had helped seemed to be turning against him. He became less trusting, more resentful, more determined to show people he didn't need them. He had reaped his success through the efforts of no one but himself. The Mi Vida was a giant. He had made a fortune on his other leases. More than ever, he was bent on building an independent mill. Few people thought that he could do it. He would show them.

By the mid-1950s, America was experiencing the greatest mining boom in its history. Almost six hundred producers on the Colorado Plateau were shipping uranium ore. Ore production was doubling every eighteen months. Employment in the industry topped eight thousand workers in mines and mills. Yet the number of processing mills to transform raw rock into yellowcake were not keeping up with the escalating production.

The AEC had turned the tap and engendered a flood. To spur exploration by individual prospectors and mining companies, the commission averaged one-million feet of test drilling per year. They paid out over $3,725,000 in bonuses. They tamed the Plateau with 993 miles of access roads. The Grand Junction office received more than three thousand visitors and processed an excess of six thousand pieces of mail each month. AEC geologists assured the inquirers that thousands of square miles on the Colorado Plateau remained to be explored.

But the government made it tough for an individual to build a processing mill. Since Uncle Sam was the only customer for uranium, it was up to the federal government to decide if there was enough ore in a particular district to warrant building a new mill. Any individual considering getting into the milling business had to be "Q-cleared" by the F.B.I. to prove that he was dependable and had not been connected with any subversive organizations. Only after such approval could persons have access to the classified information they would need.

Steen went to Grand Junction in 1954 to discuss the possibility of his mill with the AEC. The current manager of the AEC compound was Sheldon A. Wimpfen. Steen had known Wimpfen at Texas School of Mines and Metallurgy, after which Wimpfen had spent twenty years working in various phases of the mining industry: exploration, mine operation, administration, editing mining journals. Finally, he became assistant director of the AEC Division of Raw Materials, under Jesse Johnson in Washington, D.C.

In 1952, Johnson sent Wimpfen to Grand Junction to evaluate the uranium procurement program. Wimpfen was unimpressed. He wrote a "White Paper" outlining his assessment of the situation. He reported that the Grand Junction office was not well organized. People were all doing things their own way, with no common basis. There was an Exploration branch that reported to Phil Merritt in New York. And there was a Raw Materials office under the jurisdiction of George Gallagher in Washington, D.C. Wimpfen felt that the exploration and procurement programs should be consolidated under a single operations office at Grand Junction. The manager would report directly to Jesse Johnson.

Wimpfen also felt that the AEC was ineffectual in its incentive program to encourage prospecting and mining. He was convinced that uranium production could be doubled within a year. The circulars offering bonuses and other encouragements to prospectors weren't doing the job. Prices for uranium should be raised. More buying stations should be built. Additional lands should be withdrawn from the public domain for exploration and mining.

Washington was impressed with Wimpfen's "White Paper," so impressed that Johnson transferred him to Grand Junction to establish and manage the

Sheldon A. Wimpfen, manager of the AEC Operations Office, was delighted that a colorful person like Charlie Steen made the first uranium bonanza. *Sheldon A. Wimpfen.*

operations office he had suggested. Wimpfen was given unlimited authority to acquire uranium ores and concentrates and could execute procurement contracts up to $400 million. He could hire and fire employees, as needed. He was authorized to make capital investments up to two million dollars. All without asking Washington.

Wimpfen put his plan into action. By the time Steen approached him about building a mill, the uranium boom was well under way. He was pleased that his former college associate had hit it big, for Steen's sake, and for the AEC program.

"I couldn't have been more delighted because he was one of our first millionaires," Wimpfen said. "That was what we needed . . . that flair, publicity attached to someone who was on his uppers. We need guys like that. He's a departure from the norm and that's the kind of guys that civilization makes advances on. We're not going to make much progress with the ordinary pedestrian-type individual."

Wimpfen was interested in Steen's proposal and encouraged him to go forward with it. As ore production was multiplying, mill capacity was becoming critical. Besides, he was a believer in more involvement of private enterprise.

"I pointed out to him that in the history of mining very few of the individuals that found the deposit were able to hang onto it and carry it to a successful conclusion where it became a corporate entity with a full vertical line . . . mining through processing to sale of the finished product. I encouraged him to do it."[13]

With the AEC's blessing, Steen and Melich, "both country boys inexperienced in mill construction and operation and having less knowledge of 'Big Time' financing, plunged into the stream. . . ."[14] They went to Salt Lake City to talk with Melich's father-in-law, E. H. Snyder. Snyder, the president of Combined Metals Reduction Company, had many years of experience in metallurgy and operations in the lead-zinc industry. He agreed to go into partnership with Utex Exploration Company and form the Uranium Reduction Company (URECO) to build and operate the mill.

It was decided that the proposed mill would utilize a radically new metallurgical process. The system used in the older mills, which were designed to treat vanadium-uranium ores, were not as effective for ore of the Big Indian District. A better choice was the new resin-in-pulp process that had been developed in AEC pilot plants. It had not yet been used in a commercial plant.

The heart of the resin-in-pulp process were tiny, tapioca-sized beads of yellow resin. After raw ore was crushed to the proper size for handling, it was acid-leached into an unfiltered slurry. Then specially sized ion-exchange resins contained in wire-mesh baskets were jigged slowly up and down in a flowing stream of uranium-bearing pulp. The beads became coated with uranium. Finally, the uranium was stripped from the resins in a chemical bath.

The mill was planned for a rated capacity of fifteen-hundred tons of ore per day. This would require commitment to the mill of a two-million-ton ore reserve.

Before Steen could get AEC approval of the project, it was necessary to prove that the uranium tonnage in the Big Indian Wash District and surrounding areas justified construction of the facility. E. J. Longyear Company was hired to undertake a preliminary engineering and metallurgical research study. The firm concluded that the mill was feasible. Steen's mines alone were calculated to have a total of 967,000 tons. The analysts anticipated that there would be an excess of one million tons from other areas in the district within the next few years.

The AEC was satisfied with the report.

Next, URECO negotiated a construction contract with Foley Brothers, Incorporated. Foley, who was associated with New York investment bankers Kuhn and Loeb and Company, agreed to set up the initial financing. From this beginning, Steen and Melich proceeded to raise nine

Charlie Steen checking ore samples underground. *Western Mining and Railroad Museum.*

million dollars for a new company using a new process, a company that was entering a field heretofore dominated by the federal government.

At first, Steen flew to Indio to see if he could interest Floyd Odlum in putting money into the project. Odlum, who was in the process of building his uranium empire, wasn't receptive to the idea. So Steen and Melich flew to New York. There they were successful. They borrowed $3,500,000 from Chemical Bank of New York for construction and working capital. They sold New York Life Insurance Company $6,200,000 of the company's first mortgage bonds. The balance of $2,300,000 in debentures was placed with various investors and stockholders. In this way, through conventional institutional and equity channels, they financed the first large independent uranium mill in the United States.

The loans and sales were contingent on URECO's guarantee of competent management. Through Ed Snyder's connection, American Lead, Zinc and Smelting Company signed a contract to assume responsibility for finances, construction and operation of the mill.

With everything set to go, Steen and Melich decided that the best location for the plant would be alongside the Colorado River on the north side of the bridge. Steen made a long distance call to Hank Ruggeri, Melich's partner in Moab, and told him to get the property.

"Hell, that's easy," said Ruggeri, somewhat sarcastically, "I don't even know who owns the land."

"You can go up to the courthouse and find out about that," Steen replied.

"How much am I supposed to pay for it?" asked Ruggeri.

"Whatever you can negotiate," came the answer.

"You mean I have carte blanche and I simply go out and negotiate a deal and buy that?" Ruggeri asked.

"Yeah," said Steen. "And I want it bought before midnight tonight."

Ruggeri hied himself to the courthouse, looked up the titles and hurried out to the owners' farm. The farmer and his son were butchering a pig. They had dunked the dead animal in a tub of hot water and were scraping the hair off the carcass.

Ruggeri asked them if they would be interested in selling the land in question. The two men shrugged and went about their chore. Ruggeri hung around and watched them, acting as if he was not in a hurry. Finally, the father suggested he wait until they were through with the pig and then come to their house.

When the farmers had washed up and had something to eat, Ruggeri knocked at their door. He repeated his query about selling the land.

"Who wants to buy it?" the old man asked.

"I'm not at liberty to divulge the name," said Ruggeri. "But it would be for cash."

"What kind of a deal do you propose?" they asked.

"I don't know," said Ruggeri. "I don't know whether it's for sale or not."

"Well, guess we'd sell it for about sixty thousand bucks," they said.

Ruggeri visualized the barren sandy patch.

"That's ridiculous," he said, "but I'll ask the buyer and be back."

The young lawyer drove back into town and called Steen. He told him the offer.

"Hell, pay it," Steen said.

"Christ! For that damn blow sand down there?" Ruggeri said. "This is ridiculous, Charlie."

"Do you realize that every day that mill is not in operation it's costing me two or three thousand?" Charlie said. "I'm talking about what goes in my pocket after I pay the taxes and everything else. You go down and pay the price. Whatever it is."

So Ruggeri returned to the farm and offered $40,000. The farmers took it. The contract was signed that night.[15]

By June 1955, everything was ready to go. The financing was obtained, the plans were drawn, the ground was broken and construction of the mill began.

Then came word that the Ute Milling Company, a wholly-owned subsidiary of Atlas Corporation, was negotiating with the AEC for a uranium mill to be built at La Sal, Utah, about thirty-eight miles from Moab.

Steen was livid. Wimpfen had told him on many occasions that ore from the Big Indian District would not be diverted to any other mill. With the Mi Vida reserves, URECO could operate at the rate of about five hundred tons per day. This would only generate enough cash flow to pay off the lending institutions and break even. There would be no profit. They needed the ores from other mines in the area. There simply was not room for two mills.

Melich suggested they talk to Odlum about making a deal, but Steen would have no part of it.

Melich persisted. "Well," he said, "there's no need of both of us sinking in this business. If we can't make a profit, why stay in it?"[16]

Steen relented. So Melich and Bill McCormick, who was a Utex stockholder, went to Indio. They offered Odlum the whole package. The mill, and, subject to approval of directors and stockholders, the Mi Vida. They would ask $7.50 a share for the mill stock and sell the mine for about fifteen million. At face value, Odlum liked the proposition. But a couple of months later, he changed his mind. He balked at the necessity of buying out all of the individual Utex stockholders. He kept trying to lower the price, but Steen refused and the deal collapsed.

A short time thereafter, however, Melich got a call from Odlum. He wanted to take another look at the proposition. In trying to negotiate with the AEC, his people were finding the bureaucracy difficult to deal with. He invited Steen, Melich and McCormick to come to New York for further discussions.

One evening, the three men and their wives were invited to dinner at Odlum's apartment. It was their first encounter with New York society, and immediately unsettling. As Odlum's wife, Jacqueline, and Emma McCormick entered the dining room, a chandelier crashed to the floor, barely missing the women. The evening went downhill from there, as the six Utahns felt ill at ease with the formal proceedings. Jacqueline's attempts to put her guests at ease unfortunately backfired. When Steen talked at the dinner table of his early days of poverty, she launched into tales of her own childhood experiences as an orphan. Soon a verbal wrestling match developed over who had been poorest.

Finally, the hostess tackled the subject of "keeping up" intellectually. "It's very important for everyone, every woman especially, to read something at least an hour a day," she said.

Emma McCormick, who managed her husband's mercantile store, laughed.

"I can't even read the paper," she said. "My feet hurt so bad, the first thing I do is get my feet in a hot tub and relax. To heck with reading for an hour."

The fingerbowls at the conclusion of the meal were the final straw. The bowls were placed on doilies at each guest's place and then fresh strawberries were offered to be served on the dessert plates. Charlie privately asked his hostess what he should do with the doily. Jackie responded with a little demonstration on how to remove the fingerbowl and doily to the side. Everyone was embarrassed. Steen seethed.[17]

From that night on, Melich handled all negotiations with Odlum.

An arrangement was made wherein Atlas' wholly-owned subsidiary Hidden Splendor Mining Company purchased thirty percent of URECO's common stock. Odlum agreed to put substantial funds into the company immediately and on an "as needed" basis for the future. He also committed the reserves under his control to the mills. This meant that, in addition to the Mi Vida, URECO would process ore from Hidden Splendor, Lisbon, Mountain Mesa Uranium Corporation, LaSal Mining and Development Company, Hecla Mining Company and Radorock Resources, Incorporated. Operations started on October 4, 1956.

Formal dedication of the URECO "coffee grinder"[18] was delayed until September 14, 1957. Then, under a balmy autumn sky, a large crowd gathered in the shadow of the russet and gold sandstone cliffs. Rows of folding chairs were lined up in front of the pennant-festooned mill. The whole town was there to celebrate the occasion. Honored guests were United States Senator Arthur G. Watkins, U.S. Representative William A. Dawson, Utah Governor George D. Clyde, and Jesse Johnson, director of the AEC Raw Materials Division.

Mitch Melich welcomed the guests.

"These ceremonies for the dedication of the Plant are the fulfillment of an idea—or a dream—," he said. ". . . ours has been a continuous struggle to mold together into one ball a business venture which today is most significant, not only to the economy of Moab and the entire State of Utah, but to the defense and general welfare of our nation."[19]

Next, Charlie Steen took the podium to give thanks to "the Precious Few who conceived and fought from the beginning for the building of this

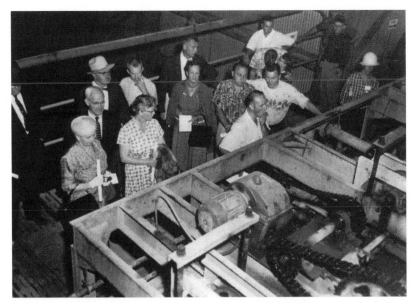

Tour of the Uranium Reduction Company mill. *Mitch Melich.*

mill—fought against the hardships of pressure, lack of money, prejudices, vitriolic abuse, and a critical whispering campaign that implied lack of faith in their integrity and in their ability to make this mill-dream of a geologist-turned-prospector come true.

"This mill symbolizes to those precious few success against almost unsurmountable odds," he continued. "No historian will ever relate the complete details of the battles that were fought in the offices and board rooms of bankers, lawyers, and public officials in Moab, Salt Lake, Grand Junction, New York, Washington, and other places at all hours of the day and night. These precious few, along with me, will carry the hidden scars and the memories of this war as long as we live."

After thanking (and issuing a few good-natured jibes to) Bill McCormick, Mitch Melich, Ed Snyder, and others, Steen concluded his talk by conferring on Jesse Johnson the degree of Doctor of Uranium.

"Greetings to all men from the University of Hard Knocks and Raw Deals," he read, "by its faculty of prospectors whose sunblistered brow, bunioned feet, seatless pants, crock haircuts, and insanely glittering eye show that they have qualified as bone-fried desert country-type prospectors and uranium hounds, thoroughly tested by blazing suns, freezing winds, reddish sandstorms, hungry scorpions, and the tall tales of crossroad and county seat barroom liars and promoters; prospectors who have blistered their rumps

Mitch Melich commends the Uranium Reduction Company mill as a significant contribution "to the defense and general welfare of our nation." *Mitch Melich.*

riding burros and jeeps, have gone without baths and women, have trekked over deserts, climbed buttes, swum rivers, run rapids and jumped arroyos—not to mention a few claims; who have sat on cactus, killed buckskin out of season, rattled at rattlesnakes, eaten stewed rabbit and porcupine, and in desperation tried a bobcat roasted in its own fur jacket; drunk skillet coffee, and smoked other men's cigarette snipes; men who have located, mined and milled uranium and who have drilled dry holes on their own without using widows' and orphans' money, and discovered that multi-million dollar lawsuits are actually filed.

"By virtue of the authority of having been graduated from all of the above experiences with heads bloody but unbowed and still unable to resist the 'click-click' of a probable Jackpot on a Geiger counter, we do hereby confer on 'Jesse Concentrate' Johnson the radioactive degree and title of Doctor of Uranium and do declare him entitled to all the rights, privileges, and appurtenances attendant on such a degree because of his having had comparable training and having shown the ability to survive in the jungles of federal bureaucracy under fire and still coming out with uranium concentrates."

The proclamation was signed "Charles A. Steen, Chief Prospector of the Royal Order of Ragged-Assed Prospectors, E. H. Snyder, Pioche-Type 'Claim Jumping' Desert Rat, William R. Square Deal McCormick, Treasurer of Hardly Able Miners, and Mitchell Maybe Melich, Barrister at Claim Jumping."[20]

After receiving his diploma, Jesse Johnson concluded the ceremony with dedicatory remarks. He congratulated Steen for making "the first important uranium discovery made by drilling in an area having no outcrops of commercial ore. This led to the general recognition of the importance of geology, particularly geologic structure, in uranium exploration."

As for the mill, Johnson stressed that the financing of the facility also had made history.

Charlie Steen introduces Jesse Johnson, director of the AEC Raw Materials Division, at the dedication ceremonies of URECO. *Mitch Melich.*

"To my knowledge, this was the first large uranium operation anywhere to be financed by New York banks and life insurance companies. When this feat had been accomplished, uranium mining and milling had become financially respectable."

The dedication concluded with a standing ovation by the audience. Everyone was thrilled with this fine new plant. The AEC was happy because the mill represented the successful conclusion of its research in pilot plants, and an increased flow of yellowcake. The politicians were happy because the mill created a fine new tax base for the city, county and state. The merchants and local citizens were happy because the mill represented a large payroll and stable employment.

There was, however, one person who was not very happy: Rose Shumaker, Steen's mother. She had, in her opinion, a valid reason for her objection to the mill. Steen had brought Ed Snyder into the partnership. Snyder was a close associate of President Herbert Hoover. In Rose's mind, the president was responsible for the Great Depression, and she didn't want to have anything to do with him or any of his associates.

Rose had been transformed from the jolly woman who mortgaged her boardinghouse in order to grubstake her son. Now, with a new home, a

Barbeque at Deadhorse Point following dedication of the URECO mill. Note the mariachi band in the foreground. *Mitch Melich.*

private airplane, furs and jewels, she was enjoying the pleasures of wealth. After divorcing Shumaker in the 1940s, she remained unmarried until about 1956. Then, in a bar one evening, she met a handsome younger man named Roy Johnson. Immediately smitten, Rose loaned Johnson $30,000 to build an ice plant in Moab. Later, she bought herself a solitaire diamond and married the man.

It wasn't long before Rose wanted to sell Johnson some of her Utex stock. Steen and the board of directors said she couldn't. Relations between mother and son became tense; they couldn't be together without calling one another names and making threats. The situation worsened as 1957 drew to a close.

Then on Christmas night, the feud came to a head. A host of friends had gathered at the Steens to celebrate the conclusion of a memorable year. As festive guests clinked glasses and Christmas lights twinkled, the doorbell rang and a United States marshal stepped through the door. He handed Steen a summons. As Steen read the notice, he blanched. Then he offered the marshal a drink.

The summons stated that Roselie Shumaker, plaintiff, was suing Utex Exploration Company, et al., for two million dollars. Rose was asking damages for "alleged conspiracy and overt acts on the part of the defendants to deprive plaintiff of the value of her stock in Utex Exploration Company, a corporation, and to prevent her from selling her stock. . . ."

Rose accused Steen and his partners of concealing information from her regarding the value of the stock, the properties and ore reserves. They wouldn't even let her duly authorized agents examine the records, she said. Worse yet, Steen was having her followed. Tapping her telephone line. Cutting off her bank credit.

This was all true, Steen stated in a court deposition.

"I told Mr. Ludlow [the comptroller]," Steen testified, "that for some time we had been bothered by leaks of confidential information concerning our business to undesirable people and strangers who were not part of the business."

Steen became suspicious about Rose's husband Roy Johnson after Roy and Rose went to New York with the Meliches and McCormicks to see the World Series. When they arrived in the city, Rose complained to Melich that she couldn't cash a check drawn on her hometown account, because "no one knew where Moab was." She asked Mitch to use his influence so that she could get five hundred dollars. Melich called his contact at the Chemical Bank and got approval for the check. But, when Rose appeared at the bank, she said she wanted $10,000. The banker gave her the money. Soon afterwards, Roy Johnson started boasting that he had placed a substantial bet on the Dodgers.

When Steen heard about this, he hired the Burns Detective Agency to check on Johnson. In a deposition for the ensuing lawsuit, he revealed that the investigators discovered "that Roy Johnson was a convicted murderer, a lifelong criminal out on parole; that he had been in prison or on a fugitive list most of his life; that she [Rose] had written a letter to the Parole Board and asked permission to marry this criminal.

"That he had kidnapped in Texas . . . a jeweler and his wife," Steen went on, "and taken them out in a thicket and tied them with wire and threatened to kill them.

"That he had been sentenced in Texas for ninety-nine years; upon his sentencing, they found he was an escaped convict from Oklahoma and returned him to Oklahoma where he was let out on parole later."

Steen told the court that Johnson's parole officer thought that the man was working a confidence game on a gullible woman.[21]

Steen was numbed and hurt by Rose's suit. This was the final betrayal, the hardest to take. When one of his attorneys expressed surprise at the depth of his anguish about the suit, Steen sent him a copy of the play *The*

Silver Cord, by Sidney Howard. The story told of a widowed mother whose domineering love for her sons almost destroyed them.

"After you read this, you'll understand," Steen said.[22]

The case dragged on for about a year. Finally, after much discussion and negotiation, the suit was dropped. Steen bought his mother's Utex stock for $1,200,000.

12 THE BUBBLE BURSTS

THE NAMES WERE APPEALING. Absaraka. Aladdin. Apache. Arrow. Atomic. Black Jack. Jolly Jack. Lucky Strike. King Midas. Newspaper readers followed daily listings of uranium stocks and figured on all of the money they could have made if they had had the nerve to buy. Many stocks went up thirty—even sixty—times the offering price in days. Soon the temptation to gamble a few bucks was irresistible. Those who had lagged behind took the plunge.

And it seemed they were in good company. The big boys were getting into the picture. How could you go wrong when Atlas, United States Vanadium and Vanadium Corporation of America, Anaconda, Homestake Mining Company, National Lead, Vitro, Kerr-McGee Oil Industries, Climax Molybdenum—to say nothing of the Santa Fe Railroad and New Jersey Zinc Company—had succumbed to the uranium frenzy? Your ordinary working guy couldn't get out there and dig the stuff himself. But he could have the fun of fantasizing about the millions he might make on a ten-dollar investment.

By the end of June 1954, eighty-one uranium firms were listed with the Utah Securities Commission or the SEC. Only a handful of the companies had as much as a whiff of ore, a fact that didn't trouble the stock promoters at all. They couldn't have cared less if a prospect had been staked in the barren Navajo or Entrada sandstone formation. As long as it was within a hundred miles of a proven mine, the product was a potential winner. After all, no one really knew where the underground trail of uranium led. Likely as not, a pod of ore would turn up in an unproven area. Look at Charlie Steen!

The scant number of stock promotions based on the legitimate possibility that ore would be produced formed a convenient backdrop for more speculative ventures. Sale of stock to finance the exploration and development of uranium mines was a vital impetus to the industry. But as time went on, entrepreneurs paid less and less attention to geologic integrity and more to the demands of a gullible clientele. Official-looking circulars and prospectuses began to appear, with "full disclosure" of a mining claim's

Penny stocks caught the imagination of thousands of novice investors. *Western Mining and Railroad Museum.*

geology, the corporation's financial statement, lists of officers, plans for developing the mine, distribution of stock and attorneys' reports.

The prospectuses were artfully prepared. The facts were "laid out on the table for all to see." The disclosures required by the SEC were adhered to. "These securities are offered as a speculation," the headline on the cover of the prospectus might say. Other pertinent information was tucked inside. For example, Standard Uranium detailed "Certain Adverse Factors" on page three.

"In the event that all of the shares offered hereby are sold and the Underwriters acquire the 143,000 shares at one cent per share," the prospectus revealed, *"the public will have paid $1,787,000, or approximately 90% of the total cost for 30% of the outstanding shares of stock, whereas the officers, directors, promoters, finder, underwriters and their associates will have paid $209,805, or approximately 10% of the total cost for approximately 70% of the outstanding shares of stock."* (Emphasis added.)

The circular then went on to divulge that, if Charlie Steen and Bill McCormick exercised their option to sell 200,000 shares of their stock back to the company for one dollar per share, the public's investment would shoot up to 99.5% of the total for approximately 31% of the outstanding shares. On the other hand, if not enough shares were sold to meet the installment payments, the claims would "revert to the optionors and all payments theretofore made [would] be forfeited as liquidated damages."

To complicate matters further, it was noted that there were "unresolved problems concerning the title" of the claims ". . . therefore, should any claims or any part thereof prove invalid, the Company would not have any right to recover any part of the purchase price paid."

Many of the "goat pasture" prospects had far less substance or indication of mineralization than Standard Uranium. Yet the disclosure documents spoke of the issuer's *option* to buy a number of claims and *his intent* to "explore and develop same."

The buyer with a check in his hot little hand didn't seem to care. He bought a few thousand shares every time he bought a house, a used car, or a load of groceries. Few prospective buyers took time to open the offering circulars, much less read the fine print.

The market started to weaken, however, after the orgy of speculation during the "seven days of May, 1954." The brokers became nervous, more cautious. Jack Coombs even tempered his enthusiasm in his newspaper column, "Uranium Topics":

> Buying uranium stocks is not a hit or miss proposition. Too many people these days are plunging into the uranium market like it was a rummage sale or a bargain basement circus.
>
> Before you invest in a uranium stock, investigate the issue.
> Look at the prospectus of the firm and do a little studying.
> Ask yourself some questions.[1]

A few days later, Coombs' yellow light flashed brighter. His column urged:

> Don't put your life savings in uranium.
> Don't hock the car—mortgage the house—withdraw your money from the bank.
> And don't borrow money from Uncle Louis!
> At the present time, uranium is an infant industry.
> It's just been born.
> And any infant industry is speculative.
> When automobiles first came out, the stocks were highly speculative—so were utility stocks.
> At one time 'smart money' scoffed at folks who invested money in the new-fangled telephone.

We strongly advise people to confine their uranium investments to money they can afford to lose.

After all, you wouldn't hock your house and your future to shoot dice at Las Vegas.

Uranium is here to stay—But a person with common sense can't ignore the fact that uranium issues at the present time are highly speculative.

That's why it's doubly important to check the background of your uranium investments.

After all, it's your money. [2]

Other warnings followed, and were flagrantly ignored by the public. Utah's governor, J. Bracken Lee, issued a statement cautioning people about blind investments in uranium stock. The next day, the penny over-the-counter market escalated.

Some of the more legitimate and conscientious brokerage houses stamped uranium stock orders with the admonition, "This Is An Unsolicited Order." Buyers still fell all over each other to purchase the latest issues.

Moab's *Times-Independent* reported that, of more than six hundred producing uranium mines on the Plateau, less than ten percent were making big money. "The remainder are putting from $50 to $75 per day into their owners' pockets—before expenses, that is—if a prospector or mining company sells stocks on the market it is a pretty sure bet that the company is not yet producing ore." Still the speculators paid no heed. [3]

Milton H. "Mickey" Love, the "one man band" of the Utah Securities Commission. *Robert Love.*

"People protectors" in the government began to express concern, however. The burden of worry fell on Milton H. "Mickey" Love, a mild-mannered, compassionate man known as the "one man band" of the Utah Securities Commission. Love, operating out of a miniscule office in the State Capitol Building with an assistant and two secretaries, was responsible for the state's real estate and securities registrations. With a law degree and some business experience, he had taken on the job in 1949 "with fear and trepidation." His first five years were relatively uneventful. Then came the uranium boom.

Love told the *Denver Post,* "We had never experienced anything like it. We had to feel our way along, with nothing to guide us but common sense, drawing up rules that seemed fair to all. We went along with two purposes in mind: first that we wanted the uranium industry developed, and second that we wanted the stock-buying public to get a fair deal."

Mickey Love had a system. Every night, after a hectic day, he carried a stack of stock registrations home with him. Into the wee hours of the morning, he would pore over the documents checking the merit of each claim. Then, next day, it was back to his desk to deal with the string of promoters and attorneys trailing down the hall with still more applications.

"Here is the list of rules," Love would tell each applicant. "Your registration complies with this, or it goes back to the bottom of the pile."

Time was of the essence. The promoters clamored to get their deals together. The bulk of registrations were for quick, intrastate offerings under $300,000, which needed state approval but were exempt from SEC filing by "Regulation A." Love's desk groaned under ever-rising stacks of paper.

One day Dick Muir and Ed Dumke appeared in Love's office wanting to register a "Reg A" issue called Southwest Uranium Trading Company. The two men could see that their application was sure to end up at the bottom of the heap.

"We have options on certain property," Muir told Mickey Love. "When do you think you'll be able to give us the go-ahead on our stock issue?"

Love measured about two inches with his thumb and index finger. "That's about a month," he said.

Then he walked the measure up the stack. "Well, you'll come along in about three months."

But Muir and Dumke were desperate.

"So, trying to figure my way out of it," Muir said later, "I thought, 'Well, if we can't do it in the State of Utah, maybe we can do it in some other state.'"

Muir called the State Capitol at Carson City, Nevada.

"I'd like to speak with your securities commission," he told the operator.

"Who?" the operator asked.

"The Nevada State Securities Commission," Muir repeated.

"Just a minute, please," came the answer.

Muir heard a flurry of buzzes and paper shuffling at the other end of the line. Finally, the operator returned.

"I don't think we have a securities commissioner," she said.

"That's very interesting," said Muir.[4]

Muir and Dumke caught the next plane to Las Vegas. The following day, they opened a branch office on the famous hotel strip, where they shared a partitioned second-floor cubbyhole with Jack Coombs, who had beat them there by a day or two. While chorus girls bartered for G-strings and sequined bodysuits in a costume shop downstairs, Muir and Dumke financed Southwest Uranium Trading Company and split a $40,000 commission in less than twenty-four hours.

Mickey Love found he had to concoct new rules as fast as he could write them. With few statutory precedents to guide him and no time for public hearings, he grappled with the problem almost single-handedly. When uranium stocks first hit the marketplace, he thought something as unfamiliar and new would be hard to sell to the public. Since the United States was so desperate to develop a domestic industry, he decided to help the brokers by allowing them to take commissions of twenty to twenty-five percent on uranium stocks sold intrastate. But the buying spree that followed soon convinced him that he must halve the allowable commissions to twelve-and-a-half percent.

"We first permitted the higher charge because we thought salesmen would have to pound the hot pavement to sell stock," he told the press. "Anyone can sell uranium stock on this market without moving out of a chair."[5]

Love required quarterly or semi-annual audited financial statements from the corporations. He notified brokers that three copies of all prospectuses were to be filed in the Securities Commission office. He assigned a flat figure of $1,000 (later $500 and then $ 300) on each claim to prevent value doctoring of worthless acres of sagebrush and dry washes that were being staked.

"A lot of people who had claims would say, 'No, this claim is worth a lot more,'" Harold Bennett, director of the Utah Department of Business Regulation and Love's boss said later. "Our experience got to the point that they were all 'motherlodes,' and no matter what the limit was—had it been $10,000—a lot of these claims would have been $10,000."[6]

Love had to get tough on the offering of unregistered shares.

"This department considers even the reading of an offering circular constitutes an offering," he warned. "The distribution of or the availability of offering circulars, the inclusion in advertising lists, or the posting of such companies on quotation boards viewed by the public will be presumed to be an offering in violation of law."[7]

And, naturally, the sale of Regulation A stocks out of state was forbidden. But Utah brokers had daily wireless contact with traders in New York, Newark, Los Angeles, San Francisco, Portland, Seattle, Spokane, Denver, St. Louis, Chicago, Phoenix, Tucson, Albuquerque, Fort Worth, Dallas, Houston, Miami and just about every other major city in the United States. Stock dealers were supposed to register issues with the state in which they sold securities, but local traders rarely observed this formality.

"We do it [sell unregistered interstate stocks] until we get a letter threatening an injunction," an anonymous broker told *Salt Lake Tribune* business editor Bob Bernick. "Then we quit. It isn't worth the candle to go through all the red tape in forty-eight states to sell a couple of hundred thousand shares."[8]

Love's mission was like attempting to hide a battered four-foot table under three-and-a-half feet of cloth. He'd get one problem covered and another would be exposed. Things were just moving too fast. Where there had been nine applications for stock registrations in 1953, there were 149 in 1954. Total registrations since the beginning in 1952 represented the sale of 837,852,000 shares of stock. Approximately twenty million dollars had been raised in primary underwritings from the Utah public. There were times when the daily volume of uranium shares in Salt Lake City topped the trading on the New York Stock Exchange.

The Salt Lake City uranium market was unique. The majority of securities dealers were young, inexperienced risk-takers motivated by greed and a hunger for action. They didn't quite know what they were doing, but they were having a ball doing it. Without realizing it, much of their activity bordered on fraud.

"In the beginning, people would come in and leave checks for different issues and when I had thirty or forty thousand dollars, we'd incorporate the company with all these people as original incorporators," said Jack Coombs. "As I got into it, I discovered that none of these issues were registered and found out that we did have a securities commission that you were supposed to file with. Anybody that bought an unregistered stock and sold it to a customer was liable to that customer. That customer could have sued him and got his money back."

Coombs learned his lesson the hard way. Ezra Gull, a former Utah securities commissioner, got five of Coombs' salesmen to raise about $50,000 for a uranium company. Coombs didn't know anything about it. Then one day a customer came into his office and said, "I bought the stock, it's unregistered, and I want my money back." Coombs had to write him a check for the full amount.[9]

Ralph and Ray Bowman were another pair of fledgling broker-dealers who got caught in a whirlwind of buying and selling. Once, after doing some fast-paced business in the East, they learned to their chagrin that their Salt Lake office had been working against them.

"The Bowmans get Mr. White from Climax Molybdenum to head up Uranium Corporation of America," Coombs remembered. "They leave the man $250,000 to run the company and they get Abram Whatner, who owned the Baltimore Colts, in the deal. Whatner is hot for the deal and gets all of his friends in it. The stock goes from twenty cents to $2.50. Whatner decides to hire an engineer to check the reports from White as to the reported ore tonnage of several million tons, all surface. The engineer reports back to Whatner that there is no ore. So Whatner proceeds to get out of the stock. At around the $2.00 level. Ralph [Bowman] is in New York selling Uranium Corporation of America to anybody who will buy. He sells off several hundred thousand shares and gets back to Salt Lake City and finds that Johnny Thomas, his brother-in-law and trader, has bought all the stock. John is sitting out in Salt Lake supporting the market!"[10]

It was a simple matter to get a Utah security-brokers license. All you had to do was file a form at the State Capitol and pay twenty-five dollars. There were no tests to pass. No special educational requirements. The Utah Code stated: "If the commission shall find that the applicant is of good repute . . . it shall register such applicant as an agent." In other words, just about anyone who wasn't currently in prison could sell over-the-counter stock.

There were twenty Salt Lake City brokers with one hundred salesmen in 1952. By March 1955, eighty brokerage firms employed four hundred and sixty-seven salesmen. Mickey Love figured that there was one broker for every two uranium offerings.

". . . some of these brokers have obtained licenses primarily to peddle a single issue to friends and neighbors," he said.[11]

Most of the trading firms worked on low budgets, with a handful of employees. Bookkeeping was haphazard. The action was so frantic that the clerical help, mostly untrained, couldn't keep up with the buy and sell orders of unsophisticated investors who gambled fifty or a hundred dollars on a few thousand shares. Sometimes a broker processed up to a million and a half dollars worth of business in a week. The clerks would churn out stock certificates in increments of ten thousand penny-shares each. It was easy to make mistakes.

Then someone would come along and plop down $25,000 for a stock. The price of the issue would zoom. Or, if that customer sold, the market would plunge.

"Because of the nature of the situation," said Noland Schneider, a trader with Coombs and Company, "you could never get your position through your auditors for your account. By the time you got your position determined, it was determined as of sixty days ago. So your trading positions were always slightly off the true facts."

One of the biggest fiascos involved the brokers' attempts to adhere to the Securities Commission Regulation T. This regulation limited a customer's period of credit to seven days. Keeping a time track on the hordes of customers was next to impossible.

"Somebody calls and orders a stock," said Schneider, "and they have a week to pay. Maybe the stock goes down substantially. So they'd say, 'We didn't order it,' or 'That was a mistake. I only ordered one thousand shares, not ten thousand.' So you could get involved in lawsuits if you wanted to. Every brokerage house in Salt Lake City lost a tremendous amount of money from people reneging on the orders that they'd placed before the stocks went down."[12]

"We came to the point that everything was bogged down because nobody knew where they stood," said Al Bain, a trader at J. A. Hogle. "You had millions of shares sold and millions of shares bought and customers on the street trading between brokers and not putting up a dime."[13]

Nationwide publicity about Salt Lake City becoming "The Wall Street of Uranium Stocks" soon attracted the attention of the National Association of Securities Dealers. This agency provided self-regulation of over-the-counter dealings in the industry. Membership in the NASD was not mandatory for broker-dealers, but non-members could not trade with any of the established offices who were affiliated with the organization. Consequently, the majority of firms joined, and tried to maintain good standing. If a broker was thrown out of NASD, he would be finished. He would have to pay full price to buy a stock and then raise the cost of the shares to sell them to the public. Competition would break him.

In June 1954, NASD staff agent Kenneth Cole landed in Utah to take a closer look at the uranium boom.

"I was one of nine men from all over the country working for NASD who was sent to Salt Lake City to examine all broker-dealers," he said. "I thought it would take two weeks. . . . We opened an office in the Boston Building on August 1st."

Cole was soon astounded by his findings.

"We'd find the principal broker was a taxi driver, a hash slinger, or something," he said. "They didn't know if securities needed to be registered or were exempt. There was no attempt at fraud; just the frenzy of an era and ignorance."[14]

Noland Schneider. *Noland Schneider.*

Fast on the heels of NASD, another watchdog came to Utah. Walter Winchell's broadcast in May 1954 that aired Federal Uranium's ore strike had alerted the SEC. The agency suspected that someone had "tipped off" Winchell in order to make a run on the market, and that there might be other unorthodox activity in the uranium business. Four federal agents arrived in Salt Lake on June 30th, and three weeks later, they opened a branch office "to provide closer surveillance of trading in uranium stocks."[15] Rumors of a pending federal crackdown buzzed up and down the street.

The "federal crackdown" came in the person of Alex Walker, Jr. Walker was an attorney, not an accountant. In January 1955, he was transferred to Utah from the SEC office in Washington, D.C., to supervise and regulate the muddled machinations of a bunch of young, green wheeler-dealers. He was alone, with one secretary, in a job he wasn't trained for that required a staff of at least ten people.

"It just seemed to be the last frontier as far as I was concerned," Walker said later. "I was born and bred in Pittsburgh and went to school in the East. I'd never been here before. The reason I was sent out here was that they looked over the staff in Washington and . . . I hate to admit it . . . I was a bachelor and they felt that it wouldn't work too much of a hardship on me to be sent to Utah."[16]

The young attorney received a quick initiation into Utah's securities scene. Making his rounds of the brokerage offices, he found every type of business operation imaginable. There were the conservative, long-established J. A. Hogle and Company and Merrill-Lynch, with their prestigious walnut-and-leather offices on Main Street. There was Doxey-Layton Company trading certificates over steamy pots of chili in the kitchen of the Grabeteria. Plato Christopulos, J. Walters, Jr.'s protégé, contributed to Walker's dilemma by keeping his record books in Greek "to fool the competition."

As if dealing with such a hodgepodge of brokers weren't enough, Walker also had to send triplicate reports to SEC headquarters of every meeting, every phone call, every decision he made.

But by June, Walker had done enough investigation to become concerned. He found that many uranium common stocks were being sold to the public in compliance with the SEC disclosure requirements, but that promoters and underwriters were getting unconscionable options and warrants on the sales. The insiders were enjoying riskless potential profits at the expense of public stockholders by diluting the customers' investment and share of earnings. In addition, most of these issues were underwritten on a "best efforts" basis. There was no firm commitment that there would be enough money raised for exploration or drilling.

"Often underwriting fees and commissions account for twenty-five percent of the offering price," Walker told the *Salt Lake Tribune, on* June 6, 1955. He warned the broker-dealers that his office would be keeping an eye on them.

The state securities office also tightened the reins. In September, Mickey Love clamped down on the insiders incorporating uranium companies through private sales to friends and relatives. He decreed that issuers of stock must put in escrow eighty-five percent of an offering in cash until three-fourths of the stock was sold.

"The cash requirement will work to deter those who wish to 'bail out' poor investments, made privately, through later sale of properties via a public solicitation," Love said.[17]

A few weeks later, two members of the House Subcommittee on Commerce and Finance flew into Salt Lake City. Representatives Arthur D. Klein of New York and John B. Bennett of Michigan announced that they had come on a fact-finding mission to look into the uranium marketplace. Bennett was sponsor of proposed legislation to extend the requirements of full SEC registration for stock offerings to the public to cover underwritings in amounts under $300,000, thus eliminating the Regulation A exemption. Klein was opposed to Bennett's proposal.

At 10 A.M. on Tuesday, September 13, Bennett and Klein, Clarence H. Adams, an SEC commissioner from Washington, D.C., and a crowd of state securities officials, stock broker-dealers, promoters and attorneys crammed into Room 230 of the Salt Lake City Post Office and Federal Building. For two days, Representative Bennett beat his legislative drum before a less than enthusiastic audience.

Starting with a few strokes to the Utah Securities Commission, he commended Mickey Love's eighteen "rather stringent" regulations governing brokers and underwriters. He lauded the state's control of the Regulation A exemption for intrastate offerings. Then he launched into a dissertation about the inequities to stock buyers in other states who didn't have such governmental supervision.

"Don't you think that the people of the other forty-eight states are entitled to as much protection as enjoyed by the people of Utah?" Bennett asked.

The Utahns replied, "Yes. But not through the SEC." The attendees argued that the existing federal laws were adequate. Other states could adopt Utah's rules for issues under $300,000 if they wanted to. But federal policing would stifle the raising of adequate venture capital to maintain the uranium industry. And elimination of the Regulation A exemption would price many small businesses out of the market. An issuer paid only one or two thousand dollars to make a Regulation A filing. Full SEC registration could run as high as twenty to twenty-five thousand.

Besides, "You can't legislate morals," the participants told Bennett. "You can't protect people from themselves."

Representative Klein agreed with the Utahns. He argued that buying stocks was "no different from shooting craps or playing the roulette wheels of Las Vegas. . . . If people want to gamble in this fashion, let them do it."

"The fact that the security is a gamble doesn't mean that the Congress of the United States should be deterred from protecting the public," Bennett countered.

Lyman Cromer, president of the Salt Lake Stock Exchange, suggested that Bennett had "fixed opinions" about uranium stocks. Bennett rejoined that the brokers had a "one-track mind" and weren't interested in protecting the public.

"Maybe the commission should think about protecting the brokers from the public," Cromer replied. "I had to write off $34,000 last year—money owed me by people who wouldn't pay for the stock they ordered."[18]

The two-day congressional hearing didn't accomplish much. The threat of more rigid federal "watchdogging" only added to the problems already facing the uranium market.

While production figures of uranium mining had leaped fifty percent over those of 1954, by mid-1955, stock sales were not keeping pace with the industry. The market was saturated with new issues. Prices of shares kept going up and speculative money was sapped. The public's buying fever was breaking. People were more thoughtful. They could see that uranium stocks were a one-time shot, not a long-term investment. What's more, no new bonanzas were discovered.

The entry of big business into the picture put a further damper on the excitement of speculating. Smaller operations were being absorbed by corporate mergers, or being pushed out of business entirely by the competition. The buying public lost interest when there was no longer the challenge of pitching in with the little guy.

Then the AEC Grand Junction office put a lid on disclosure about ore grades in the mines, the reserves, and production figures. The intent, according to the manager, Sheldon Wimpfen, was to protect national security.

"The gathering and subsequent publication of ore reserves and ore production data on this individual basis very rapidly leads to the accumulation of ore production information that is a measure of our domestic uranium ore supply," Wimpfen explained. "We do not wish to make information regarding our overall uranium supply readily available to unfriendly nations. . . ."[19]

But for potential investors, the lack of information made all of the companies alike. Besides, it made buying shares too much of a gamble. Stock buyers, becoming more sophisticated, refused to throw out money blindly.

And then the "wolves and vultures" started coming in, Noland Schneider remembered. These were "the old promoters from Texas, Canada and the East, who were looking at this thing not on really developing property, but coming in and raising money on anything and then dissipating [it] in other ways."[20]

The market was infested with crooks and con men. Several brokers suffered losses after receiving counterfeit certificates. Unsuspecting salesmen peddled stock to customers they didn't know. Later, they discovered that the strangers had used phoney names.

One incident involved an organization known as Trading and Investment Company and a man who claimed to be a Canadian citizen. The man opened an account with a San Francisco stock firm and named a Los Angeles attorney as a reference. He said that he wanted to liquidate approximately two hundred shares of Trading and Investment Company stock, and professed that there was a market in Salt Lake. A check was made and the man's credentials seemed to be in order, so the San Francisco brokerage wired Marilyn Coon, a trader with Muir, Dumke and Light. She was told that the Whitneys, Dempsey-Tegler and J.A. Hogle all knew about the offering and had placed bids.

"Cookie" Coon wasn't familiar with the stock. She called Whitneys. They informed her that the stock had just been issued within the past few days. The market was $130 bid at $150 asking price.

Coon started to sell the stock. She sold Dempsey-Tegler twenty-five shares at $140 per share. Whitneys bought three packages of about fifteen shares each at $150, $170 and $180. Hogles purchased a block of stock. Other securities firms in Los Angeles and Dallas sold the issue in their cities. In about four days, approximately one hundred and seventy shares had been sold.

Then Cookie Coon got word that Hogles had issued a cease and desist order on Trading and Investment Company sales. Three weeks later, Whitneys refused delivery on the stock. Fraud was suspected, and later proven.

A subsequent court case revealed that the Trading and Investment Company books and capitalization had been sold for $4,500 in Las Vegas. The new owner had transferred the stock certificates, placed buy orders in Los Angeles and Dallas, and then put in the sell order in San Francisco. Five days later, the San Francisco brokerage paid him. By then, Whitneys had refused delivery of stock. The brokers tried to go after the buyer. He was nowhere to be found. His Canadian bank account had been closed. There was no one at his alleged business address. The court rested on the ruling, "Buyer beware."[21]

That November, the Utah Securities Commission threw another curve at the broker-dealers. In order to remain in business, they would be required to have a minimum of $10,000 in clear, quick assets on hand. Mining claims wouldn't count. Mickey Love wrote letters to all of the brokers requesting them to send him reports on their financial condition.

Then Alex Walker, Jr., surfaced from his sea of official paperwork. Deciding to expose brokers for infractions of SEC regulations, he ignored the bucket-shop operators and pounced on the largest and oldest brokerage firm. He would make an example of J. A. Hogle and Company.

One of Hogle's employees, Walker discovered, had sold fifty thousand shares of Arrow Uranium, an intrastate Reg A stock, to five Idaho residents. Walker suspended the company's license, thus causing the prominent, established brokerage house to suffer a great deal of unwarranted embarrassment. In a subsequent hearing the accusation was dismissed on the grounds that the employee had mistakenly sold the shares "as an accommodation to a customer."[22] Even so, Hogle's temporary suspension agitated the investment community.

The SEC indictments multiplied, and eventually few brokerage houses escaped the bureaucratic purge. The Bowman brothers had their license revoked for illegal transactions in selling Reg A stock interstate. Eighteen firms, who had neglected to send Mickey Love the reports on their financial condition (including Coombs and Company), had their licenses revoked for not being financially sound. In a little over eighteen months, the uranium stock boom was over. It ended as abruptly as it had started. Few of the major participants came out with the hundreds of thousands of dollars that they had made so quickly.

Jack Coombs and most of the others involved in the penny over-the-counter market contend that the stock dealings played a critical role in establishing the domestic uranium industry. But virtually all of the brokers and promoters were put out of business by the SEC and NASD.

Ralph Hendershot, financial editor of the *New York Herald-Tribune,* agreed.

"The free world has licked the threat of a shortage of uranium supplies," he wrote, "but we can't resist the temptation to point out that this happy state of affairs has been brought about through no help from the more 'respectable' people in the financial district. And the SEC and the rank and file of our responsible people in business have made no notable contribution either. Too much emphasis is placed on the protection of the investor and not enough on the need of taking risks to keep the nation safe and progressive. We have the strange situation of one branch of government encouraging the search for uranium and another branch keeping the strictest kind of curbs on those helping to finance it."[23]

Probably of all the people involved in the uranium boom, it was the players in the over-the-counter penny stock game who fared the best. A promoter is as resilient as a rubber ball. He might be a loser today and a winner tomorrow. There is always a new deal around the corner. While numerous broker-dealers were felled by the SEC—many right down to bankruptcy—most of them bounced right back with other business enterprises.

"It sometimes seems that I have been prospecting, in a way, ever since I left Moab," said Dick Muir. "Sometimes you have to wade through a lot of dirt before you hit paydirt. Sometimes the values improve, but, in general, I've been going through plenty of gravel in search of jewelry rock. Still looking, though."[24]

Muir's career was far from over after the closing of Muir, Dumke and Light in 1956. He spent thirteen years in San Francisco as a managing partner with Schwabacher and Company. In 1967, he moved to Las Vegas as director of planning and development of Howard Hughes' Summa Corporation. Later, he was responsible for the entertainment division for Summa's fabulous hotels on the Strip.

Today, Muir has returned to his family's produce shipping and real estate developing businesses in Salt Lake City. In fact, he returned to the job he postponed for three weeks in order to check out the uranium scene thirty-five years before.

"It was a fantastic experience for everyone connected with the business," said Jack Coombs. "It was big money, a tremendous cash flow. But, when the boom ended, I didn't go out with flying colors—but on a stretcher."[25]

After losing a small fortune in the brokerage business, Coombs closed his eight offices and moved into the insurance field. At New York Life Insurance Company he became a member of the Million Dollar Round Table, and earned a Chartered Life Underwriter degree. But you couldn't take the stock business out of the boy. At New York Life he continued to trade stocks and put money into deals.

"I loved the securities business," he said. "It was exciting, interesting, fun—and also full of heart breaks. As a money raiser for new issues and start up ventures, I had found my niche in life. The experience, knowledge and contacts gained in the uranium boom was my MBA. The curriculum is not taught in school. Our graduating class of game players in the '50s became the promoters of the late sixties."[26]

13 LEETSO—THE MONSTER THAT KILLS

"The Navajo word for 'monster' is Nayee. The literal translation is 'that which gets in the way of a successful life.' Navajos believe that one of the best ways to overcome or weaken a monster as a barrier to life is to name it. Every evil—each monster—has a name. Uranium has a name in Navajo. It is leetso—meaning 'yellow brown' or 'yellow dirt.' Aside from its literal translation, the word carries a powerful connotation."[1]

Esther Yazzie and Jim Zion

KERN AND McRAE BULLOCH thought nothing of it at the time. They were in camp, trailing their herd of 2,510 sheep from their Nevada ranch to the lambing sheds in Cedar City. It was four o'clock in the morning. Dark. Cold. The light from their gas lantern flickered as they lit the small wood stove in the sheep wagon and prepared breakfast. It would be a long day of riding, and they would have to start before daylight in order to travel ten miles or so while keeping the large flock in tow.

Moving the sheep was a thirty-day job that the men enjoyed. They loved those animals and knew how to care for them. It had been a dry year on the winter range, so they had supplemented the natural vegetation with "mush," gruel of cottonseed meal, corn and barley, with a little salt added so they wouldn't eat too much. They carried a supply with them in troughs on the front of the wagon. They also brought along a large cattle truck to hold supplies or any sheep that might get weak or sick on the long trek.

Several mornings while they were eating breakfast that spring of 1953 the sky would suddenly light up "just like day."

"Well, they shot another bomb off," they would say.

"There would be a towering mushroom cloud come up just like this—bip!—a big one," McRae later remembered. "It was red and pink at first, then it kind of changed colors and was kind of a haze. We didn't know. We didn't pay any attention then, have any idea."[2]

Then on the morning of May 19 the pre-dawn sky illuminated like a massive burst of sheet lightning. In slow motion, an immense cloud spread into a dense gray fog that floated toward the Bulloch camp. The haze was heavy with dust and smoke. McRae's arms reddened with a burning sensation.

The Bullochs were herding the sheep in preparation to leave when two men drove up in a government vehicle. As the visitors got out of the car the sheep ranchers were surprised to see that they were wearing protective coverings over their hands, feet and faces. They carried a Geiger counter.

"Boy, you guys are really in a hot spot," one of them said. "How many people are here?"

"Two."

"You'd better try to get these animals out of here. You're in an awfully hot spot."

Then one of the monitors went to the jeep to radio into headquarters. Pretty soon he returned and reported his orders. "Well, they said if there's just two men forget about them and go up to the Lincoln Mine [three or four miles away] and put those people under cover. There are two or three hundred people there."

"So they left us," McRae remembered. "And we just stayed. You couldn't just get that herd of sheep and corral them out of there. We didn't know a thing about it [radiation] at that time. They didn't explain what a hot spot was. It didn't even dawn on us that our government would do anything to hurt us."[3]

The ranchers continued their long, slow trek. A few days later they noticed little scabs forming on the lips and noses of the sheep.

"They were eating the brush and they ate snow a lot of the time to get water," McRae said. "They'd skim their lips and the front part of their nose because they cropped the vegetation and sagebrush real low."

As the days went on, the brothers saw strange white spots on the backs of the black sheep and the black horse that they rode. The marks appeared to be burns that had scabbed over and sloughed off wool and hair. They thought it strange that so many animals would be affected.

A week later the sky lit up again. It was Shot Grable, a 15-kiloton bomb fired from a 280 mm. gun. In another ten days 61-kiloton Climax was dropped from an aircraft. Again the fallout clouds came.

"It always drifted our way, never toward California or Vegas," McRae remembered.

The sheep weakened. Lesions on their muzzles worsened. Many shed patches of wool. Numbers of the ewes miscarried, some of them delivering lambs with two legs or none at all. The cattle truck started to fill up with the sick and those too weak to walk. By the time they reached the lambing yards in Cedar City, fifty head were dead in the truck. The ranchers were dumb-struck.

"Well, we trailed on into Cedar City—I guess it was 200 and some odd miles," Kern later testified.

"When they started to lamb, we started losing them, and the lambs were born with little legs, kind of pot-bellied. As I remember some of them didn't have any wool, kind of a skin instead of wool. . . .

"And we just started to losing so many lambs that my father—[who] was alive at the time—just about went crazy. He had never seen anything like it before. Neither had I; neither had anybody else."[4]

During the spring and summer of 1953, of 11,710 sheep grazing within an area 40 miles north and 160 miles east of Frenchman Flat, 1,420 lambing ewes and 2,970 new lambs died. Numbers of cattle and horses in the vicinity experienced burn injuries, as well.

Local veterinarians were called in to investigate the strange happenings, but having never seen anything like it, they were baffled. They contacted the state veterinarian and a representative from the Department of Agriculture who examined some animals and reported to the Utah Department of Health that it was a most unusual case. They allowed that radiation was a possible cause.

The state health director then turned to the Public Health Service, which dispatched veterinarians Monroe A. Holmes of the PHS Communicable Disease Center, William G. Hadlow, a pathologist, and Arthur H. Wolff, a radiologist, to the scene. They were met in Cedar City by Major R. J. Veenstra, an army veterinarian, and R. E. Thompsett, a private vet doing contract work with the AEC at Los Alamos.

The researchers toured the ranches to get a firsthand look at the struggling flocks. Lambs were stillborn or deformed, and many that survived were undersized and too weak to nurse. The ewes had little or no milk and died during lambing or shortly thereafter.

They slaughtered one test sheep and passed a radiation detection device over it. They noticed extreme readings on the thyroid, which indicated high accumulation of radioactive Iodine-131. "This is hotter than a $2 pistol," one of the investigators said. The team also identified the lesions on the heads, mouths and backs of the animals as beta burns, similar to those found in radiation experiments conducted on livestock at the Health Research Center in Los Alamos.

A preliminary report noted "the concentration of radioactivity in these thyroid glands as of June 9, 1953 exceed[ed] by a factor of 250–1,000 the

Shot Grable. *U.S. Department of Energy photograph.*

maximum permissible concentration of radioactive iodine for humans as stated in the National Bureau of Standards Handbook 52."[5]

The sheepmen told the veterinarians, "You know, it's unfortunate that you don't have the animals that were really affected, because they're dead. You're looking at animals that survived, and even they have these unusual indications that we've never seen before."[6]

But a final report submitted by Holmes was inconclusive. Based on the fact that there were too many variables and more hearsay than absolute proof, the veterinarians blamed the situation on a combination of fallout and malnutrition.

Individual reports by Veenstra and Thompsett gave more credence to the radiation theory, however. Veenstra declared radiation was "at least a contributing factor." Thompsett went even further, opining that "the Atomic Energy Commission has contributed to great losses." Their conclusions were not popular with the AEC. They were nudged out of the investigation. and their reports were immediately classified.

Thompsett had previously promised Dr. Stephen Brower, the Iron County agricultural agent in Cedar City, that he would send him a copy of his findings.

"[But] they were taken off the case," Brower later testified. "In fact, Dr. Thompsett, who said he would give me a copy of the report and provide a copy of his report to the livestock men indicating the readings and the appearance of the animals definitely were similar to an experimental radiation done [on] animals, told me later his report was picked up—even his own personal copy—and he was told to rewrite it and eliminate any reference to speculation about radiation damage or effects."[7]

With Thompsett and Veenstra out of the picture, further study was turned over to the Division of Biology and Medicine. Lt. Col. Bernard F. Trum and John Rust, veterinarians in the Oak Ridge, Tennessee, Agricultural Research Program, were in charge. Their investigative tour of southern Utah resulted in a report more to the AEC's liking. Dry range conditions. Lack of feed. Malnutrition.

On the heels of the new findings, Dr. Paul Pearson, chief of the AEC Biology Branch of the Division of Medicine, went to Cedar City to call on Dr. Brower.

"Dr. Pearson told me . . . that the AEC could under no circumstances afford to have a claim established against them and have that precedent set," Brower later testified. "And he further indicated that the sheepmen could not expect under any circumstances to be reimbursed for that reason. . . .

". . . there was a clear mandate that under no circumstances would they do research . . . that involved radiation."[8]

On October 27, Gordon Dunning, the division's health physicist, met with government and scientific consultants and found them still unable to agree on the cause of the sheep deaths. Dunning, determined "to continue at any cost the nuclear weapons testing program," urged his associates to release a statement absolving the AEC of blame for the sheep disaster.

Despite his inability to get concurrence he released a final report on January 13, 1954. ". . . it is apparent," the statement said, "that the peculiar lesions observed in the sheep around Cedar City in the spring of 1953 and the abnormal losses suffered by the several sheepmen cannot at this time be accounted for by radiation or attributed to the atomic tests at the Nevada Proving Grounds."[9]

The press release did not mention that there had not been complete agreement among the participants, and further studies would go forward.

The ranchers were given a token offering of a $25,000 range nutrition study that might provide some definite answers.

In the meantime, complaints of skin burns, nausea, dizziness, headaches, eye problems, nervous disorders and other unusual physical conditions were increasing dramatically among the human population. People were eyeing the Nevada tests with suspicion The AEC continued to discount radiation as a possible cause of these disorders and attributed the symptoms to hysteria, gastrointestinal disturbances, hyperthyroidism or just too much sun.

With the conclusion of the Upshot-Knothole Series the continental site was given a year's rest and the AEC turned its attention back on Bikini and Enewetak. Six devices were exploded on land or atop barges in a lagoon of the deserted islands.

Then on February 18, 1955 the tests returned to Nevada with Operation Teapot. The new fourteen-shot series debuted with a public relations blitz of films, booklets, and speakers targeted at downwinders who were becoming increasingly worried about the effects of radiation since Dirty Harry and the other Upshot-Knothole detonations. Monitors had registered many "hot spots" and other violations of the 3.9-rad safety standard in 1953.

"Fallout does not constitute a serious hazard to any living thing outside the test site," a new booklet entitled "Atomic Tests in Nevada" announced.

James E. Reeves, the Nevada Test Site manager, went even further in a message to the downwinders.

"You are in a very real sense active participants in the Nation's atomic test program," he wrote. "You have been close observers of tests which have contributed greatly to building the defenses of our country and of the free world. . . . Some of you have been inconvenienced by our test operations. At times, some of you have been exposed to potential risk from flash, blast or fallout. You have accepted the inconvenience or the risk without fuss, without alarm and without panic. Your cooperation has helped an unusual record of safety . . . I want you to know that each shot is justified by national or international security need and that none will be fired unless there is adequate assurance of public safety. We are grateful for your continued cooperation and your understanding." [10]

In the meantime Gordon Dunning developed a simplistic formula ". . . of 'reduction factors' that could be applied mathematically to shrink the radiation dumped on the downwinders below the 3.9-rad threshold. . . .

they enabled their author to wave a mathematical wand and expunge the hot spot islands identified by the monitors and to blot out the dangerous doses measured by the beta-burns."[11]

But it was becoming very difficult for the people to swallow the government's propaganda. Communities north and east of the test site were facing some alarming changes. An unprecedented number of children were developing leukemia. Diagnoses of cancer were becoming more frequent. There were several miscarriages, stillbirths, and delivery of deformed babies. Local newspapers were publishing more and more articles questioning the culpability of the atomic tests.

Then the AEC was formally called on the carpet. The day before Operation Teapot commenced, David, Kern and McRae Bulloch, along with fellow ranchers Douglas Corry, H. C. Seegmiller, Myron Higbee, Nelson Webster, Lillian and Douglas Clark, Lambeth Bros. Livestock, T. Randall Adams and Dee Evans, brought suit under the provisions of the Federal Tort Claims Act against the United States government, alleging that thousands of sheep were killed by radioactive fallout from the bomb tests. Judge A. Sherman Christensen presided in the federal district court in Salt Lake City. The plaintiffs only asked for $177,000 in reparation for their actual losses.

Under the veil of secrecy, and with the ability to muster a cadre of expert witnesses in their defense, the AEC went to work. Potentially incriminating documents were "classified." Thompsett was persuaded to modify his report and write a letter indicating that he had changed his mind about the possibility of radiation from the bomb tests causing the sheep losses. Veenstra disqualified himself on the basis of information not available to him. The remaining persons involved united in a statement that the deaths were caused by malnutrition, not radiation.

Dan Bushnell, a young attorney who had received his degree from Stanford University in 1948, represented the sheepmen. Prior to the trial, he submitted written questions for the government lawyers.

McRae Bulloch.

"Did anyone involved in the investigation disagree with the report?" he asked.

"We are not aware of anyone who is involved in the Commission's investigation of the alleged sheep losses who *now* disagrees with the report issued by the AEC." [Emphasis added.]

"Did anyone conclude that radioactive fallout was a possible cause of the injury to the sheep?"

"We do not know of anyone connected with the AEC's investigation of the alleged sheep losses who has *now* concluded that radioactive fallout was a possible or probable cause of the injury to the sheep." [Emphasis added.]

"Both answers were evasive in that the questions were phrased generally in the past tense and the answers were pointedly limited to disagreements as of the time of the answers," Bushnell later wrote. "Such calculated answers were not an inadvertance, but rather an intentional evasiveness. Not only were they unresponsive to the questions asked but the answers themselves were untrue."[12]

The trial lasted fourteen days. The sheepmen didn't have a chance. The Judge ruled that on the factual record presented it was his conclusion that the sheep deaths were not due to atomic radiation.

"I was just a young attorney out of Stanford," Bushnell remembered. "I worked hard at it and we had a strong circumstantial case, but we had no expert witnesses because the federal government had a monopoly on anything that had to do with fission and atomic energy. Anybody in that had to be licensed by the government.

"All I could show was that there was a reasonable belief and causation for the damage to sheep. And the judge so admitted. He said I can't go against expert testimony. They had people from Harvard and people who had been to the Marshall Islands, and so forth, who said there was not enough evidence that there was proximate cause.

"Judge Christensen was a very thorough judge. He wrote an analytical decision and I didn't think that I had a chance to over-rule him, so we let it sit there. It was very disappointing to everyone."[13]

The sheepmen were devastated.

"It was a bad deal," McRae remembered, "and then the government, they just would tell so much information. And they twisted it around and lied. They figured, well, if we let the Bullochs win this then we'll have so darned many cases come against us that we just can't afford it. We were only asking enough money to cover our losses.

"We lost half our herd, and we were in debt. So we had to sell property right here in town. We had to give it away because there wasn't much of a demand. And then we finally sold that Nevada ranch because we was scared to go out there."[14]

On the heels of the Bulloch trial, seven-year-old Martin Bordoli died of leukemia.

Martin lived on a ranch that his father had homesteaded in 1880. The ranch, on the northwest edge of the Humboldt National Forest, was approximately seventy miles from the Nevada Test Site. The Bordolis were quite isolated so Martin spent five days a week with his aunt Helen Fallini and attended the small school at the Twin Springs Ranch.

The Fallinis were some thirty miles away from the test site, so the ranch and school were frequently engulfed by clouds after a nuclear blast. The chil-

Dan Bushnell represented sheepmen in lawsuit against the AEC. *Dan Bushnell.*

dren often suffered headaches and developed a burn that "wasn't exactly like the ordinary sunburn because after this you would get kind of like little water blisters or white spots."

Typical of ranch families, the Bordolis and Fallinis drank unpasteurized milk from cows in the pasture and churned their own butter and cheese.

"We raised our own garden," Martha Bordoli Laird later testified. "I canned everything. We canned all our own fruit. We put up our own vegetables. We ate many deer. About the only thing we did buy is we would go to town and we would buy sugar and things like that."[15]

In 1955, Martin developed stem cell leukemia, a disease local doctors knew little about and had never treated. The boy died ten months later. The attending physician noted that radioactive fallout *might* have caused the illness.

Although no AEC personnel had ever visited the ranch nor warned the people to stay indoors, shower and take other precautions during testing, two representatives called on Mrs. Bordoli following Martin's death. After offering condolences, one of the men assured her that the boy could not have been affected by the amount of radiation he received.

"So he said he had children," Mrs. Bordoli Laird testified later. "I told him that I would gladly take his children to our ranch and baby sit them

Shot Sedan was one of the dirtiest underground detonations. *Raymond E. Brim.*

for the rest of the summer while they were setting off those bombs. He said 'My God, woman, don't wish that upon me.'"[16]

The summer Martin died, the AEC initiated a program of issuing film badges to many school children and other downwinders to check amounts of external radiation they received after a detonation.

Mysterious plights befalling the sheepmen and occurrences of leukemia heightened the growing fears of downwind residents, and possibly gave pause to the AEC. Once again stepping up public relations efforts, the commission held the 1956 tests in the Pacific, before returning to Nevada for Operation Plumbbob.

A record thirty shots was planned for the 1957 series, a fact that the AEC down-played to minimize public concern. They countered the worrisome number by emphasizing that only small weapons would be tested. Despite their protestations the tests were conducted under the shadow of antagonistic media and downwinders, plus convocation of the first full congressional hearing on the subject, *The Nature of Radioactive Fallout and Its Effects on Man.*

The AEC attempted to convince Congress that radiation exposure from fallout was a necessary evil of nuclear testing. But, with new evidence exposed, the agency was criticized for whitewashing these dangers and taking the "body in the morgue" approach, in other words, waiting until there was concrete evidence of the problem before doing anything about it.

After seven days of listening to a battery of scientists and AEC commissioners, the congressional subcommittee's findings were inconclusive. It

Heavier particles from Shot Sedan cascaded to the ground as the cloud continued to rise. *Raymond E. Brim.*

was felt that there was not enough knowledge about the actual effect of fallout on humans, the threshold of tolerance to radiation and the impacts of low-level exposures. In conclusion, the subcommittee recommended that the AEC accelerate its research program.

The AEC was off the hook for the moment, but then a new roadblock hit when the eminent scientist Dr. Albert Schweitzer, Albert Einstein, Pope Pius XII and other humanitarians called for an end to nuclear testing, warning of the dangers of radioactive fallout to unborn generations of children. A few months later a battery of international scientists petitioned the United Nations to stop the atomic testing that was increasing the incidence of cancer and other diseases worldwide.

Finally in August 1958, President Dwight D. Eisenhower announced that the United States and Great Britain would sign a unilateral agreement for a moratorium on nuclear testing, providing that the Soviet Union would do likewise. The agreement would commence on October 31.

Accordingly, the AEC rushed to launch one last operation before the deadline. In its desperation to complete the proposed Operation Hardtack II, the scientists hurriedly sped up the planned schedule, sometimes firing three to four shots a day. Thirty-six detonations were conducted between September 12 and October 31.

Shot Sedan left a crater 1,280 feet in diameter and 320 feet deep. *Raymond E. Brim.*

In his book, *Justice Downwind: America's Atomic Testing Program in the 1950's,* Howard Ball reported, "The flurry of tests in October, 1958 led to radiation scares in Salt Lake City and in the downwind area, especially in the Washington County area of St. George, which had 'the highest reported level of fallout in the nation.'"

The respite lasted three years before the Soviets broke the agreement and resumed atmospheric tests. The United States countered soon after, but staged atmospheric operations on the Pacific islands, firing only underground blasts from shafts and tunnels in Nevada.

Contrary to government predictions, the underground detonations produced off-site radiation. One of the dirtiest was Shot Sedan fired on July 6, 1962. The 104-kiloton device, designed as an experiment in earth-moving for peaceful purposes such as digging canals, was buried 635 feet underground. In a dramatic explosion that shot 7.5 million cubic yards of dirt and stone in the air, its cloud, twice the predicted size, spewed heavy fallout forty miles from ground zero and spread radiation out into the Atlantic Ocean. The thermonuclear device left a crater 1,280 feet in diameter and 320 feet deep.

All above-ground tests were stopped in 1963 when the United States and Russia signed the Limited Test Ban Treaty. But underground operations continued. Off-site radiation was detected in twenty-six events during the next seven years.

14 THE AMERICAN EXPERIENCE

"It's started," Dr. Victor Archer thought.

The new medical director of the uranium miner study laid the letter on his desk and sighed. It wasn't much. Nothing conclusive. Not enough to prove that the "European Experience" was being repeated in America. But the report Archer received from Uravan, Colorado, that day in September 1956, read like a portent.

Dr. David J, Berman had admitted a patient named Tom Van Arsdale to St. Mary's Hospital in Grand Junction. Van Arsdale, a fifty-one-year-old hardrock miner from Nucla, Colorado, had spent over half of the past sixteen years working in uranium mines. He had received a physical from the Public Health Service field examiners in 1953 and was part of the uranium miner study cohort. He had lung cancer.

Although the thirty-four-year-old surgeon and radiation specialist had only recently taken over Dr. James Egan's duties at the Salt Lake field station, Archer recognized that there were indisputable parallels to be drawn between Van Arsdale and his Old World counterparts. The average age of the German and Czechoslovakian miners who died of lung cancer was forty-seven. Van Arsdale was fifty-one. Most of the European workers had died approximately seventeen years after their first exposure to uranium. It had been sixteen years since Van Arsdale's initial contact. Typically, the foreign miners died within months of their diagnosis. Accordingly, Van Arsdale's days might be limited.

Archer knew that it was important to learn all he could about the case. He would try to discover if the man's cancer had been caused by his exposure to radiation in the mines. Regrettably, the only reasonably sure way to determine the fact was by autopsy. On September 28, 1956, Archer dashed off an urgent letter to Dr. Lynn James at St. Mary's Hospital.

"If Mr. Van Arsdale should die of his disease, I would like to urge you to make every effort to obtain an autopsy," Archer wrote. "If an autopsy is performed I would like very much to have the opportunity to be present. If you were to send a collect telegram to me . . . I would plan to fly to Grand Junction (flights are twice a day), or to drive (whichever would be quicker)."

A few weeks later, Archer received a sample of Van Arsdale's urine from Dr. James. The specimen contained 34.4 picocuries of polonium-210 per liter—radio-lead, the final product of disintegrating radon daughters. The highest value the laboratory had previously found in the urine of uranium miners was 4.8 picocuries of polonium per liter.

Two months later, Tom Van Arsdale was dead. An autopsy, the first necropsy contracted on a uranium miner, was performed by Dr. Geno Saccamanno of St. Mary's Hospital. The diagnosis was death due to an oat cell (small cell) carcinoma of the left lung. The "oat cell" type of cancer, fast-growing and inoperable, was the same as that in the majority of uranium miners who died of lung cancer in Joachimstal, Czechoslovakia and Schneeberg, Germany.

Van Arsdale left a widow in poor health with small children to support. Her husband had workman's compensation insurance, so Mrs. Van Arsdale contacted an attorney in Grand Junction to apply for payment of benefits. As there had been no legal precedent in awarding compensation to uranium miners for radiation sickness, Dr. Archer was asked to testify at a court hearing regarding the matter. While he had no firsthand knowledge of the case, the doctor was wanted as an expert witness on radiation. It was hoped that his explanation of the "European Experience," the current miner study, and his interpretation of the results of Van Arsdale's physical examinations and autopsy would prove that the miner's death might be attributable to his employment in uranium mines.

As Archer had not been present at the postmortem examination, he wrote a letter to Mrs. Van Arsdale requesting permission to study the report. The letter stated that the autopsy might "contribute considerably to the solution and correction of the health hazards in uranium mining."

But Archer was not free to appear in court without the approval of Pope Lawrence, coordinator of the health study. On October 10, 1957, Archer wrote a memo to Lawrence. He explained that his personal connection with the Van Arsdale matter was tenuous, but added, "It is not likely that a better documented case will come along for some time."

Then, anticipating the Public Health Service's sensitivity to publicity about the miner radiation program, he attempted to allay Lawrence's fears.

Due to the fact that Mrs. Van Arsdale's law firm had "a continuing relationship with Union Carbide Nuclear [formerly U.S. Vanadium and Van Arsdale's employer] which they do not wish to sever," Archer was convinced that his testimony would in no way inflame the mining company. A similar case for a uranium miner who had died in 1954 had been "turned down flatly, as having no foundation in fact for the claim that the man's lung cancer was caused by his occupation. . . ." There was no possibility of the case being reopened.

"With this as an indication of past performance and with the obvious desire of the Workman's Compensation Fund, Mrs. Van Arsdale's present lawyer, of Union Carbide Nuclear, and of the Public Health Service to avoid undue publicity, it appears to me that the chances of keeping the publicity under control are very good," Archer wrote.

All that Mrs. Van Arsdale's attorney sought was to bypass certain statutes of limitation that would prevent the widow from pursuing her case later. The lawyer hoped to convince the Colorado Industrial Commission to hold the claim in abeyance until definitive information from the Public Health Service study was obtained.

"It seems to me that a fairly strong case can be presented to the effect that the man's occupation may have contributed to his development of lung cancer, but that it may actually have been a 'spontaneous' cancer or may have been due to smoking," Archer explained. "In other words, a case which *suggests* a relationship but does not *prove* one." (Emphasis added.)

Archer went on to state, "Against the obviously emotional appeal of helping a widow get what she may deserve, we must balance the possible effects of Public Health Service testimony at this hearing upon the Public Health Service itself, upon the Occupational Health Program in general, upon our miner study in particular and upon the career of Dr. Archer."

But he felt that no individual or agency would suffer disfavor or embarrassment as a result of his testimony. The hearing "is intended as a 'holding' action, rather than an all out attempt to convince anyone that here was a man who died of radiation-caused lung cancer," Archer wrote.

As for the mining companies, "even if the objectives of the hearing are achieved, the mining industry would have little reason to take alarm," he added.

"Management . . . is sufficiently enlightened to realize that there may be a radiation health risk involved in uranium mining, and that they do not wish to be cast in the light of obstructing efforts to determine what that risk is."

Archer considered his involvement in the case to be essential for progress in the miner study. "Certainly, from the scientific viewpoint, this case is premature," he concluded. "I would much prefer that it had waited for a few more years until we had some good data from the Uranium Miner

Study. However, it is from deaths such as this that our statistical case will be built. We must be prepared to cope with an increasing number of them."

Archer was allowed to testify. The case was declared "not clear." Mrs. Van Arsdale received no financial compensation for the death of her husband. But her attorneys did succeed in keeping the door open for future pleading.[1]

The annual field examinations of the mine workers under Dr. Egan, Victor Archer's predecessor, had always consisted of a health check-up, history, and X-rays. In 1957, Dr. Archer, hoping to reach over four thousand uranium miners, decided to enlarge and improve the study. He invited a pathologist from the National Cancer Institute in Washington to do some experiments with sputum cytology.

An offshoot of the revolutionary Papanicolaou's test for early detection of cervical cancer ("the Pap Smear"), this new technique for identifying malignancies analyzed sputum coughed up from the lungs. Only a few labs were doing the process at that time. Archer figured if the test could be successfully conducted in the field, the practice could strengthen future diagnoses in the miner study.

Samples of sputum from approximately thirty miners were collected in the fall of 1957. Robert D. Johnson, another Nucla miner who had been a participant in the health study since 1951, was one of the workers examined. It was his fourth on-site physical. He was given a check-up, an X-ray and a sputum test on July 15, 1957. Two months later, Dr. Archer notified Johnson that his laboratory findings indicated that he needed further medical attention. Dr. O. P. Gableman, of Grand Junction, was also alerted that "Urgent repeat sputum smears [are] recommended."[2] Johnson's lab findings showed "suspicious looking" cells. The patient had a "History of chest pain, chronic cough, shortness of breath. . . ." However, his chest X-ray film "was interpreted as being within normal limits."[3]

Johnson was forty-two years old. A hard-rock miner since 1936, he had worked in uranium mines continuously for the past nine years, mostly in Union Carbide Nuclear's Long Park # 1. This was a mine where Duncan Holaday had measured from fifteen to one hundred thirty times the recommended working level of radon per liter of air.

Robert D. Johnson died of lung cancer in November 1958. An autopsy showed "excessive amounts" of radio-lead in his bones, kidneys, liver and muscles. His widow applied to the Colorado Industrial Commission for a workman's compensation settlement. In November 1961, Mrs. Johnson was awarded $11,486. This was the first case in which the effects of radiation as a working hazard were judged.

The sputum cytology diagnosis on Robert Johnson had been correct. But Archer was alarmed at the overall results of the 1957 experiment. The NCI pathologist contended that approximately sixty-five percent of the examinees had cancer. Archer was skeptical about the accuracy of such high readings. He called Dr. Geno Saccamanno and asked if Saccamanno would go to Washington with him to personally examine the slides.

"We both went back," Saccamanno said later, "and I looked over those slides and I didn't know if they had cancer or not because the preparation of the material was so poor."[4]

The standard technique for making the slides was primitive at that time. The sputum sample was examined for deeper colored spots, which were removed with an applicator stick and then studied under a microscope. As the spot was but a small part of the whole sample, it contained only a few of the total cells. Therefore, an accurate diagnosis of lung cancer was only nominally successful.

But Archer persisted. Anything that might facilitate early cancer detection was worth a try. He proposed to conduct sputum tests on the entire miner study population in 1958. It would be the first time this type of examination would be tried as a mass screening procedure.

Archer asked Dr. Saccamanno if he would take part in the project and apply for a grant to finance research to refine the techniques of sputum cytology. Saccamanno accepted the offer. Better still, he succeeded in formulating a technique of taking sputum samples in a way that caused less nausea for the patient, and of sustaining and examining the material more satisfactorily. He developed a fluid fixative to preserve specimens collected at the mines until they could be processed in the lab. Then, instead of preparing a slide with a small piece of the sputum, Saccamanno homogenized the entire sample in a blender and then centrifuged it to decant off the mucous and leave just the cells. Concentration of all of the cells made for a more reliable diagnosis.

Marvin Kushner, Professor of Pathology, NYU Medical Center, and Dr. Oscar Auerbach, chief of Medicine, Veteran's Hospital, New Jersey, collaborated with Saccamanno on the lab work and diagnosis. The sputum tests were added to the yearly physicals. (Saccamanno's procedures are now widely used throughout the world.)

By the end of 1958, the lung cancer death count of uranium miners participating in the Public Health Service study had reached four. Inconclusive proof, to be sure, but enough to make Victor Archer and his colleagues cringe. It wasn't simply a hypothetical scientific study anymore. Real human

Dr. Victor E. Archer organized the first mass screenings of sputum samples for early detection of cancer. *Victor E. Archer.*

beings were dying. They were American miners, not distant workers in Germany and Czechoslovakia.[5]

While Archer's miner study was shifting into high gear, Duncan Holaday's activities had been curtailed. Since 1955, when "the Bureau of the Budget had stated that they did not approve of [more] funds being spent on the uranium study,"[6] Holaday's environmental phase of the project had been limited to assisting state agencies. And only upon their request.

During that period, however, Holaday had been invited to join the AEC Division of Biology and Medicine on a junket to South Africa. There, he inspected the uranium workings, took air samples for radon counts and studied the findings of the local mine inspectors.

Holaday returned to Utah with glowing reports about the Belgian operators of the African companies. In the late 1940s, they had sent inspectors to the United States to confer with radiation experts at the AEC Division of Biology and Medicine and the Oak Ridge Health Physics Department regarding health and safety underground. They "did their best to find some advice somewhere,"[7] as to what levels of radiation would be acceptable, but it was early in the game and no one really knew the answer. Finally, the AEC pulled the number of one thousand picocuries of radon per liter of air "out of their hat" as a standard that might be safe. The Belgians returned to Africa and began to implement radiation standards in their mines.

". . . out in the middle of the Congo," Holaday said . . . "a laboratory for measuring radon samples [was set up] and [they] had a steady collection procedure going around with people collecting air samples in their working places. And they have the levels quite a ways below one thousand picocuries. . . ."

Not so in the United States, however. The AEC was doing even less than before. In February 1957, the Mining Division of the Grand Junction office stopped its routine inspections of thirty mines that were owned by the federal government and privately operated under standard mine lease agreements. The legal department decided "that if anyone in the Atomic Energy Commission was responsible for health and safety practices in mines, it would be the inspection branch of the Civilian Application Division and that the Raw Materials Division should have nothing further to do with this work."

When Duncan Holaday learned of this decision, he wrote a letter, dated May 8, 1957, to Dr. Charles L. Dunham, chief of the AEC Division of Biology and Medicine in Washington, D.C. Holaday told the chief that an experienced mine safety inspector had been doing a commendable job and that the Colorado mines involved "were not only the best studied mines but were also in the best condition of any of the operations of this type . . . the most difficult to ventilate."

Holaday explained that the mines were in small, isolated, flat-bedded deposits. Costs of control were high. Extended planning for mining operations was difficult.

"I am afraid that the operators of these small mines will be happy to seize upon any excuse to cease spending money on control methods, and without bothering to ascertain all the facts, will assume that the AEC itself is no longer particularly interested in any potential hazards from radioactive gas and dust and that, therefore, they need not exert themselves particularly," Holaday wrote. He urged the chief to intercede.

Even so, the mining of uranium continued, indeed increased. In April 1958, just two days before the government bonus offer expired, Lisbon Uranium Company collected the AEC's $10,000 bonus for delivery of twenty short tons of high grade uranium from a new working. Cumulative bonuses paid out for the fiscal year totaled $2,040,118. Approximately four-and-a-half-million tons of ore were produced.

Some of the mines that worked large ore bodies were doing a fairly reasonable job of controlling atmospheric concentrations of radon. But many mines in the Four Corners area were not regulating their mines as well. Arizona, Utah and Colorado mine inspectors did little more than take air samples. None of them felt that there was sufficient data to warrant imposing effective control measures. Instead, they tried to accomplish their goals through education of the mining companies, and by exhorting them to lower radiation levels.

Union Carbide Corporation now employed a full-time safety engineer and a ventilation expert. Climax Uranium Company's engineer spent half of his time improving ventilation. But three hundred older and smaller workings, employing approximately fifteen hundred men, were still averaging fifty times the recommended working level of radiation. Duncan Holaday was hog-tied as far as doing anything about it.

New Mexico was one exception. There, in 1958, the mine inspector finally put some teeth into the regulations. The state recommended a standard of one hundred picocuries—or one working level—of radon per liter of air. If a mine was found to have three working levels, immediate corrective measures were demanded. Should the working level reach ten or higher, the area would be closed until proper ventilation was installed. To show that he meant business, the mine inspector shut down the biggest mine in the Ambrosia Lake District when it exceeded the standard.

"This was the requirement that made the mining companies sit up, because they stop producing ore from that area and, therefore, the money stops coming in," Holaday said.[8]

But, as 1958 drew to an end, conditions were far from improved. Now four miners in the study cohort were dead of lung cancer. Thousands of their co-workers, working under environmental conditions that were woefully unsatisfactory, were pushing production to an estimated five million tons of ore per year.

Finally, on October 27, after a year of prodding by Holaday and his associates, Surgeon General L. E. Burney wrote a letter to Dr. Dunham. He reported that "there has been little, if any, improvement in the mines over the past two years," and pleaded with the AEC Division of Biology and Medicine to take "more intensive action."[9]

Noting that the Occupational Health Program of the Public Health Service, the Bureau of Mines and Department of Interior were "enlarging the educational program for mine operators and the miners themselves," Burney went on to say:

"It has occurred to us that the Atomic Energy Commission may be in a unique position to accelerate the adoption of more adequate control measures. The United States Government is the sole purchaser of ores produced in the mines. If the Federal Government were to require the supplier mines to conform to the recommended health standards promulgated by the official state health agencies involved, an effective measure would be available which could bring about prompt changes in working conditions."

This was what Merril Eisenbud had suggested in 1948. It was the same tune that Duncan Holaday had been playing for the past eight years.

Dunham responded by stating that "it was doubtful that the Commission, under its present law, could exercise regulatory authority over the mines." He suggested that the Public Health Service convene an "inter-agency committee to decide what the role of the federal government should be."

Accordingly, twenty-six representatives from the Department of the Interior, the Department of Labor, the AEC, and the Public Health Service gathered on May 20, 1959. After hours of deliberation and subsequent channeling down to a smaller action group, it was concluded that "the role of the Federal Government is not well defined." The AEC could not assume jurisdiction over the mines "both by its licensing regulations and the Atomic Energy Act as amended."

The Department of Labor might have some responsibility through the "Public Contracts Act," the report went on, but the department "does not now have the technical staff needed to enforce the Public Contracts Act in uranium mines" and such enforcement "would not give the mines the needed technical assistance to control the potential problem."

The Public Health Service was "limited to research, investigation, and the provision of technical assistance to the States." At the request of the American Standards Association, Duncan Holaday had headed a sub-committee to formulate a standard of radiation for the mines. The experts settled on three hundred picocuries of radon daughters per liter of air. But this was no more than advice. The Public Health Service had neither the muscle nor the money to enforce the recommendation.

The Bureau of Mines of the Department of the Interior had been inspecting and reporting on the AEC's leased mines since the commission stopped doing so. However, the bureau had "no enforcement power in mines located on private property." The agency's authority only "provides to the mines, upon request, technical assistance in safety and health problems."

So the buck was passed back to the states. Again.

In 1959, five more miners died of lung cancer. The Uranium Study Advisory Committee met and agreed that the incidence of lung cancer in uranium miners with three or more years' experience "significantly exceeded the number expected among the population based on the mortality experience of a nonuranium mining control group." The advisors reported that the "lung cancer deaths were probably the first indication of what could be expected and that further cases would occur."

They did. Nine miners died in 1960.

As a result of the inter-agency conference the year before, Arthur S. Fleming, the outgoing secretary of Health, Education and Welfare, called a meeting of the four western governors whose states were concerned with uranium mining. On December 16, 1960, the dignitaries gathered in Denver to be briefed about the Public Health Service study. They got an earful.

Dr. Harold J. Magnuson announced that, in the three years since sputum tests were initiated into the miner study, the percentage of "doubtful" diagnoses had gone from 1.2 percent to 12.2 percent, and "positive" findings jumped from .09 percent to 3.3 percent. Lung cancer was occurring "at five times the expected rate among the experienced mining population."

The Occupational Health official went on to state that environmental studies taken in 1959 of three hundred seventy-one mines employing 3,619 miners revealed that twenty-two percent had atmospheres measuring between one and three times the recommended working level of radon. Twenty-three percent registered three to nine times the suggested standard (a two-percent increase over the preceding year), and twenty-two percent (four percent over the 1958 figure) had concentrations of more than ten times the working level. One mine had a count of 47,000 picocuries per liter of air. Yet it was "believed that in virtually all uranium mines radioactive contamination can be controlled, and that the means of control is practicable."

"If preventative action is not taken immediately," Magnuson warned, "it is possible that, on the basis of the European mortality rate of one percent per year, some thirty miner deaths from lung cancer may be expected annually in the years ahead."

Governor Steve McNichols of Colorado was quick to blame the federal government.

"State attempts to improve working conditions in the mines have been handicapped by the government's reluctance to furnish information on the dangers of radiation," he said.[10]

McNichols also charged the AEC with forcing "the price for uranium so low that small mine operators cannot afford proper safeguards and ventilation for the miners."

The consensus among small operators was that they would have "to shut down sooner or later if the Atomic Energy Commission and the Public Health Services have their way."[11]

These smaller operators were subject to undue pressures. The AEC forced them to sell their ore only to selected mills and at lower rates than they could get from independent processors. The Public Health Service, "on the basis of sketchy information and because anything concerning uranium is sure-fire publicity," were forcing them "to install such costly equipment that it leaves no alternative but to close down or sell out to the cartels." The independents felt that if new regulations based on "scare-head contentions" were necessary for them, they should also be enforced on *all* existing base metal mines in the Rocky Mountains.

But despite the federal agencies' games of musical chairs and the emotional opposition waged by small mine operators, the governors agreed that the states must take some aggressive action.

The bureaucratic gears started to grind. A week after the governors' conference, G.A. Franz, deputy commissioner of the Colorado Bureau of Mines, issued a warning memo to all mine operators in the state.

"This has been the first definite information of this type given your Bureau of Mines...," he wrote.[12] He urged compliance with the American Standard of three hundred picocuries and vowed to enforce it. By June 1961, Franz had five additional trained mine inspectors in the field.

Utah's mine inspector, assisted by the U.S. Bureau of Mines, launched a program to survey the state's uranium operations and institute a control program. New Mexico funded an engineer from the Division of Occupational Health to assist its mine inspector. Arizona's inspector, denied the authority to regulate exposure to radon and its daughters in the mines, had to be satisfied with educational work only.

With these small steps underway, the field station in Salt Lake City analyzed its position. Funding was tighter than ever. The program would have to be modified. There would be no more physical examinations of workers at the mine sites. Instead, an annual sputum cytology program was instituted, under the direction of Dr. Geno Saccamanno. The samples would be taken during the yearly census at the mines. An attempt would also be made to autopsy all miners who died, no matter what the cause of death.

Dr. Victor Archer would head the epidemiological study. His staff would keep track of a cohort of miners through the census and by conducting follow-up research using questionnaires, work histories, newspaper articles and obituaries.

Duncan Holaday assumed administrative duties as deputy chief of the Division of Industrial Hygiene. His environmental efforts were limited to assisting with mine surveys in Utah. He also ran experimental tests on respirators and materials to seal radon from the mine drifts.

From all of the talk, the meetings, the memoranda and "enforcement of radiation standards," it would seem that the goals of the uranium miner study had been reached. Duncan Holaday allowed that some of them had. Most of the mines now had control procedures. Miners who had been exposed to significant amounts of radon daughters had been identified and were under observation. Other nations had followed America's example in taking action against radiation hazards in the mines. But by 1964 Holaday felt that many of the items of action outlined in 1950 still awaited attention. He wrote:

> It is now fourteen years since the uranium study was started. . . . At times, the project has been quite active; for other periods it has been quiescent. The study was undertaken with the belief that all that was required was the evaluation of environmental conditions in the industry and comparison of the results of these studies with the data on human experience which was available in the literature [The European Experience"]. Measures to control the exposures of the workers to toxic materials could then be recommended. This belief was a delusion. It required ten years and the accumulation of a number of deaths to convince the authorities that real hazards existed in the uranium mines. The contrast between this neglect of the miners and the protection given to workers in other parts of the nuclear energy field is amazing. At no time has it been possible for the Division to secure enough support to conduct more than a fraction of the basic work that needed to be done.[13]

Sadly, Holaday realized that it was the classic case of "too little too late." With twelve years since Charlie Steen's strike had triggered the uranium rush, there had been time for signs of cancer to appear in scores of miners who had been exposed to underground radiation. No one knew how many others, still without symptoms, had contracted the disease and would eventually die because safety measures had not been adopted.

And now, when there was some degree of control of radiation hazards, the AEC decided that "it is no longer in the interest of the Government to expand production of uranium concentrate."[14] After producing almost nine million tons of ore per year, valued at $250 million, the uranium market was saturated. There were seventy-one million tons of reserves— enough to satisfy the needs of the United States through the next four years. Since 1963 when the United States and the Soviet Union signed the Limited Test Ban Treaty, military requirements of uranium for nuclear weapons were diminishing. Private enterprise, now entitled to purchase uranium oxide, wasn't developing power plants and other peacetime uses of nuclear energy as quickly as anticipated. So the AEC announced that, from 1962 to 1966 it would only buy "appropriate quantities of concentrate derived from ore reserves developed prior to November 28, 1958.

Many of the smaller operations closed up shop. Ostensibly, uranium mining was on hold. The boom was over.

15   SENATOR STEEN

IN 1958, PEOPLE WEREN'T calling Charlie Steen "that crazy Texan" anymore. It was hard to ridicule a guy who had parlayed his geologic hunches into a bonanza and triggered a modern-day prospecting rush. Steen's Uranium Reduction Company mill, which processed some seventeen hundred tons of ore per day, had the largest payroll in Grand County. Utex Exploration Company was one of the area's biggest property owners. Besides its mines, the corporation held twenty-one rental homes, eight executive residences, and acres of commercial and residential land. Moab Drilling Company dominated the field in local mine exploration activities.

And Charlie Steen had shared his good fortune—with the business community, the schools, the new hospital, the churches. To say nothing of the annual party Utex threw every year, which got grander and grander.

Public honors had been showered on Steen. On March 1, 1957, Orval Hafen, president of the Utah State Senate, read a resolution of appreciation to "our distinguished fellow citizen, Charles A. Steen," who has "rendered valuable service to the people of the State of Utah through his untiring efforts to develop the uranium mining industry in the State. . . ."

The proclamation lauded Steen's "determination and strong devotion to a purpose." The senate saluted him as "an outstanding pioneer in the exploration and search for uranium." He was applauded for being "instrumental in establishing the uranium processing industry in the State. . . . enlarging employment opportunities, and enhancing the material well-being of Utah's citizens."[1]

The citation concluded by thanking Steen "for his accomplishments in the field of mineral development, for his demonstrated willingness and desire to serve his fellow man, and for the inspiration he has been to the citizens of the State of Utah and the nation as a whole."

Already something of a local legend, Steen had also been idealized in bits of doggerel published in Moab's *Times-Independent*. One editorial "chatter column" turned the famous Davy Crockett song into a "Ballad to Charlie Steen."

> Charlie—Charlie Steen,
> King of Uranium.
> Went to school at Texas mines,
> The professor said, 'You might as well resign.'
> So he packed his gear, and off he came
> To find that yella' stuff, and stake his claim.
> Charlie—Charlie Steen,
> King of Uranium.
> He went out huntin' every day,
> Ate rabbit stew, and that ain't hay.
> Finally his geiger went bizerk
> And now old Charlie's richer than a Turk.
> Charlie—Charlie Steen,
> King of U-r-a-n-i-u-m.

Mitch Melich, an experienced politician himself, reasoned that the timing was right for Steen to make a bid for the senate seat.

"Charlie, I've been thinking," Melich said one day. "The state senate race is open in our district this year. I think it would be a good thing if you got into politics."

"I'm a geologist, not a politician," Steen replied.

"Yes, but with the discovery of your mine and all you've done for the community, I think you could win it," Melich persisted.[2]

Melich went on to elaborate on his logic. Steen had all of the prerequisites for getting into office. He had the name, he had the money, and southeastern Utah really mattered to him. Besides, Melich added, civic involvement would be good for the "Uranium King." Finally his arguments prevailed.

"Hell, I'm paying eighty-five percent of the taxes in Grand County," Steen said, "guess I just as well have a say as to how they're spent."[3]

Melich knew that Charlie's election would depend on a well-programmed campaign. The new senator would represent neighboring Emery and San Juan counties as well as Grand County. His opponent in the primary election would be George Hurst, a prominent Mormon from Blanding, eighty-two miles south of Moab. Hurst had served as a Republican member of the legislature a number of times. His roots and his religion would pose a formidable challenge.

Melich and Steen enlisted Jennings Phillips from Salt Lake City to help them plan their strategy. Phillips, a seasoned politicker, was a lobbyist for the Safeway grocery chain.

Charlie Steen, front right, with, left to right, William R. McCormick, president Standard metals, Mitchell Melich, president of URECO, and Roy F. Hollis, president of Atlas Minerals. *Mitch Melich.*

The organizers knew they would have to find some way to combat Hurst's Mormon stronghold. They decided to go after the American Indian vote. Steen's committee learned that a fellow named Rusty Musselman was running for commissioner of San Juan County. Musselman lived in Bluff, Utah, about thirty miles south of Blanding. The aspiring commissioner claimed he had succeeded in convincing about five hundred Navajos from the Indian reservation in Bluff to register and vote for him in the primary.

"So we decided that if Rusty had this Indian vote, we've got to have that Indian vote if we're going to beat this Mormon out of Blanding," Melich said later.[4]

Steen's organizers concluded that the best way for the Indians to meet Charlie was to have a big party on the reservation. They hired a committee of Navajos to take care of refreshments. All Steen's cohorts had to do was show up and spotlight their candidate.

On the appointed day, Melich and Phillips flew down to Blanding with Charlie in his twin-engine plane. A car met them at the dirt landing strip and drove them to Bluff. There, they turned onto a gravel road for another short drive to the Aneth Oil Field, where the party was to be held.

But when they got there, the three men were horrified. Over four hundred Navajos had gathered for the big "pow-wow." The catering committee was hovering over some open kettles boiling up with the meat of several lambs. Nothing else. There wasn't enough food for twenty-five people, much less hundreds of expectant Native Americans.

While Steen mingled among the crowd, Melich and Phillips hopped in the car and drove to the nearby trading post. There they bought the entire inventory of soda pop, whole cheeses, canned hams, anything that looked like a party. When their purchases had been sliced and packaged, the two men returned to Aneth and spread out the feast. After much dancing and "speechifying," the rally ended on a happy note.

A reporter from the Cortez, Colorado, radio station later reported the event. "Here are two men running for state senator from that district," he said. "One, George Hurst, the Mormon. The other, Charlie Steen, the American."

The night of the primary, Steen was too upset to sit around waiting for the results. He went flying. When he came back to earth, he found that he had won a place on the ballot. As for the Native Americans, Charlie wangled a scant twenty-five votes from them. But Hurst only garnered a dozen.

The real battle would be the November election. Steen's advisors told him that he would have to change his image in order to get the votes. He had to counteract his reputation for being aloof. Mix with the locals, smile, kiss babies, tell jokes . . . *clean* jokes, and control his temper.

For the past year, Charlie had staged a running argument with the Moab Chamber of Commerce over economic development. In a typical surge of anger, he had "popped off" against the chamber's plans to "make a metropolis out of this desert center." Grand County was in a tug-of-war with San Juan County to get a twenty-million-dollar potash plant constructed within their boundaries. Plans were afoot to build a ski resort in the nearby La Sal Mountains to attract winter business in Moab. The city boosters wanted to woo summer trade by constructing paved roads to open the Arches National Monument and other scenic backlands to tourists. There was also a proposal to establish a summer school for students in geology and the arts in cooperation with the Texas Technical College.

Charlie didn't mince words in letting people know that he considered most of these projects impractical for the small, isolated community.

"You're going to have to eat your words," the election committee told Steen. If he wanted to get elected, the senate hopeful must diplomatically admit that he had been wrong in some of his past statements. Say something like, "I can understand now that the new picture is a little different

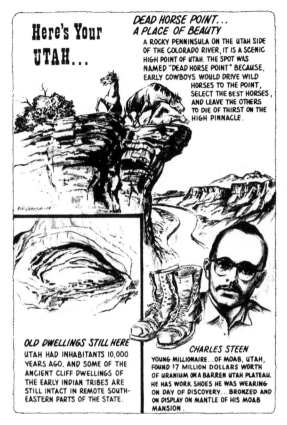

Steen's discovery and the subsequent boom turned
sleepy Moab into a tourist mecca. *Western Mining and
Railroad Museum.*

than I'd thought, and that the new half-million-dollar Arches road, the
river marathon, the Texas Tech summer art and geology group and the ski-
ing plans will certainly pour red blood into our tourist endeavors."

As for the potash plant, the strategists suggested he say, "Well, hell, even
Eisenhower can't call them all right, so let's get on the stick and be sure our
potash mill is located *north* of the boundary for Grand County tax revenue."

"Use the same light-hearted delivery you used in your dedication
speech for the mill," the planners told Steen.

The committee was sure that Charlie could sell this I-was-wrong-this-
time approach. The voters would conclude, "Maybe Charlie pops off too
damn much once in a while, but if a man as big as he is now can admit he's
wrong on a point or two *in front of thirty business men in his own town,* he's
big enough for my vote."[5]

Consequently, Charlie Steen became a full-time campaigner. He shook lots of hands. He confessed to past mistakes and pushed for the potash plant to be located in Grand, not San Juan County. He rallied behind ski area developers. He offered use of the Utex laboratory and library to the proposed summer program for Texas Technical College. Advertising himself as "The Man Who Didn't Move Away," Steen vowed "that the Grand-San Juan-Emery area has tremendous possibilities for growth."

"I will fight for new roads," he promised. "I will push the development of new water supplies. I will strive for more aggressive advertising and promotion of our tourist attractions. Equally important, I will insist on a tax structure that will encourage incentive. These measures will benefit ranchers, businessmen and wage-earners alike."[6]

However, the Uranium King was not completely tamed. Just days before the election, his temper got the best of him again.

The Ex-students' Association of the Texas Western College (formerly Texas College of Mines and Metallurgy) had invited Steen to attend the annual Homecoming Banquet on November 1, 1958. The association wanted to present Steen with the Outstanding Student Award. Charlie accepted, and, nattily-dressed, took his place beside the college and association presidents and other officials at the banquet. After the dinner and preliminary ceremonies, Charlie was introduced and stood to accept his award. The large crowd of faculty and alumni rose to applaud their famous colleague. When the ovation ended, the 1943 graduate began.

"I know the proper way in which to accept this award. I am expected to say 'thank you' and sit down," Steen said.

"However, inasmuch as I did not seek this award, and as Dean Thomas reminded me last night that I was the only son-of-a-bitch he knew who had made a career at being one, and was a success as a result, you need not expect the proper response."

There was an uncomfortable stirring in the room.

After this, Steen began to berate the Chamber of Commerce.

"In their desire to attract more students," he said, "a feat which would mean more money for the people they represent, they destroyed a brand name—Texas College of Mines—that was known the world over through the efforts of men who had dedicated their lives to building the school and their students who have achieved recognition in every country of the free world for their ability and proficiency in Mining and Engineering fields. A graduate of Texas College of Mines was able to obtain a job for which a graduate of a teachers' college would not be considered."

Then, with no little disdain, Steen listed some of the courses currently offered in the college catalogue: Techniques, Methods and Materials of Instruction in Rhythm; Coaching Basketball; Coaching Football; Real Estate Brokerage; Radio Listening as an Aid to Elementary Teaching; Public Relations; Drama; Storage and Warehousing; Jewelry Construction; Baton Twirling.

Steen went on to suggest a few more "worthwhile subjects [that] should have their proper place in the new scheme of things. Beer Guzzling, a course in how to chug-a-lug beer out of a gallon pitcher without getting a permanent crease on the bridge of your nose . . . Mexican Relations. How to go to Juarez and keep enough money to get back across the Bridge . . . The Art of Tobaccos. How to chew tobacco and not dribble it on your chin."

Protesting that he was "a product of the Texas College of Mines and Metallurgy" and "refuses to wear the badge of the neither fish-nor-fowl institution that now exists in El Paso," Steen's rancor reached its zenith near the end of his speech.

"This son-of-a-bitch previously mentioned at thirty-nine years of age is a living legend of the Uranium Boom that he helped create, a boom that raised the U.S.A. from a 'have not' nation to a number one position in Uranium reserves of the world. Whether he dies a multimillionaire or a broken-down, ragged-ass, prospecting tramp, his place in the mining history of our country is secure. History, if true, will show he graduated from the Texas College of Mines."

The audience sat in stunned silence.

Steen concluded by stating that he would accept the trophy, but "replace the name that I do not recognize with Texas College of Mines." When he took his seat, no one applauded. Some booed. His companions at the head table left in a huff.[7]

The next day, Paul H. Carlton, president of the Ex-students' Association, rushed a letter to the college administration and faculty.

" . . . we come to Homecoming to honor our school and you people who devote your lives so that we may enjoy happier, more meaningful ones; so that we may live with our fellow men in a peaceful, understanding world," Carlton wrote.

"I am sorry that I did not have the insight to get up at the time and refute the irresponsible abuse and blasphemy," he apologized.[8]

After receiving a copy of Carlton's letter, Charlie responded. ". . . 'fess up', Paul: Was it lack of insight or lack of guts?"

Defending his actions, he wrote: "As for 'living with our fellow men in a peaceful, understanding world,' what sandpile have you buried your head in since you got out of college? We are living in a world in which two systems are locked in mortal conflict to determine which kind of a world our kids are going to inherit, a race which we shall have to work hard not to lose since our institutions have been turning out a majority of knuckle-heads instead of men and women who are trained to think and weigh the issues of our society."[9]

Despite festering criticism, Steen refused to retract any part of his speech.

"I realize in retrospect that never have I created a controversy with more sincerity," he stated later.[10] "I retract no part of it and apologize to no one who chooses to take offense. As a graduate of Texas College of Mines, I am entitled to express my beliefs, and I happen to have enough financial independence to be able to voice my views, as no one can hire me, fire me, or buy me."[11]

Fortunately, the mutiny in El Paso didn't affect the election. A few days later, Charlie Steen became a Utah state senator.

Mitch Melich gave Steen some parting words of advice before he departed for Salt Lake City and Capitol Hill.

"I said to him, 'There are two things you shouldn't do when you go out there. One is that you shouldn't introduce a bill for liquor-by-the-drink.' Which he did."' (Utah law restricted the sale of alcohol to the purchase of package goods at state-owned liquor stores. The predominant Mormon Church forbids *any* alcohol consumption.)

"'And,'" Mitch continued, "'you shouldn't introduce a bill for parimutuel horse racing.' Which he did."[12]

During his tenure in office, Steen served on the Industry, Labor, Mines, Revenue and Taxation, Judiciary and Sifting committees. He succeeded in getting a tax repeal and a county agent for his local constituency. But his foray among his staid, Mormon colleagues was rocky.

"I was a wild radical in college," he once boasted.[13]

He was similarly unconventional in the Utah state senate. Contrary to Melich's caveat, Senator Steen worked for a bill to lower the legal age for purchase of cigarettes from twenty-one to eighteen. (The Mormon Church bans smoking.) He introduced legislation to permit Utah restaurants to serve wine with meals, and he sought a law allowing the purchase of liquor-by-the-drink. All were defeated.

"I then thought of introducing a head tax on virtue," he later quipped. He reckoned that "quite a lot of money could be collected from those who couldn't furnish a signed affidavit that they neither smoked or drank."[14]

"I got ten thousand letters over this liquor-by-the-drink business," Steen said. "One of them said, 'If you don't like it in Utah, why don't you move to Nevada?' And so I did."

Despite rumors that there was a movement afoot to run him for governor, Steen resigned from the senate during the last year of his term.

"I think I would have run pretty good for Governor," he told *Cosmopolitan* writer Murray Teigh Bloom, "a lot of Republicans were ready to back me. Then one week while I was fishing off Baja, California, I did a lot of thinking. I felt I had been pretty futile as a state senator, and I realized I would be almost as limited as Governor. At best, I was only a partly-accepted outsider. I suddenly found myself saying, 'What are you doing in Utah, anyhow?'"

On March 11, 1961, Steen handed in his resignation from the senate and moved his family to Nevada. Three months later, he bought three Washoe Valley ranches about twenty miles south of Reno. He planned to run five-thousand-head of cattle and breed Arabian horses.

". . . what I liked best about Nevada was the protective coloration," he told Murray Teigh Bloom. "In Moab, I was the only millionaire. In the Reno area, there were at least one hundred and thirty others, some a lot richer than me."

And Steen was fascinated by the legends of the Comstock Lode. Stories about the pick and shovel miners who had struck fortunes in gold and silver a hundred years earlier. One of the most famous of these prospectors-turned millionaires was Sandy Bowers, an illiterate teamster from Missouri who had gone west to prospect. There he struck a mother-lode and fell in love with his landlady Eilley Orrum Hunter Cowan.

When she met Sandy Bowers, Eilley had two marriages behind her, one to a Mormon bishop whom she left after she found him practicing the secret polygamous rites of "Celestial Marriage." Fleeing to Washoe Valley, Eilley married another follower of Brigham Young. She divorced him after he was "called" back to Utah when President Buchanan threatened to send the military to prevent the Mormons from seceding from the Union.

The resourceful divorcee stayed on in Nevada. She bought a log cabin close to the "diggings" and let it be known that she would take in boarders, "but not sleep with them."[15] She filled her rooms with sourdoughs such as "Old Pancake" Comstock, "Old Virginny," Pat McLaughlin, Pete O'Reilley and Sandy Bowers. When Bowers hit paydirt, she accepted his proposal of marriage.

Eilley Bowers boasted that she could divine future happenings by gazing into a crystal ball. Deep inside the transparent globe, she saw herself as

"Queen of Washoe." Dressed in royal purple and carrying a sceptre, she sat on a golden throne before her bowing subjects.

If such was her calling, Eilley decided, then she and Sandy must have a castle. They took a grand excursion to Europe to furnish a mansion. They returned with mirrors from a Venetian palace, skylights of richly-etched Bohemian glass, lace curtains costing $1,200 each, books bound in Moroccan leather, and paintings in gilded frames. Bullwhackers freighted rare marble over the Sierra Mountains. San Francisco silversmiths hammered doorknobs and hinges from bullion cast in Bowers' mill. And on a dais in the recess of a deep window sat a gilded throne embossed with golden fleur-de-lis.

But the King and Queen of Washoe were not destined to enjoy a long reign. Sandy developed a chronic cough and died in 1868. Eilley frittered away her inheritance and, by 1873, she had lost it all.

It was on a wooded hillside next to the historic Bowers mansion, once reputed to be "the most luxurious dwelling between Chicago and San Francisco," and currently a museum, that Charlie and M. L. decided to build their dream home. "If I go broke like Bowers did," Steen jested, "they can sell tickets for two mansions, practically side by side."[16]

Living on a ranch and having horses was a novelty for the Steen family. A certain amount of anonymity was a pleasure. "We didn't make the papers until a year after we moved here and we made our friends before the news broke," Charles, Jr., told Murray Teigh Bloom. "We were no one special—people had forgotten that the name Steen meant anything—and that was a big help."

Shortly after their arrival from Utah, Charles, Jr., announced that he had entered the Steen family in the annual Admission Day Parade in Carson City on October 31st. The event was only three days away. There was little time to prepare an entry. But the kids were all excited. So Steen, an indulgent parent, instructed the hired hands to wrangle some horses out of the hills. He told everybody to shine up their boots and freshen up their Western attire. He and M. L., the three older boys and eight of the ranch cowboys would ride smartly in formation. Ten-year-old Andy would lead the crew with his pony-cart. They would show these Nevadans some Texas style.

On the big day, the Steen family found itself near the end of a lengthy lineup. While floats and clowns, marching bands and other family groups mounted on magnificently-groomed Arabian horses marched up the street, the ill-prepared Steens waited in the noisy crowd for the starter's signal. Their half-wild range horses, unused to the hoopla, became skittish. The four impatient boys clamored to begin the ride.

Finally, their starting call came. Little Andy pulled his pony and cart into the procession. Then, suddenly, he surprised his parents by hurling handsful of candy, smuggled into his ponycart, out to the spectators. Squealing children dove for the candy, in and out of the pack of spooked, shying horses, around and under the ponycart. Steen, forcing a smile through clenched teeth, attempted to keep his thirteen prancing horsemen in line.

"Our entire progress, then, consisted of starts and stops as Andy was mobbed with sweet-toothed kids while I kept up a running soto voice [sic] recital of what I would do to the whole bunch if we *ever* got out of that interminable parade," Steen later told Dr. Wesley Hoskins, chairman of the Arabian Horse Show.

To make matters worse, every time the group approached a reviewing stand, the announcer would blast, "And here is the entry of Charles Steen, the Uranium King!"

"At that point, the 'Uranium King' would have been happy to drop through a convenient manhole," Steen said.

Thoroughly scorched by the incident, Steen vowed that the next year would be better. He located a restored Spanish calèche, circa 1875, in California. He imported hand-tooled saddles from Chihuahua, Mexico, and silver-ornamented harnesses from Spain. M. L. found an antique white gown in Philadelphia. Steen bought a costume of a Spanish grandee. With a trained team of four horses pulling the Venetian red calèche and sixteen outriders in full Spanish costume, in 1962, the entry of "Charles Steen, the Uranium King" garnered first place in four categories and took the Sweepstakes Award.

Steen judged that the move to Nevada had been fortuitous. Not only had he begun to feel smothered by the Mormons in Utah, but M. L. had grown unhappy in Moab. And taxes had become unbearable. Even before he closed his cliff-top home and moved to Reno, Steen decided to sell the Mi Vida Mine and Uranium Reduction Company. He figured that changing his residence to Nevada could save him up to half a million dollars in Utah state income taxes.

On May 29, 1962, Steen and his partners entered into a stock purchase agreement with Atlas Corporation to buy eighty-five percent of the Utex stock for $12,890,000. Steen's personal take was approximately eighty-three percent of the total. To ease the capital-gain tax bite, Steen took a 29.9 percent downpayment and agreed to collect the remainder in installments over a four-year period. In addition, Atlas was to pay Steen a share of certain state tax liabilities on the mine if it turned out that the taxes were not really owing.

M. L. Steen wearing an
antique gown in the
Nevada Days Parade, 1962.
Charles A. Steen, Jr.

The negotiations were complicated. The reorganization called for the dissolution or merger into Atlas Corporation of Utex, Hidden Splendor Mining Company and the separate shares of Uranium Reduction Company held by Atlas and Utex. The investment firm founded by Floyd Odlum, and now directed by David A. Stretch, A. Payne Kibbe, Edward R. Farley, Jr., and Walter G. Clinchy, officially became a company principally engaged in natural resource, manufacturing and other industrial activities. The sale essentially marked Steen's departure from the industry he had helped create.[17]

But Charlie was not deserting the mining trade. He didn't opt for the security of tax-exempt bonds, which would lead to what he called "the Three B Syndrome of bridge, bourbon and boredom."[18] He still kept looking for "that big ore body."

"The same type of people who said I was crazy to expect to find a million dollar uranium ore body would now say that one big district is all a prospector should expect to find in his lifetime," he said. "A limit on a prospector's discoveries is a self-imposed limit. Certainly, if he stops looking, he stops finding mineral deposits. I still believe a big ore body is easier to find than a small one. If you seek the big ones, the little ones come along as a matter of course."

But this time around, Steen diversified. The Charles A. Steen Company still managed the MiVida Mine. Steen was president of various mining companies that were developing gold and silver prospects in Utah, New Mexico, Nevada, British Honduras and Mexico. He had oil and gas interests. He also headed Basic Chemical Corporation, a manufacturer of rock dust used to

The Steens' restored Spanish calèche won the Sweepstakes Award that day. *Charles A. Steen Jr.*

reduce the hazard of explosion in mines, and hydrated lime for water purification systems, agricultural and construction uses. He owned a marble quarry.

"Ordinary income doesn't mean a thing," Steen told *Wall Street Journal* staff writer, James E. Bylin. "I go for capital gains—two or three years of development and then sell out."[19]

But Steen's banking advisors cautioned him against spending too much in risk ventures. They told him to get into something more stable. So Charlie switched his policy. He bought a San Antonio concern that built and repaired prop airplanes for executives. He rounded up twenty-five hundred head of cattle for the ranch. His herd of twenty-four Arabian horses contained the largest band of brood mares in the world. He picked up a Reno flying service, put $150,000 in a pickle factory, purchased a California citrus grove, and he invested heavily in Nevada real estate.

In the mid-1960s, the Steens moved into their 27,000 square-foot home next to the Sandy Bowers mansion. A driveway threaded through stands of giant Ponderosa pines to the hilltop house of native stone, wood and glass, topped by a copper roof with five pagoda-like domes. Inside, a massive circular foyer with a sweeping curved staircase welcomed guests. Off to one side was an indoor swimming pool that was forty feet long and nineteen feet deep. The dining room featured a twenty-place, elliptical table centered on a marble pad that was surrounded by a moat. The 13,500 square-foot living room had a freestanding fireplace in the middle and was furnished with teak cabinets, suede upholstery, marble and copper fixtures and a cupboard holding five thousand dollars worth of liquors. A Dean Fausett oil painting of Deadhorse Point, a scenic site near Moab, hung on a wall. In a place of honor were Charlie's bronzed boots and a gold-plated gasoline lantern, the one that Charlie and M. L. had used throughout their prospecting days.

". . . it cost $285 to have it gold-plated," M. L. said later. "And the man said, 'What kind of nuts are these people? You can go out and buy a lantern like this for six dollars!'"[20]

It took almost three years to complete the Steen mansion. Then, on March 26, 1966, the monument to Charlie Steen was christened with a gala housewarming party. The invitation, carrying a photograph of the "Casa de Cisco—1952," Steen's tarpaper shack, announced:

> Charlie, M. L., Johnny, Charles, Jr., Andrew and Mark Steen have moved into a new shack in Washoe Valley, Franktown, Nevada. This shack is better than the one we left in Utah; we enjoy indoor plumbing and an electric light, but our Coleman lantern is still around (if the Power Company should disconnect service for nonpayment of the overdue bills.)—Our shack was under construction for three-and-a-half years. The whole family scrounged during the week for used materials. Charlie worked on our shack weekends because he had to promote and prospect on weekdays!— We have a good sized water hole, so please bring swimming suits for mixed bathing—no towels—we will supply you with feed sacks (dividend from Arabian horses.)
>
> It's a Shack-Warming Fiesta. Let's Celebrate!
> Place—"Hacienda de Washoe—Franktown Nevada.
> Date-Saturday, March 26, 1966.
> Cocktails—5 o'clock,
> Rim Rock Vittles (Buckskin, biscuits, and gravy)—7 o'clock.
> Dancin,' Drinkin,' and Cavortin'—'Til we run out of moonshine, buckskin, and night!

Only two years later, Steen's quip about the Power Company cutting off electricity became a sad reality. At eleven A.M., on February 27, 1968, while Steen was in California on business, six agents from the Internal Revenue Service marched into his plush, circular office building adjacent to the Reno Airport. One of the men stood in the hallway and read a statement announcing that the office was being seized due to delinquent taxes. While the federal agent instructed the employees to "get their coats, hats, and purses, and leave," the other five men fanned throughout the facility with padlocks and seizure notices. They padlocked Steen's private office that had been papered with stock certificates from defunct uranium companies, and locked his eighty-five filing cabinets. They confiscated Charlie's six-thousand-volume geology library and his prized mineral collection, "which couldn't be replaced for $500,000." The government men even took Ringo, a pet spider monkey.

"They're supposed to send you a letter ninety days in advance of a seizure, saying you owe so much back taxes," a bitter M. L. said later. "We got the letter a month and four days after they had locked us up."[21]

The government claimed that Steen owed $1.8 million in back taxes. The action was taken because of alleged illegal deductions Steen had taken on some of his ill-fated business ventures. The citation also stated that $459,000 received in 1966 from Atlas Corporation for uncollected Utah sales taxes in the Mi Vida deal disqualified Steen's capital gain from being taxed in installments. The installment contract, alone, had an ascertainable value in 1962, the government contended. Steen should have reported the entire gain that year.

As for ill-fated business ventures, Steen rued his decision to diversify. "Everybody kept advising me, 'Diversify, Charlie, diversify,'" reported *Los Angeles Times* writer, Charles Hillinger. "So I diversified. I got into things I couldn't hack. I got spread out so damn thin. Everything I touched went sour.

"For example, I got into the aircraft business, but I was manufacturing piston planes for businessmen when jets came in. It was like getting into the buggy whip business when the Model T was introduced."

Steen dropped $3.5 million on the aircraft factory.

"I bought a large cattle ranch with 2,500 cows," he told Hillinger. "The price of beef dropped to new lows.

"I was losing $30,000 a month on the ranch when I traded it for an orange grove. The orange grove didn't work out either. I lost $700,000 on it."

And somewhat sheepishly, Steen added, "A fellow came to me. He was making gourmet pickles in his basement. He said if I put up the money, it was a sure thing.

"First year, the cucumber crop froze. Next year it failed again."[22]

Steen lost $250,000 on pickles. A mine strike sent millions down the drain.

"I got outside my specialized field of knowledge where I could make my own decisions," Steen told the Wall Street Journal. [23]

With the I.R.S. seizure and consequent publicity, other creditors came out of the woodwork. Approximately three hundred debts, ranging from

$450,000 to five dollars, totalled some five million dollars. Steen's newly-refurbished yacht, three twin-engine planes, and his office building were repossessed. Efforts for bank loans failed. Steen's credit dissolved into mountains of unpaid bills. The I.R.S. attached his income-producing properties.

Desperately, Charlie tried to float the sinking ship. He pawned M. L.'s jewelry. His mother, whose Utex interest he had bought in 1962 after a reconciliation, pitched in her jewels, as well. By taking shelter under Chapter Eleven of the Federal Bankruptcy Act, he staved off repossession of their home only a day before the mortgaging bank was poised to kick them out.

By November 1969, Steen told Charles Hillinger, "Here I sit in this big house that M. L. and I designed, without a damn dime to our name.

"We're sitting in the middle of all this luxury—my wife, four sons and I—eating canned beans and stale bread just like we did in that tarpaper shack seventeen years ago before we struck it rich."

That December, Dobson Benedec, an Omaha painter and art gallery operator, read about Steen's plight. Noting the federal aid and charitable organizations serving the nation's poor, Benedec asked, "Who helps the destitute millionaire?" He organized a food drive and sent a truckload of coffee, beans and meat to the Steens.[24]

Despite the unfortunate turn of events, Charlie maintained the philosophy he had espoused in 1963 when Murray Teigh Bloom asked him what he would do differently if he had it to do over.

". . . life isn't like that," he replied. "You can't go back and make one or two changes to get a neater, happier pattern. You've got to buy the whole package as it is: blemishes, sorrows, mistakes, and all. If you ask me on *that* basis, I'd say: no changes, no regrets."[25]

Since the I.R.S. raid in 1970, Charlie and M. L. had spent agonizing months in court battles—scrambling to cash in on any assets not seized by the government, struggling to retain their home and personal possessions. And all the while Charlie, recognizing the error he had made by diversifying his business interests, took every possible chance to get back into the thing he knew best, prospecting for ore.

"If there's something I've got out of this whole bankruptcy mess, it's that it's got me out into the field again where I belong," he told the *Salt Lake Tribune* on February 24, 1970.

Steen headed back to Utah, where it had all started. Attempting to disguise himself, he would go unshaven for a week at a time. He moved around, staying in a different motel every night, and he used an assumed name. His sons, who went with him, "thought it was great hiding out like the law was after us or something."

But Charlie was too well known in mining circles to go unrecognized, even with his low profile.

"... we stopped at a café, and someone said, 'Hi, Charlie!' I couldn't believe it. I guess my disguise wasn't too good," he said.

Steen always denied that the resumption of prospecting had anything to do with the bankruptcy case.

"I'm going to get out of that without any new find," he said. "My only ambition now is to leave all my kids multimillionaires."[26]

In many ways, Charlie was back where he started. He was broke, using beat-up equipment and growing gaunt from the hard work of prospecting.

But in early 1970, it looked as if he was in luck again. When he filed three hundred and fifty claims near Escalante, Utah, one test in an oil exploration hole that had been thought unsuccessful showed radioactivity that "goes right off the log." Steen was able to get financial backing from a Canadian firm to drill the prospect to twenty-four hundred feet.

Associated Press reporter Marty Thompson asked Steen what he would do if he struck it rich again.

"I'm going to buy some new clothes and a Mercedez Benz automobile," he said "I only had twenty-five percent of that $60 million [the Mi Vida]. This time I've got a helluva lot more of it."[27]

But three months later, Charlie had to admit defeat.

"There's uranium," he told the Associated Press, "but it's not commercial. I'm done for the time being."[28]

The following year, he found a promising copper prospect in Deep Springs Valley, near Bishop, California. After staking approximately three hundred claims with the hope of intercepting a large deposit of low-grade ore, he purchased a small coring rig and hired a couple of untrained helpers.

Then, on Monday, September 6, 1971, while they were drilling a claim, Charlie accidentally punched the wrong button on the control panel for raising the rotary table. The drill began to whirl and a wrench fastened to the vertical pipe swung around and hit Charlie on the side of the head, knocking him unconscious. When he came to an hour later, he had difficulty speaking. His vision was blurred. The hired hand carried him to the truck, and they bumped over twelve-and-a-half miles of dirt road to the paved highway. By the time they arrived at the hospital in Bishop, Steen's head was bruised and swollen. When his condition had not improved two days later, he was rushed to the Washoe Medical Center in Reno. There a neurosurgeon removed a blood clot from his brain. Charlie was not expected to come out of the operating room alive.

M. L. and the four boys waited in the intensive care unit. Charles, Jr., had come with his wife and children from British Columbia, where he was living. As the long hours stretched through the day, the family members were allowed to sit with Charlie briefly, one at a time. None of them expected ever to talk with him again.

Thirty days passed, and Steen remained in a coma. The prognosis was still poor. Then, finally, he rallied. The family was told that he would live, but it would not be an easy recovery.[29]

"When my family brought me home," Steen said later, "they didn't know if I would be an idiot or what. I couldn't speak a word of English, but when I got home, I started singing in Spanish to my dog."[30]

At first, he could scarcely talk at all. There was no association between his words and his intent. If he needed to go to the bathroom, for instance, all he could come out with was, "Gold and silver." It was a stressful time for both M. L. and Charlie.[31]

"Recovery has been slow," he told a reporter in 1973. "When you ask me a question, it may take four or five minutes, but I'll answer. I know what I want to say, but the words won't come out. I'm in good shape physically. I can out-walk my sons. There's no damage to my mind, but it's hard to talk—especially when I get excited."[32]

Steen's hardhat, bearing the inch-deep dent where the wrench had struck, was placed alongside his bronzed prospecting boots as a memento of the accident in which he tricked fate once again.

His brain injury and the death of his mother in 1971 made the ensuing bankruptcy proceedings all the more traumatic for the Steens. There were endless hearings, submission and rejection of revised plans, hiring and firing of attorneys, resignation and replacement of bankruptcy referees, and aborted sales of properties and business enterprises that might have bailed Steen out. In 1974, Charlie and M. L. were evicted from their Reno mansion. The ranch house was taken by the court to help pay an estimated six million dollars in debts. (Two years later, the million dollar 21,000-square-foot home sold for less than $300,000.)

Despite all his troubles Charlie Steen was still able to muster some of the old optimism. "I'm not bitter," he told Salt Lake Tribune reporter Clark Lobb in the summer of 1975. Then, stressing that his thoughts lay more with the future than the past, he said, "It's not a matter of thinking I'll make another big find; it's knowing it."

Eleven years to the day after the I.R.S. proceedings began, Steen's bankruptcy ended on May 9, 1979.

The I.R.S. claim had been raised to $29,282,244. Steen's assets were to be liquidated for $3,394,746. After paying secured creditors, attorneys' fees and the I.R.S., there was nothing left to pay the $1,167,453 owed general creditors.

Without hesitation, Charlie's sister, Maxine, gave him all the money she had earned from Utex stock. Carolyn and Holly Seeley, friends from the early Cisco days, immediately responded to Steen's call for help with $15,000, and asked no security.

The support of the Seeleys was "a contrast to a lot of people I made money for that wouldn't offer to give me any help when I needed it," Steen said later, with some bitterness.[33]

"You know, the government owes Charlie Steen a lot of thanks," John J. Gibbons, one of Steen's attorneys, told Hillinger of the *Los Angeles Times.* "The United States was buying uranium from South Africa at $46 a pound at the time Charlie ran onto Mi Vida.

"Two years later, the government was paying Charlie and others $8 a pound.

"Charlie saved the United States a billion dollars."[34]

But Steen knew the past was over.

"All I want to do is climb out from under," he said. "If I come out of this without a red cent, that's okay. I'll go prospecting and make it again.

"But next time if anybody suggests to Charlie Steen to diversify, I'll flatten him."[35]

The Steens were allowed to retain half interest in twenty acres of timberland in Washoe Valley, their home in Moab and the old Cash Mine near Boulder, Colorado.

"It's always been our ace in the hole," Charlie said of the Gold Hill property that Utah Exploration Company had purchased for $100,000 in 1958.[36]

He opened the mine in 1983. Young Mark Steen, the project manager, revealed that the mine would be a joint venture between Cosmos Resources, Inc., of Vancouver, British Columbia, and the Steen family.

The Cash Mine, discovered in 1870, had been one of the most productive operations in the district. Early assays of the ore averaged over one ounce of gold and ten to twelve ounces of silver per ton. The underground workings had been idle from 1919 to 1946, when William E. Brewster drained water from the four levels and rehabilitated the shaft. Brewster struck gold on two of the levels and continued mining until 1953. Then, he stopped work and let the mine flood.

Steen did some development work on the Cash Mine after he purchased it and, by 1964, the facility was ready to go into operation. An estimated

$600,000 had been spent in blocking out over nineteen thousand tons of ore. Reserves were calculated at over 128,000 tons, and about 150,000 tons of waste dump material could be brought up to ore grade, it was thought. But Steen opted to wait.

"I didn't need the money [in 1964]," Steen said in 1983, "and $35 per ounce gold and $1.10 silver were too low. I decided to speculate on precious metals rising in the future. At the time, I hoped gold would reach $100 per ounce."[37]

A small-scale sorting and concentrating mill was operated by one man between 1967 and 1971. It wasn't until the early 1980s that actual rehabilitation work on the mine, re-working of the dump and construction of a fifty-ton-per-day processing mill was begun.

But today, due to a family tragedy of another kind, the mine is dormant. A rift between the four sons and their parents split the family into two camps. Mark, manager of the Gold Hill property and spokesman for his father and brother John, is ostensibly in charge of the estate. (M. L. Steen died on July 14, 1997.) They live in Longmont, Colorado.

Charles, Jr., and Andy contended that Charlie retained no interest in the Moab holdings or Gold Hill operations. The estate, they claimed, belonged to the four Steen sons in return for a $175,000 note drawn by their father on a trust fund established for them years before. The opposing pairs of brothers also accuse each other of wrong doings.

Time has not yet brought a new bonanza to Charlie Steen, and family disputes persist. If one were to ponder whether he was a "victim" or a "survivor" of the uranium boom, possibly more than anyone else, he would be both.

"Certainly there aren't many survivors [of the uranium boom]," said Edward R. Farley, Jr., former chairman of Atlas Corporation. "Some people in the industry never made it, never achieved any economic success—which was the name of the game. Some people like Charlie (and there were a lot of others) made a lot of money and, in one way or another, got rid of it. I assume they're losers, too, even though they were big winners. . . . They undergo a lot of hardships and take a lot of risks to try to find that strike and then their [venturesome] nature doesn't change. So, when it's all reduced to dollars, they get rid of it anyhow."[38]

16   A WIDOW FIGHTS BACK

Tex Garner didn't know he had less than twenty-four hours to live.[1] It had been one week since the doctor, talking of "cancer" and "one to fifteen years," dismissed him from St. Mary's Hospital in Grand Junction and sent him home to Moab. Tex was a bit tired, but there wasn't much pain in his lungs and the scar from his liver biopsy was fading already. He thought he was making good progress. In no time, he'd be pitching a few innings and showing the kids how he could walk on his hands all around the front yard. The old Garner optimism was returning.

"Fifteen years?" he would say, negating the possibility of anything less. "Let's see. I'm forty-seven now. Sixty-two is not a bad age to die."

It was a pleasant time of year. The September sunshine was taking on a bit of crisp and the leaves were just beginning to blend with the red and gold cliffs surrounding Moab. Tex had been underground in the uranium mines for so many years that he had almost forgotten the brilliance of fall—summer's last gasp before the brown chill of winter. He spent his days resting under the peach trees that shaded the red and white mobile home on Cliffview Drive, below the mansion where Charlie Steen had lived before he moved to Reno. Sometimes Tex would pick a plump fruit and watch the RVs filing along the highway leading into town. Fall brought out hordes of retirees who had postponed their wanderings until vacationing families returned their kids to school. It would be less crowded, then, at Arches National Monument, Deadhorse Point, and other geologic wonderlands where tourists flocked after mining roads in uranium country had opened up the backlands.

September was also playoff season and baseball was Tex's passion. For years he had pitched in the local leagues and he had even had a chance to go pro when he was younger. Now his favorite team, the St. Louis Cardinals, was favored to play against the Los Angeles Dodgers for the 1963 National League title. On the 13th, the Cards had downed the Milwaukee Braves for their seventh straight win. The next night, they would meet the Braves again.

Tex usually listened to the games in the car. After supper, he would tell his wife, Eola, that the radio reception was better outside. She would smile knowingly as he ambled out to the blue and white Buick (his other passion) and stretched his six-foot-two-inch frame over the front seat. Sometimes, he would fold his long pitcher's arms behind his head and bend his knees to make himself fit. On balmy evenings, he would dangle his excess length out of the window. Every now and then, Eola would step outside to see that all was well. Tex would rise up in the seat and tell her that Ron Taylor had retired three batters or Ken Boyer had fumbled a grounder. She would nod her head and then return to her television program or help the two younger kids with their homework.

On the evening of Saturday, September 14, 1963, the Cardinals won again. Tex was elated when he came back into the trailer. He vowed that if the Cards made it to the World Series, he and Eola were going to see it. Then, admitting that he felt a bit "under the weather," he said he thought he'd forego the prizefight telecast that night and go to bed early. Eola said she was tired, too, and joined him.

At two in the morning, Tex awakened with excruciating abdominal pains. By eight o'clock, he was in the Moab hospital. At five-thirty Sunday afternoon, Douglas G. "Tex" Garner was dead.

Tex's untimely death shook the small community. The Methodist Church was crowded on the day of the funeral. Boys in the Pony League baseball team that Garner had coached served as pallbearers. One of the youngsters, devastated at losing his coach and good friend, placed Tex's baseball cap on the casket and said goodbye just before the coffin was lowered into the ground.

The finality of Tex's death did not come easily to Eola. He had always seemed so indestructible, energetic, easy-going—and impractical, sometimes exasperatingly so.

Eola was conservative and serious-minded. She was always thinking of putting aside for the future, so they could buy a nice home and furnish it with pretty things. Tex had lived vigorously and relished each day as it came.

"Money was made to spend," he would say, "and I'm going to enjoy myself."

As long as he could play ball, drink a couple of beers with the guys, and buy a new car every few years, he wasn't interested in a closet full of clothes or a fancy house.

Eola's adjustment to Tex's sudden death was also hampered by the fact that he was hardly ever sick. He'd had some cartilage removed from his right knee after a service injury some twenty years earlier. He'd had an appendectomy way back in 1948. More recently, he had suffered a duodenal ulcer. But only eighteen months ago, he had taken out a New York Life Insurance policy. The doctor who examined him had said that if he stayed as healthy as he was then, he'd live to be ninety.

As for the well-known hazards of hard-rock mining, they didn't seem to phase Tex. Five years before, he had been walking through a dark tunnel in Standard's Big Buck Mine when he plummeted down a fifty-five foot shaft. His quick reflexes in grabbing at protruding rocks saved him from splitting his pelvis on a jagged ledge, or worse, getting killed. He came out with a sprained ankle, a slight concussion, and a few cuts and bruises.

The Public Health Service doctors, who had conducted triennial examinations for their uranium miners study since 1957, always pronounced Tex a healthy specimen. Neither Tex nor Eola dreamed that the insidious radon he had ingested had been gradually turning to lead since his early milling days with the Manhattan Project.

Tex first got into the uranium business in 1939, primarily because he was a baseball player. At eighteen, the young pitcher from Mason, Texas, had been picked up by the St. Louis Browns, and sent to a training camp in California. But he soon balked at the strict rules and tough regimen, and cancelled his contract and went home.

Shortly thereafter, Sandy Sterns, Tex's friend and superintendent at the vanadium mill in Uravan, Colorado, got in touch with him. Sterns told Tex that if he would come and play baseball, he'd give him a good-paying job in the plant. The superintendent, an avid baseball fan, needed a pitcher for his team. Tex accepted. And it was in Uravan, at a night game in 1940, that he was introduced to the slender, five-foot-nine-inch high-school football queen from Estelline, Texas, who was visiting her sister. Tex and Eola Eddins were married three months later.

Within a year, Sterns was transferred to the mill at Rifle, Colorado. He still needed a pitcher so he talked the Garners into following him. The young couple rented a modest house on a corner lot filled with fruit trees. Eola was ecstatic. When the owners offered to sell the home to them at a low price, she begged Tex to buy it. But baseball fever had hit again. Scouts for the Cincinnati Reds had seen Tex play. They made him an offer.

It was 1942, and America was at war. The vanadium mills were working the old tailings piles for uranium. Nobody knew why. The workers were told it was part of the "war effort" defense work. It was important enough, however, so that mill employees were deferred from the draft.

The last portrait of Tex Garner and his family. Seated, left to right: Eola, Pamela and Douglas "Tex." Standing: Ronald Keith, Jimmy Douglas and David Wayne. *Eola Garner.*

Sterns warned Garner that if he quit his job to accept the Reds' offer, he'd lose his deferment. But Tex had to give himself that last chance at baseball. He set out for Ogden, Utah, and the team's training farm, but he arrived too late. Having missed the session, he signed with the semi-professional Industrial League in Brigham City, Utah, and then he was drafted.

Garner first joined the air force, and then transferred to the corps of engineers. The last place he expected to be stationed was Uravan, Colorado. But after boot camp at Fort Bragg, North Carolina, and brief duty with the engineering department at Oak Ridge, Tennessee, that's where he landed, totally unaware that his work would have anything to do with an atomic bomb.

The Manhattan Project had constructed an experimental plant in Uravan to process uranium from the vanadium tailings. Garner and several other experienced mill workers were drafted and sent to the small company-town to work the operation. Thus, military service scarcely affected the Garners. Tex was assigned to his old job as a leach operator. He pitched for the local baseball team, and Eola bore their first two sons.

After the war, when the AEC made its call for stockpiles of domestic uranium, Garner left mill work and went underground. As was typical of hard-rock miners, he moved from place to place as new ore deposits developed. He leased the Yellow Bird and a few other claims near Uravan from Vanadium Corporation of America. He worked in the Radium Group at Slick Rock, Colorado. He mined high grade uranium at the Basin #1 in Cottonwood Wash, Utah, and leased some diggings from Kerr-McGee close to Cove, Arizona. Then he returned to Colorado at VCA's Hideout and King Incline #2. From there he ended up in Utah at Standard's Big Buck and Hecla's Radon mines, where he worked twelve hours a day, five days a week—drilling, blasting, mucking, hauling loads of ore—right up to the time that he fell ill for the first time.

The two older Garner children—Ron, who was nineteen, and Dave, twenty—thought their dad was kidding when he collapsed that Sunday afternoon in July 1963. Tex had always been a practical joker. When he came out of the bedroom and fell to the floor, choking and gasping for breath, the boys thought it was another Academy Award performance. They knew there was nothing wrong with their father's health. He'd been pronounced okay by the Public Health Service examiners six months before. He had just been treated in the hospital for his ulcer. The doctors had found nothing else wrong with him. Besides, he was keeping a full schedule working in the mine, coaching a "pony" baseball team of youngsters and playing ball, himself, on weekends.

But when Garner didn't jump up and call off the joke, his sons realized it wasn't an act. Thinking Tex had suffered a heart attack, Eola telephoned the family doctor. He had them come to his office and examined Tex but found nothing. Garner was just over-tired, the doctor said.

Three weeks later, in August 1963, Tex took his pony league to Spanish Fork, Utah, for a playoff. He returned home in a few days with the good news that the Moab kids had taken first place. But Eola noticed that Tex had lost weight, at least ten pounds while he was away. Alarmed, she called the doctor again, and Tex was referred to another physician at St. Mary's Hospital in Grand Junction. There he was put through a series of tests. This time the X-rays revealed a small cancerous growth on the lung.

The doctor at St. Mary's called Eola at home to tell her the news. They would say nothing to Tex right now, he advised. And there would be no lung operation. It would be better first to perform a biopsy on the liver to determine if the malignancy had spread.

A few days after Tex entered the hospital, he and Eola were chatting with friends who had come to visit when a nurse poked her head into the room and asked Eola to step down the hall to take a phone call.

She could tell immediately from the doctor's voice that the news was bad. The cancer had metastasized, he told her. It was in the lungs and liver, perhaps other organs, or bone. The prognosis was very poor.

Eola leaned against the wall to catch her breath.

"Don't tell him," she told the doctor. "He's so optimistic that I don't think he'll ever know what's wrong with him and I'd rather that he didn't."

The doctor promised that he wouldn't say anything. At least for the time being.

Eola hung up the phone and tried to compose herself. She couldn't let on about what she had heard in front of Tex and his visitors. Somehow she was able to mask her feelings by the time she got back to the room.

"What did he say?" Tex asked as she walked through the door.

"He said you could go home in a week and a half or two weeks," she said.

"I told you there wasn't anything wrong with me," he said. "When I get home with your good cooking, I'll get better."

Everybody laughed with relief.

He was more serious when she arrived at the hospital a few days later, however. He told her about an article he had seen in the *Denver Post.*

"You know what?" he said, "I read in the paper this morning that working in the uranium mines could cause lung cancer. And that really scared me. I'll never go back into another uranium mine."

Eola didn't know how to respond. Tex didn't know about the diagnosis yet. She changed the subject.

But that weekend, when she took the children to see their father, Tex was not himself As she leaned over to kiss him, he said abruptly,

"Go check me out."

"Have you been discharged?" she asked, surprised.

"Yes," he answered.

"The doctor said you wouldn't be discharged for another week, or so," she said.

"I'm going home!" Tex insisted. "Just go check me out."

Then he drew her to him and started to cry.

Tex heard about his cancer the morning he was released from St. Mary's Hospital. The impact was severe, but Tex just didn't have it in him to stay down for very long. When the doctor said he could possibly live for fifteen more years, that was the number Tex settled on. As far as he was concerned, it would be no less. Only when he was rushed to the Moab hospital a week after he returned home, did he accept the fact that death was near.

The spasms had come on fast and strong. Even after hospitalization, Tex was unable to find relief. He would ask the nurse to raise the bed's backrest, and then moments later tell her to put it back down. The effects of pain killers would wear off shortly after they were administered, and he would call for another one. Tex Garner knew that he was dying.

"I got this lung cancer from working in the uranium mines," he told Eola. "You won't have enough money to get along when I'm gone. You've got to go after workman's compensation for you and the kids. Sue the mining company, the state or even the AEC if you have to. But promise me that you won't give up till you get it."

Eola promised. The doctor had already talked to her about it before Tex had left St. Mary's Hospital. He had told her that some widows of uranium miners in Colorado had been awarded workman's compensation by the State Industrial Commission. It had been proven that their lung cancers were caused by radiation in the mines. Eola knew that Tex's five thousand dollar life insurance policy would not go far in supporting and educating her two younger children. The older boys would need some help as well. An additional sum of money would ease her load.

Certainly Tex had earned it. Twenty-three years of work in the mills and mines. If his cancer was caused by radiation, someone should pay. The mining companies. The state. The AEC. Whoever was responsible for the hazardous conditions.

Eola had been told that an autopsy was necessary to establish that a death was due to excessive exposure to radon. It was hard to think about such a thing while her husband was still alive, but she had faced the fact that he would be gone in a matter of hours. She asked their family doctor to make the arrangements. He was not licensed to perform a post mortem, and the Moab hospital didn't have the facilities, so the examination would have to be done in Grand Junction.

When Garner died late that afternoon, an autopsy was scheduled with Dr. Geno Saccamanno at St. Mary's Hospital for the following day. The pathology report revealed that Tex Garner had oat cell carcinoma of the lung and abundant tumor tissue in the liver, spleen and adrenals. A subsequent analysis of body tissues by Dr. Victor Archer showed exorbitant amounts of radioactive polonium and lead.[2]

"The average range of lead-210 reported in the literature for fresh bone of normal people (non-miners) is between 15 and 54 picocuries per kilogram," Archer wrote. "The normal body content of Po-210 [polonium] is slightly less than that of Pb-210 [lead]. . . ."

Garner's bone samples registered 448.64 picocuries per kilogram of Polonium-210 and 1852.50 picocuries of lead-210. The kidneys contained 978.10 picocuries of polonium and 224.80 picocuries of lead. The liver 213.74 picocuries of polonium and 167.90 picocuries of lead. When the radiation readings in mines where Garner had worked were tabulated with his years underground, it was figured that he had been exposed to 1,870 working-level months of radiation.[3]

Eola wrote a letter to Charles J. Traylor, an attorney in Grand Junction. Traylor had successfully handled workman's compensation cases for several widows of uranium miners. Eola asked if he would represent her. He said he would.

A few days later, Eola drove to Grand Junction to meet with her lawyer. When she parked her car in front of the old public-service building on Third and Main Street, she was not favorably impressed. As she zig-zagged up the rickety back stairs to Traylor's suite of offices on the second floor, her misgivings intensified. The electrical system was so dilapidated that the automatic typewriters wouldn't work if the swamp coolers were operating. Consequently, pigeons flew in and out of open windows in the poorly-ventilated structure. Traylor's office looked into a brick wall a couple of feet away. There was a musty odor of age.

"Surely anyone who is a good attorney wouldn't have this kind of office," she thought.

She was somewhat reassured when Traylor told her that he was handling three other compensation cases in Colorado. And then, after reciting her husband's work history since 1940—a long succession of mills and mines in Uravan, Naturita, Dove Creek, Slick Rock and Cortez, Colorado, and the Big Buck and Hecla's Radon Mine in Utah—Traylor promised that he would figure out a way to file for her not only in Colorado, but in Utah as well.

"We'll hold one state in abeyance while we're trying the case in the other," he said.

With compensation in Utah, Colorado, or maybe even both states, Eola's odds would be improved. There was a hitch, however. There was no precedent of a similar suit in Utah.

On October 23, Traylor wrote a letter to Brigham Roberts, a legal colleague in Salt Lake City. He asked Roberts if there was a workman's compensation expert in Robert's office who might collaborate with him.

"The Colorado Supreme Court has ruled that this type of injury is compensable and we probably have three or four cases dealing with this problem," he wrote. "We are not aware of any cases now pending before the Utah Commission, or one where they have ruled that there is a causal relationship between working in the uranium mines and lung cancer. . . . It is entirely possible that if we break the ice with this case in Utah that there will be numerous other cases available to your firm in Utah."

Two days later, Roberts replied. Richard C. Dibblee would work with Traylor, he said. But he had conferred with a friend in the Safety Division of the Utah Industrial Commission and been summarily told that the state was aware of the presence of radon gas. Roberts was informed that the mine supervisors had kept "a close check" and "the safety rules provide a maximum amount of this gas allowable and, if this maximum is reached, the mines are closed down."

It didn't appear that the Utah commission would be too receptive to Traylor's argument.

Traylor decided to hedge his bet. Because the statute of limitations on a workman's compensation claim filed under the Occupational Disease Act was six months, he decided to register an accident claim in Utah as well.

"It is our idea that if we file under both that the Commission will have to make a decision somewhere along the line and it should prevent any statute of limitation from running," he said.[4]

But in November 1965, Hecla Mining Company, which had been named in the suit as Garner's last employer, denied liability. According to law, the medical issues then had to be referred to the Occupational Disease Panel of the Utah Industrial Commission, with the ensuing tangle of red tape.

A full autopsy report had to be be filed. Certain portions of Garner's vital organs needed to be analyzed for radioactive lead content. Expert witnesses, such as Duncan Holaday, Dr. Victor Archer and Dr. Geno Saccamanno would have to be subpoenaed. A few of Tex's co-workers would be asked to testify. To complicate matters, nobody really knew how to proceed. This would be the first such case heard by the Utah commission.

"To begin with," said Otto Wiesley, Utah Industrial Commission chairman, "we'll have to establish that the kind of radioactivity claimed does cause cancer of the lungs. Then, the panel will have to satisfy itself that the particular cancer involved resulted from exposure."[5]

Almost admitting to the inevitability of the suit's defeat, Wiesley added, "The very absence of any such claims convinces us we've done a good job at keeping radioactivity dangers in Utah mines down."

Then began a process that dragged on for three years. First there were hearings before a committee of physicians on the Occupational Disease Panel. In August 1964, the pleas were denied. The plaintiffs immediately filed an objection and the next year they were granted further hearings before the group of doctors. Once again the panel voted against the plaintiffs. Finally, the case was taken before the Utah Industrial Commission itself.

On May 16, 1966, Chairman Otto Wiesley opened the hearing. Witnesses for plaintiffs and defendants went over their testimonies once more. Mine operators were quizzed about ventilation practices. Did they rely on natural ventilation from mine entrance portals? Were there fans in the air pipes? Did surface air holes intersect with underground tunnels?

Miners were asked if they drilled "wet" or "dry." Were they ever present during blasting? How long were they required to remain outside after the dynamite blew? Did they ever hear about radiation being measured in the mines?

Dr. Victor Archer was questioned, at length, about his epidemiological study and its conclusions. He testified that the study had established that lung cancer occurs more frequently in uranium miners than among other people who contract the disease. Furthermore, he said, the small cell, undifferentiated (oat cell) type of cancer occurs more often in the uranium miners. And it was proven that the greater the dose of radiation the miners received, the greater the chance they would contract lung cancer.

As for Tex Garner, the doctor went on, he had spent approximately fifteen years mining underground in exposures well over the recommended working level. The lead content in his bones was about thirty-four times the amount of that in a non-mining control group. Garner's cancer was the small cell, undifferentiated type. In view of the surgeon general's recent proclamation that smoking could be dangerous to a person's health, the fact that Garner had smoked a pack of cigarettes a day might have contributed to the problem, Archer stated. But he added, "the odds are greater than ten to one, perhaps as high as a hundred to one, that his lung cancer was caused by radiation." [6] Letters from four other doctors were introduced as evidence that there was a cause-effect relationship between exposure to radiation in uranium mines and the development of lung cancer.

Dr. Elmer L. Kilpatrick, chairman of the Occupational Disease Panel, appeared as the final witness. Taking the floor, he admitted that the three doctors on the panel had never participated in surveys or studies concerning the relationship of radiation exposure to lung cancer. They concurred with certain of Dr. Archer's findings and agreed that Garner had certainly been exposed to an excessive amount of radiation. Kilpatrick even allowed that the incidence of lung cancer in uranium miners appeared to be higher than with the general population.

But Garner was a cigarette smoker, he pointed out. As far as the panel of non-smoking Mormon doctors were concerned, "If the miner was a smoker, it was the smoking that killed him." At any rate, they were not entirely convinced by Archer's scientific findings. The panel upheld its earlier decision. It was their opinion, Kilpatrick stated, that "there was a high 'possibility' as distinguished from a 'probability' that the cause of death was over-exposure to radiation.

"I wouldn't use the word 'probable,'" Kilpatrick emphasized, "I would use the word 'possible.' And we consider the term 'beyond a reasonable doubt' is a pretty good phrase to use."

"You mean in Utah, in order to recover, you doctors have to say it is beyond a reasonable doubt? Is this your understanding?" Traylor asked.

"Yes."

"I see," said Traylor. "And you are not willing to say that beyond a reasonable doubt Mr. Garner died of lung cancer as a result of overexposure, is that correct?"

"Yes."[7]

The Utah Industrial Commission filed an order denying the Garner claim. Furious at the decision, Traylor applied for another hearing.[8] Proving a case "beyond a reasonable doubt" was a standard for criminal matters, Traylor objected. Such proof was not mandatory in industrial cases, which only needed to be supported by "a preponderance of the evidence."

"The Medical Panel required an unreasonable amount of proof that the deceased died of an occupational disease," he raged. He further accused the Mormon doctors of being "biased and prejudiced against the applicant."

The Industrial Commission refused to schedule a re-hearing. Traylor promptly transferred his appeal to the Utah Supreme Court.

Eola Garner was deeply discouraged. She had spent three years working on the case, making arrangements for Tex's former co-workers to appear as witnesses; spending hundreds of dollars to pay their transportation, room and board when they travelled to Salt Lake City; facing repeated postponements of hearing dates and waiting months for decisions.

In the meantime, she had had to table her own plans for the future. She wanted to move to a college town where her children could get a better education. She longed to open up a dress shop. But she couldn't follow through on either plan without some kind of a financial settlement.

"I really need to win this case," Eola Garner wrote Traylor, "as I have children to put through college, and I have spent so much money on it [the case]. To win it would help me get established in a business. . . ."[9]

Heavy smoking by miners contributed to the danger of lung cancer from under-ground radiation. *Museum of Western Colorado.*

She started to write people in state government in hopes of enlisting their support for her cause. She requested pamphlets about lung cancer from Dr. Victor Archer, and sent that information to representatives Sam Taylor and Ed Drake from her district. She also wrote to Utah Governor Calvin L. Rampton.

The responses were disappointing and noncommittal. The politicians regretted that the Constitutional Doctrine of the Separation of Powers precluded any interference with the Supreme Court. They were sure that the high court would see that justice was done.

While Eola Garner was waging a one-woman battle in Utah, national attention was starting to focus on the plight of the uranium miner. In May

1966, Charles Traylor received a letter from Leo Goodman of the Industrial Union Department AFL/CIO in Washington, D.C.

Goodman, who was also secretary of the AEC Technical Committee, had been concerned about the miner radiation problem for some time. Seven years before, he had urged the Congressional Joint Committee on Atomic Energy to "step up to their responsibility regarding the health of uranium miners. . . ." He had pointed out that France had resolved its radiation problems by installing adequate ventilation with an increased cost to mining operations of only one percent. Goodman contended that the U.S. Congress, "which had brought the whole of atomic science into being, had the responsibility to help protect all of the individuals who were employed in it for the national interest."

Goodman had then devoted his efforts towards convincing Utah, New Mexico, and other uranium mining states to join Colorado in compensating the families of miners who had died because of excessive radiation. Now he had another idea.

"It has occurred to me," he wrote Traylor, "that the whole question of safety falls under that portion of the U.S. Department of Labor supervised by Esther Peterson, who is a native of Utah, and that it might be desirable for me to talk to her about this whole problem of the refusal of the State of Utah to accept the Public Health Service's certification of causation as the Colorado State Fund now does."[10]

Peterson, who described herself as "a gym teacher appointed to a political position,"[11] was assistant secretary for Labor Standards under Secretary of Labor Willard Wirtz. The former instructor at Utah State University in Logan, Utah, was one of the few women holding a key position in the federal government. It was her job to administer the workman's compensation program.

But by the time Goodman contacted her, Peterson had already been touched by the problem. She had received a brief, handwritten note from the distraught wife of a miner who had been stricken with cancer.

"My husband has been a uranium miner in Utah, Wyoming and Arizona," the woman wrote. "I understand that the compensation is different in different states. Please tell me where my husband should go to die."

Peterson was appalled by the fatalistic simplicity of the request.

"That killed me. Wrenched me," she said later. "It really got me going."[12]

Peterson immediately requested permission from Secretary Wirtz to make a trip to Utah and assess the situation for herself. She wanted to go underground and inspect the conditions in a uranium mine first hand. She felt that she should personally talk with the miners to discover what they understood about radiation hazards and safety practices. Wirtz told her to go and take a look.

Esther Peterson was not happy with what she found. In talking to the miners and residents of Moab, she learned of the growing numbers of underground workers suffering and dying from lung cancer. She heard about widows like Eola Garner who were filing futile lawsuits for workman's compensation. She could see how the industrial and government authorities were passing the ball back and forth as to who should take responsibility for radiation control and settling insurance claims.

Peterson returned to Washington determined to make some changes.

"Maybe it's my old Mormon heritage," she said. "But it just haunted me that we must do what was right."

Reporting to Wirtz about what she had seen, she gave him a small sample of uranium ore.

"We've got to do something about this," she said. Then she outlined the sluggish activity of the Federal Radiation Council, where she had sat in on meetings in Wirtz's absence.[13]

The council, organized in 1959, had been formed as a presidential advisory body charged with recommending a general policy for radiation protection. The secretary of Health, Education and Welfare (HEW), the secretaries of Commerce, of Labor, and of Agriculture, the chairman of the AEC and the president's science advisor all sat on the council.

The threat of radiation in the mines was first brought to the attention of the Federal Radiation Council after Arthur S. Fleming, the out-going secretary of HEW, called an emergency meeting in 1960 of the four western governors whose states were engaged in uranium mining. At that conference, Dr. Harold J. Magnuson of the Public Health Service had announced that lung cancer was occurring "at five times the expected rate among the experienced mining population." Magnuson also stated that radioactive contamination could be controlled by ventilation. The usual discussions followed about which agency should regulate mine safety standards.

When proceedings of the Governors' Conference were reported to the Federal Radiation Council, the council members agreed that they must look into the matter. But it wasn't until 1965 that they initiated a study of the problem. An unofficial report, entitled, "Radiation Protection Policy: Guidance for Control of Radiation Hazards in Uranium Mining," and outlining broad guidelines for radiation control, was prepared by Executive Director Paul C. Tompkins. A primary objective of the action paper was to maintain a reasonable balance between biological risk to the miners and minimizing the expense of requisite ventilation equipment to the mining companies: health versus economics. But, with each faction buttering its own bread, a consensus among the conflicting viewpoints of the committee seemed next to impossible. Thus, the subject of uranium miners and radiation was discussed intermittently, but never resolved.

"We've got to do something to get the Federal Radiation Council to move towards establishing national radiation standards for the mines," Peterson told Secretary Wirtz."[14]

Wirtz agreed. He gave her full authority to represent the Labor Department on the committee and push for any standard she felt was necessary.

"And take a hard line," he said.

17 FULL CIRCLE

It was the Siren Song all over again. The mailbox at the AEC compound in Grand Junction overflowed with letters and postcards. "Tell me about the new uranium boom." "Where can I go to prospect?" "Where can I sell it if I find it?"

By 1966, word had spread that the long-awaited commercial uranium industry was finally gearing up. Fifteen nuclear power plants were in operation in the United States. Eight more were under construction and twenty in the planning stages. This was music to the ears of mining companies whose production had been slashed by a series of stretchouts in the government procurement program.

"Nuclear Power Plant Needs Begin to Revive Sick Uranium Industry," the *Wall Street Journal* headlined on August 1, 1966. "About half the total power output of new generating plants placed on order so far this year will be drawn from nuclear fuel, compared with about half of the new capacity ordered in all of 1965," the article reported.[1]

The Grand Junction AEC office corroborated the news. "The nuclear power industry is developing rapidly," said staffer Nielson B. O'Rear.

"The capacity of plants announced to date will result in nuclear capacity in 1970 of at least 10,000 MWe [megawatts], six times estimated year-end 1966 capacity. If the industry is to supply domestic requirements, a uranium production capacity of 27,000 tons of U_3O_8 per year by 1980 is indicated. [AEC purchases in the peak year of 1961 were 17,671 tons of U_3O_8]. . . . Therefore, the uranium mining industry has a growing market to serve and a major job of exploration, development, and the construction of mining and milling facilities to carry out."[2]

Ten years earlier, it had been a different story. The AEC, up until then the only customer for uranium, had all of the ore and concentrate that they wanted. The government's campaign to create a domestic uranium industry had been so successful that the United States had more than enough stockpiled uranium for its nuclear testing program and for an atomic defense arsenal. Despite the wars in Korea and Vietnam, there wasn't the hysterical buildup of reserves that had dominated the decade after the end of World War II.

But the futuristic "atomic age" of commercial nuclear power that everybody had talked about hadn't yet materialized. It was taking longer to perfect the technology than had originally been thought, and the federal government was left with too much of a good thing. It was necessary to slow down—but not kill—this monster industry that had been created.

In May 1956, Allan E. Jones, the new AEC manager at Grand Junction, announced a stretchout in the federal buying program "to maintain a high rate of exploration and development until that time when the power industry had greater demands for the ore." The new procurement program guaranteed a government market for five hundred tons of concentrate per year from any one operator at a flat price of eight dollars per pound, through 1966. This was a change from the old policy which had a set rate for ores and negotiated the price of concentrates.

Eighteen months later, large ore discoveries in the Gas Hills and Crooks Gap Districts of Wyoming, and near Grants, New Mexico, indicated that the reserves would grow even greater. On October 31, 1958, President Eisenhower announced that the United States and former Soviet Union had agreed on a moratorium on testing atomic weapons. With these new developments Jones said that the AEC would have to pull back again. He redefined his 1962–66 buying schedule in order to limit production to the approximate level which would be reached with existing commitments. The government would buy only "appropriate quantities of concentrate derived from ore reserves developed prior to 1958." No ores would be purchased.

"It is no longer in the interest of the Government to expand the production of uranium concentrate," Jones announced flatly.[3]

And four years later, Jones could see that the private market for uranium was still insufficient to sustain a viable industry through 1966, when the procurement program was scheduled to end. Reserves were still large enough to accommodate the new series of underground detonations taking place after the Nevada Test Site was reopened in 1961 when the Soviets resumed their testing. Therefore he established another stretch-out program for 1967 through 1970. Prices would be reduced again, and the mills would be asked to voluntarily defer some of their deliveries of concentrate until 1967–68. Since restrictions against private interests purchasing uranium had been lifted, and the federal government had announced that it would be out of the uranium business by 1970, Jones encouraged the mining companies to

The Colorado Plateau was studded with uranium mine headframes.

woo the power industry as future customers. The hope of the uranium industry lay with its commercial development for peacetime purposes.

The prospects of non-military uses of the atom caught people's imagination.

"Throughout a world beset by growing problems of overpopulation, lack of food and limited traditional resources, nuclear energy is emerging as a most powerful force for the benefit of man," reported the *Unesco Journal*.[4]

A single nuclear-powered generating unit could supply cities with nearly half a million kilowatts of electricity. Atomic power would be cleaner than coal, less expensive than oil, and its generous supply would supplement the fossil fuels, which were facing depletion of reserves.

There was talk of immense "nuplexes," complexes of industrial plants powered by a common energy center fed by breeder reactors that were self-replenishing. The future held hopes of irrigating arid lands with sea water that had been desalinated by nuclear power. Scientists promised techniques for controlling or eradicating insects, preserving foods, fertilizing land and even using the atom for medical purposes.

By 1963, it appeared that the future was here. Contracts were awarded for four nuclear power plants with a total electrical capacity of nearly two-million kilowatts. One of the generating units, located in New Jersey, claimed

Miners with uranium ore competed for AEC bonuses. *Museum of Western Colorado.*

that projected power costs were substantially cheaper than those of a coal-fueled plant of the same capacity, even at significantly low prices for coal.

But three years later, at a meeting of the Joint Atomic Energy Committee, it was revealed that by 1980 the United States would be hardpressed to provide enough uranium for its burgeoning nuclear-power industry.

Rafford L. Faulkner, director of the AEC Raw Materials Division, told the committee that government stockpiles could meet requirements until about 1979, but only if nuclear-power development continued at the present rate.

"However," Faulkner said, "mining could not be maintained at the required levels in the absence of new discoveries and would tend to decline some years earlier as individual mines approached exhaustion."[5]

Faulkner told the committee that the AEC had requested the U.S. Geological Survey to map areas most favorable for potential uranium discoveries. The Grand Junction office was also making as much information as possible available to the public as to likely ore-bearing formations. Already interest in prospecting was being revived. Newspaper headlines were urging, "Save That [Geiger] Counter: A Shortage of Uranium."

By the fall of 1966, prospectors were starting to return to the Colorado Plateau. No one really noticed it at first. In February 1967, reporter Maxine Newell wrote in the *Times-Independent,* "Moab has been geared all year for a rumored second boom, but for all practical purposes, nothing seems to be happening. The extent of uranium activities is almost as hush-hush as the use which set it in motion a couple of decades ago." [6]

It didn't take long for the activity to quicken, however. But the hopeful ore-hunters who figured that everything would be just as it was in the early 1950s—a guaranteed federal market, plenty of incentives, purchase depots close to claims—were in for a surprise. Allan Jones had to tell the applicants that the AEC no longer bought ore. They didn't operate buying stations anymore, pay for hauling, or offer bonuses. He advised the miners to check with Westinghouse, General Electric, or one of the other big companies that was getting into the nuclear power business.

But, Jones warned, chances of an individual prospector hitting paydirt were slim. Exploration for uranium had become sophisticated. No longer could hopeful miners hike into the hills with their Geiger counters and expect to detect ore. The shallow deposits had just about been mined out. It took more sensitive detection instruments to locate the deeper deposits.

"Locating an unstaked uranium deposit shallow enough to make a Geiger tick is like locating a needle in a haystack," wrote Maxine Newell in the *Times-Independent.* [7]

Deep drilling with expensive machinery was now the standard procedure. The estimated cost of finding a uranium deposit, developing the mine and building a new processing plant was twenty million dollars, and the timing averaged eight years. An AEC survey in 1967 showed that approximately seventy-seven million dollars would be spent by a number of large oil companies during the next four years just on drilling programs to prove newly-discovered ore bodies.

Major corporations, with unlimited funds for exploration, started acquiring large blocks of claims. The AEC encouraged the firms to reexamine some of the former ore-bearing areas in light of current knowledge of uranium geology and improved technology. In a number of these districts, the previous drilling had not been deep enough to test the entire

Mucking uranium ore. *Museum of Western Colorado.*

thickness of a favorable formation. Much of the work was poorly logged and inadequately evaluated. Commonly, only surface indications were investigated. When early operators failed to discover important ore after their cursory explorations, they had shifted their attention to other properties. Companies searching for other mineral resources—petroleum, coal, salts, phosphates, non-metallics, or metals—were also urged to test their exploration boreholes for radioactivity on a routine basis.

Hauling ore from a mine portal. *Museum of Western Colorado.*

"A windfall discovery of major economic value could result," said Elton A. Youngberg, assistant operations manager of the AEC Grand Junction Office.[8]

The Grand Junction AEC office catered to the corporate prospectors by conducting a series of technical workshops for professionals. Scores of industry representatives were briefed on uranium geology, exploration techniques, and mining and milling procedures. Bands of company-sponsored ore-seekers joined the individual prospectors in a search for hidden deposits. Competition between the corporate parties was often tense.

"In the chill before dawn on a Saturday last April [1966]," wrote *Wall Street Journal* correspondent Glynn Mapes from Crook's Gap, Wyoming, "a crew of twenty-five sleepy employees of Utah Construction and Mining Company drove quietly out of town in a motley assortment of cars and pickup trucks. They were sneaking off to stake mining claims, hoping literally to catch their competitors asleep.

"But a field crew of Western Nuclear, Inc. was awake, and noticed the frenzied activity of the band of strangers."

As they looked more closely, the campers realized what their rivals were doing. Losing no time, the Western Nuclear men dispatched a messenger to their main camp to gather other forces. The man returned the next morning with a group of fellow workers.

Soon both teams were feverishly staking and drilling claims. They piled rocks and hammered stakes all weekend. Few of the men had taken the pains to measure boundaries with any accuracy. By Monday morning, almost five hundred acres of overlapping monuments had been placed.[9]

Competition was even harder for the individual ore-seeker. About the most one could hope for was to lease one's claims to such industrial giants as Gulf Oil, Continental Oil, American Smelting and Refining Company, Standard Oil of New Jersey or Cities Service, who had drilling programs. If the company struck ore, the claim owner would be paid in royalties.

Hundreds of prospectors decided it was worth a try. The Grand County, Utah, Recorder tabulated 5,810 location notices filed in 1967. Two-hundred-fifty-nine of the claims were staked in chill December, a month when activity was generally slow. Forty-five hundred affidavits, proving a miner had done the required yearly labor on his property, were issued on claims that had been previously filed.

San Juan County topped its neighbor, Grand County. Miners filed 19,521 claims during the year. Five hundred and thirty-nine of them were documented in October alone.

The reawakening of the Colorado Plateau didn't go unnoticed in Salt Lake City. In December 1967, *Deseret News* reporter Jerry E. Henry reported that "the West is throbbing with a renewed uranium fever. . . . The Atomic Energy Commission (AEC) office in Grand Junction, Colo., predicts that the demand for uranium concentrate will grow to 38,000 tons by the year 1980."

Henry added that "the prospect of shoestring prospectors striking it rich overnight is unlikely." Finding new reserves would mean expensive drilling and exploration activities that would take capital.[10]

Salt Lake City's stockbrokers and promoters perked up their ears. Raising capital for speculative ventures was their game. Many of the broker-dealers who had "gone out on a stretcher" when the SEC had cracked down a decade ago were quick to sense new action. The underground buzzed with scuttlebutt about large corporations planning deep drilling on old claims that had been tabled as worthless. Prospectors were ranging the backcountry for new deposits. There had to be something to the rumor that a second boom was in the offing.

Meanwhile, the money men thought of another way to get uranium back in play. They knew of scores of small, developing companies that wanted to get financing by selling stock shares to the public. Little-known pharmaceutical outfits needed growth money. Lone inventors looked for funds to manufacture and market their discoveries. Recreational developments required capital to build resorts and sports facilities. But starting up a new corporation and filing with the SEC and state securities office was an expensive and time-consuming proposition. Developers were eager to find a way to launch their projects without all that expense and red tape.

In the late 1960s, a solution was presented in the form of defunct uranium corporations. Worthless companies with no holdings. Empty corporate structures with nothing in them. But they had one very important asset: the shell of an organization that was registered with the SEC.

"The more worthless the corporation, the better chance that it was worth money," said Noland Schneider, former trader with Coombs and Company. "Promoters found that, due to expense and time under the SEC regulations, it was much more expedient to go out and buy a stock at one-fourth or one-half cent—buy most of the stock—and then put new assets into that company."[11]

At first there was a stigma identified with using a "shell" to go public. Since the plan circumvented securities regulators and regulations, it was thought of as going public through the back door. In time, however, shells became considered as the sophisticated way to start issues. It took less time and money.

So the brokers went out and bought up the shells of old uranium companies and turned them into the pharmaceutical house, a corporation for the newly-invented product or the resort. Then the issues were put back on the market. Shares that had been worth less than a penny a few days before, suddenly jumped to ten, twenty, fifty cents, sometimes even a dollar or more.

Salt Lake City took flight once again. Not only were there plenty of corporate shells available, but Utah was one of the few states that allowed reactivation of a dormant corporation simply by paying back franchise taxes. In most states, inactive firms that were delinquent in taxes for from three to five years could not be revived.

"If you want to find a shell corporation, you go where there has been a previous boom," Schneider said. "And there hadn't been many booms in this country in the last thirty or forty years. So everybody came to Salt Lake City to pick up these old shell corporations."

Local residents ransacked basements and cleaned out the drawers of old chests. High closet shelves were explored for old uranium stock certificates.

And some were lucky. Long-time broker Lyman Cromer had a trunkful of certificates. After the 1950's market had collapsed, many of his customers sold their old stocks to him for nearly nothing so they could take a tax loss. Cromer never threw anything away. So, by 1967, he had a position in just about every uranium company that had ever existed.

Hal Cameron, former developer of Standard Uranium, went so far as to advertise in the newspapers for old uranium certificates.

"I put up about twenty thousand dollars and bought anything I could buy," Cameron said later. "I probably got ten times my money back."[12]

The Elliott brothers were astounded to find that they owned a sixty thousand dollar outhouse. After the uranium boom, Jim and Bob had papered the walls of their rustic privy in Brighton Canyon with worthless stock certificates in Lightning Uranium Corporation. Then, ten years later, the stock—once twenty cents a share—was suddenly worth nearly half a dollar.

Frantically, the Elliotts tried to steam the paper off. It wouldn't budge. They considered chopping down the outhouse and delivering the siding to a stock transferring office, but this was impractical. Finally, they dragged a transfer agent to the site and had him properly cancel and reissue the certificates for resale.

At about the same time, a Mormon bishop in Salt Lake City presented a local stock broker with a list of uranium certificates owned by one of his ward (parish) members who was ill and in need of money. The sick man had tried to sell his securities four years earlier, only to discover that they were worthless. But he was desperate now and thought he would give it another try. After the stock dealer had shuffled through the documents, he told the bishop that his friend's troubles were over. The broker offered to buy the lot, then and there, for fifty thousand dollars.

Another stockholder, Gerald W. "Andy" Anderson, wasn't so fortunate. In the late 1950s, Andy had decided to sell his old uranium paper just to get it off the books. He let one-hundred-and-fifty-thousand shares of Seagull Uranium go for an eighth-of-a-cent per share. In 1968, Seagull was going for as high as twenty-two cents per share, and Andy was out thirty-three thousand dollars.

Jack Coombs was in Montreal, dealing in insurance, when he learned about the "shell game." Since he had closed his stock brokerage offices in 1957, Coombs had done a stint with New York Life Insurance Company. Later he operated as a general agent, specializing in substandard risks, and then became a free-lance broker for a number of agencies.

Deciding it would be profitable to go into business for himself in Canada, Coombs formed B.C. Life Insurance Company in British Columbia. Finally, in 1966, he moved his family to Montreal, where he started Le Citroyen Compagnie de l'Assurance Vie.

As luck would have it, the action with over-the-counter penny stocks in obsolete uranium companies was just revving up in Salt Lake City. When Coombs learned that many of his former colleagues were back wheeling and dealing, it didn't take him long to shift gears. He began to fly back and forth between Montreal and Salt Lake City, sometimes several times a week. He dug out all of his old uranium stocks and bought up as many others as he could get his hands on. He'd find the last known board of directors of a corporation. Get what records they had retained. Locate the transfer agent for each issue and pick up his files. Then, after he had got the companies reinstated with new financials, he would turn them over to a battery of out-of-state lawyers who would locate some little business with a few concrete assets to put into the shell.

More often than not, Coombs looked to New York, Houston, Denver, San Francisco, or Los Angeles for new enterprises. There, demand for corporate shells was high. All kinds of struggling new companies wanted to go public. And all of the old big-time promoters were touting their favorites.

The big money was "not from local and hopeful citizenry as was the case in the mid-50s," wrote Robert H. Woody, business editor of the *Salt Lake Tribune,* "but by big and self-assured money from the East and elsewhere in the nation."[13]

"The action by all of the people trying to get a cut when a deal is being made is unbelievable," Coombs said. "The same people [as in the 1950s] are all in the action today with shells. Those who were knocked out of the box by the SEC have all come back to life—it is bigger than the Uranium Boom."[14]

The old, familiar faces were back in Salt Lake City, too. Dick and Frank Whitney were "doing deals." Ralph and Ray Bowman were in it, along with Lyman Cromer.

"The guys who were young in the '50s became the big promoters of the '70s," said Coombs, who had returned to Utah.

But they were smarter, claimed Don Glenn, a Salt Lake stock trader: "We all *made* money in the last boom and we plan to *keep* some in this one."[15]

There were newcomers to the broker-dealer scene as well. Paul Barracco had missed the first uranium frenzy, having emigrated to the United States from Italy in 1959. In June 1968, he opened a brokerage company with a capitalization of sixty thousand dollars. By the end of December, Barracco and Company had a staff of twenty-one employees, and was worth three hundred thousand dollars.

As with most stock firms at this time, Barracco's was a madhouse. From seven-thirty in the morning until two in the afternoon (four o'clock in New York) bells rang, lights flashed and teletypes chattered away under clouds of cigarette smoke. The traders, tuned in to earphones, frantically memorized vacillating stock prices while scribbling buy and sell orders on trading forms and stamping them with the time of the transaction. They gulped coffee, chain-smoked and could scarcely take the time to go to the bathroom.

Then at two o'clock the switchboards and teletypes went dead. For only a few seconds, however. As soon as the trading day was over, customers began calling to check on their orders and telephones rang off the hook until late in the evening.

The "shell game" caught on fast. There were about one hundred and twenty stocks listed. Some of the names were familiar: Blue Lizard, Mayday, Horsethief, Justheim, Uranus, Comanche. Some issues changed their tag. Arrow Uranium became Controlled Metals. Strategic Metals became known as Greenwich Pharmaceutical. Trans-Western went into Flying Diamond. There were some winners among them, but more losers.

What began as a legitimate effort to expedite the formation of new companies quickly deteriorated into a "fast buck" scheme. And buyers had the same cavalier attitude that they had exhibited in the first uranium boom. People just didn't seem to care about the substance of the business entities. Anything was all right so long as they could buy and sell those old shells.

The SEC office in Salt Lake City was soon inundated with paperwork relating to the re-formed companies. Director G. Gail Weggeland viewed the over-the-counter activities as a no-win situation for the customer. It was a case where the promoters get control of cheap stock, vote themselves into management, "tout the acquisitions beyond reason and cause a market rise in trading of the stock," and then sell, he said. Insiders made private placements of shares to friends and relatives, and paid themselves profits in salaries. Sometimes they would switch the companies' assets into some other corporation they held. Weggeland warned broker-dealers that they must give their customers full disclosure, or risk indictments straight from the attorney general.[16]

Bob Woody tried to analyze the phenomenon in his column prophetically entitled, "Penny Stock Mart Booms, So Let th' Buyer Beware."

"Is this resurrection real? Has it flesh and blood? Are the assets (or about to be acquired assets) there?" Woody asked.

Noting that the SEC was "taking a concerned look at the phenomenon," Woody pointed out that there had been a few success stories among smaller companies.

"But for the most," he said, "present management is unknown, the assets unstated."

"They [customers] don't want a lecture," an unnamed broker told Woody. "They want me to make a buy order."

The public didn't want to listen. Too many easy bucks were at hand. Everybody knew of someone who had panicked and sold out too soon. [17]

"I had one-hundred-seventy thousand shares of Western Standard that cost me two cents a share," said trader Don Glenn. "The SEC came out with a big article saying that all of these penny stocks were phony, so I sold it all. It went to $3.50 a share! I had forty thousand Shoni—cost me a dime—I sold it all at forty cents. It's fourteen dollars now."[18]

And everybody knew people who had "made a killing" by cashing in on their recycled stocks.

Paul Barracco told of the International Business Machines salesman who came into his office to demonstrate some new equipment.

"The man was quite poor. Had five children," Barracco said. "We got chatting about uranium shells. He thought he might have some old uranium certificates, so I told him to check them out.

"He went home that night and rummaged through the desk. He didn't tell his wife what he was looking for and she got mad at him for messing up the drawers. At midnight, he sneaked out of the bedroom and called me.

"'I found 20,000 shares of Western Standard,' he whispered. 'Is it worth anything?'

"'Four dollars,' I answered.

"The man was elated. He said, 'Sell five thousand shares. But don't tell my wife.'

"A few days later, he called me again.

"'Meet me at [a certain address],' he said. 'Bring the check. But don't let my wife know.'

"So I met him and his wife at the appointed time in front of a new house," Barracco continued.

"'Give that check you brought to my wife,' the man said.

"He had bought the house."[19]

The public was less interested in circulating other stories that supported the SEC's admonitions—like the case of Elkton Company.

Elkton's promoters advertised that the shell of an old Colorado mining corporation had been refitted with a number of assets. The company had acquired Solar Airlines, an air-taxi service, which had formerly operated in Texas and New Mexico. They had a car rental concern and an interest in Hasty House International, a national restaurant chain. The corporation owned part of a gold mine. They manufactured an industrial cleaning product and had an accounting systems-management company.

But when the SEC started to investigate the facts, Elkton turned out to be a disaster. The rental car company was inactive after operating a limited time and losing the few lease cars it had for rent. Solar Airlines had been acquired for one dollar and a promise to supply some capital and assume liabilities. But there were so many tax liens and other debts against the company that they lost all of their planes. The accounting system was nil. The gold mines inoperative. Acquisition of the Hasty House International restaurant chain fell through. The SEC suspended trading of the stock.[20]

Another sham discovered by the SEC was Micro-Biology, a firm that purported to be engaged in cancer research. The medical specialist who sat on the board was found to be a veterinarian.

The president of Silver Shield turned out to be a minor, therefore ineligible to be a stockholder, much less the head of a corporation.

Finally, in January of 1969, Don Stocking, administrator of the SEC regional office in Denver, came to Salt Lake City. The SEC and the Justice Department had been investigating the uranium stock market. They were not happy with their findings. The agents had concluded that buyers and sellers were essentially just trading paper. In terms of real assets, most of the revitalized companies were no better off than they had been years before. Stocking also announced that the investigators had even uncovered "...'persuasive evidence' some stocks are being triggered off on their fantastic rise by outside seed money from organized crime."[21]

Stocking issued a warning to the local stockbrokers. From now on, he said, the SEC would take the position that anyone quoting prices on the over-the-counter market was obligated to gain full knowledge of the assets and disposition of a company's stocks and make them known to buyers and sellers.

In addition, the SEC would assume that all shares purchased from former management of a company for deployment into market stocks were re-registered with full disclosures before they entered the trading arena. No longer could promoters consider the original offering circulars and registrations as legal vehicles for new corporations.

Stocking promised that his investigators would be watching the action closely.

"There will be," he said, "suspensions, injunctions and criminal references to the Department of Justice against those who are distributing unregistered securities or anyone aiding and abetting the process."[22]

On May 14, Don Stocking proved that he meant business. Preliminary injunctions were slapped on several brokerage houses for violating registration requirements. The action virtually closed their business.

The surviving broker-dealers knew that they would have to make changes in their operations or give up. Therefore in July 1969, a small group of traders formed the Intermountain Association of Over-the-Counter Broker-Dealers. Hoping to create a more efficient and orderly marketplace, they made plans to distribute upgraded information to their membership and clients and establish a better forum of communication with government agencies. They would develop brokerage schools and professional seminars. And, primarily, the association would lobby for legislation that would protect the investing public without handcuffing the activity of publicly-owned corporations and the brokerage industry.

Seven months later, Bob Woody reported in the *Salt Lake Tribune* that the "Salt Lake Penny Mart Reaches New Measures of Stability." After one brokerage had taken out bankruptcy, several others had been suspended from doing business due to infractions in registering stocks, and public fervor had been dampened by the SEC criticisms of the trading community. Woody observed that the over-the-counter market was "now pretty well shaken out."

"No doubt, the penny market is the place for those who want a little faster action," said broker Richard Parker. "But the high speculative flush— and the trading of shells as characterized early last year—is over. Now there are some good, solid companies with money in the bank, with property, with good ideas."[23]

Just as the uranium mining business was shifting from military emphasis and federal sponsorship to private industry for peacetime purposes, the over-the-counter penny stocks, born of the uranium frenzy, had found legitimacy.

"The market has calmed," Bob Woody wrote in his *Salt Lake Tribune* column on March 3, 1970. "And it now is considered a vehicle not only for resource development but also for the funding of new technology and manufacturing ventures."[24]

The "shell game" had gone straight.

"But the laws didn't hold water, and the cat was out of the bag anyway," Coombs said. "All the shells were gone."

In the early 1970s, Coombs got on to another concept. He had seen many instances in which a shell did not satisfy a client's need. They needed money.

"What if I had a public company with cash and then looked for a start-up deal?" he thought.

The idea was to raise capital from a number of investors and then find a viable business opportunity to invest it in. In other words, going public with a securities offering before you have any specific use of proceeds. A "blind pool."

Coombs launched the first "blind pool," Kipper Corporation in 1972. He used a shell of the defunct Capitol Furniture Company and registered with the Utah Securities Commission 300,000 shares at thirty cents, for a total offering of $90,000. When the money was raised he looked around for some kind of business to put in the shell. He found Billings Energy, a young Utah company that was trying to develop a hydrogen-powered automobile. Kipper's name was changed to Billings Energy. Later on John K. Hanson, president of Winnebago, came on board and financed the venture, along with George Romney, former Michigan governor. A year or two later the stock traded at twenty-six-and-a-half dollars per share. Other blind pools followed. Most of Coombs' ventures were successful. He was described in *The Wall Street Journal* as "promoter par excellence."

"For a number of years it seemed I was the only source for a blind pool," Coombs said in 1987. "No one was doing them and my success was becoming a legend. I always had an inventory of three or four companies with various amounts of cash from $150,000 to $500,000. My success prompted a lot of competition, and by 1982, I was pretty much out of the business.

"But now, everybody's doing it. Except what they're doing is raising little fifteen to twenty thousand dollar deals just designed for market plays—up, down and worthless."[25]

By 1984, there were two-hundred-eleven blind pool offerings listed in Utah. The Utah Securities Division became concerned. The blind pools showed "a tendency toward a work of fraud," they claimed.[26] Steps must be taken to eliminate them and to clean up Utah's growing reputation for being "The Securities Fraud Capital of the World," the agents warned.

Governor Scott Matheson agreed. He appointed a securities fraud task force and the blind pools were eliminated by law.

But by the time this happened, Coombs had turned to something else.

"It seems that most promoters that make money in the penny market put it all back," Coombs said. "They don't ever get their money out of the penny speculative issues to invest in solid deals. And that is what I did that was so different. I'd take the money out of Kipper Corporation and buy a ranch in Spokane. I'd make $100,000 here and I'd buy $100,000 in bankrupt bonds. As I started accumulating a portfolio, the profits were extraordinary. So now I've ended up with family partnerships and manage around seven million dollars. But I don't think the investment profits would have come if it hadn't been for those penny stock deals."

Not all of the penny stock brokers were able to emulate Coombs' success, but most of them fought their way back to solvency after the boom. Frank and Dick Whitney turned their brokerage house into a travel agency. Frank is now retired. Dick, who died of a massive heart attack in 1980, took a flyer with the Intermountain Smelting Company and then turned to real estate sales.

After the boom, Dewey Anderson returned to his career of promoting oil and gas leases until his death, from cancer, in 1984. Ralph Bowman continued promoting deals in Salt Lake City until his death, and his brother, Ray, died later in California. Jay Walters, Jr., who started the speculative spree, never lived to see the second boom. His license was permanently suspended by the SEC for using the mails fraudulently in delivering unregistered stock. In 1956, Walters died of pneumonia.

Most of the people involved in the various aspects of the uranium boom feel that, regardless of their personal gains or losses, this unique moment in American history served its purpose.

Few who lived through that time would have missed having the experience of the frantic fifties.

18   THE STANDARD IS SET

It was at a hearing of the Joint Senate-House Committee on Atomic Energy that he first heard about it. J.V. Reistrup, a reporter new to the beat of science, space and energy for the *Washington Post,* was intrigued when someone asked, "What about these uranium miners that are dying of lung cancer?"

Reistrup started to dig for more information, asking questions and delving into files. Finally he tracked down a copy of the 1967 revised edition of the Federal Radiation Council's action paper, "Radiation Protection Policy: Guidance for the Control of Radiation Hazards in Uranium Mining." He knew he had a story.[1]

The report, which had triggered the current subcommittee hearings, was incredible. The booklet explained the natural radioactive decay of uranium into a series of radon daughters. It told how Duncan Holaday and the Public Health Service, in cooperation with the AEC and state agencies, had been studying the problem for seventeen years. The report pointed out that, despite Holaday's conclusions "that underground uranium miners are subject to lung cancer to a degree substantially greater than the general population," and the fact that stepped-up ventilation systems would virtually eliminate the risk of radiation to the miners, no regulatory agency was doing anything about health and safety in the mines. The Public Health Service didn't have the money or the power to enforce regulations. The Atomic Energy Commission had sidestepped jurisdiction over uranium "in its place in nature." The states had exercised little authority over the industry until recently. Still the U.S. Public Health Service estimated that over half of the three thousand miners in their study had been overexposed to radiation.

Reistrup took the council report to his editor, Lawrence Stern, and proposed a story on the subject.

"Find a miner," Stern said, giving Reistrup the go ahead.

A few days later Reistrup was on the western slopes of the Colorado Rockies, talking to a fifty-six-year-old "casualty of the atomic age."

"This is probably the last spring John Morrill will ever see," Reistrup wrote from the small town of Nucla, Colorado, "because he is dying of lung cancer.

"He is dying of lung cancer because he was a uranium miner."[2]

The young newsman had never been west before. He was moved by the beauty of the mountains and the freshening of springtime greenery. He was touched by the open friendliness of the people. But the tragedy underlying the serene facade was inescapable. John Morrill, the miner of Reistrup's story, stoically facing his last days, had a child born with Down's Syndrome, and was wondering whether the disorder was also related to radiation in the mines.

When Reistrup interviewed Duncan Holaday at the field station in Salt Lake City, Holaday told him that Morrill was probably the fiftieth or fifty-first miner to be recorded as dying of lung cancer.

"There'll be more," Holaday had said. "We haven't seen them all."

Reistrup's *Washington Post* story on Morrill in March 1967 was the first of a series about the problem. The article attracted much attention. People started calling him, telling him about other miners afflicted, or already dead, from lung cancer. One of them was Tex Garner.

When Reistrup heard that Eola Garner was fighting a landmark compensation case in Utah, he wrote her, asking for more information. Her answering letter told of filing her claim four years earlier, and being repeatedly refused compensation by the Utah State Industrial Commission. Her case was now before the Supreme Court. A motion for the Oil, Chemical and Atomic Workers International to enter her case on her behalf had just been granted.

"I would have already had it [the case] through if I could have filed in Colorado," she wrote him. "It is a shame two States have such a different viewpoint on this, as it is a very serious matter and something should be done to the State of Utah to make them come up to the standards of Colorado."[3]

Reistrup discovered that Colorado was the only state that had recognized lung cancer cases as being caused by the radiation in underground mines. The state expected that there would finally be about four hundred and twenty such claims, totalling $8.5 million in compensation costs.

"It is suggested that the Federal Government assist in paying the claims, 'since these costs arise out of a national defense activity,'" Reistrup wrote.

Some investigative reporting among his personal contacts "on the hill" turned up other interesting information. He found that there had been a bureaucratic coverup.

"More than a thousand uranium miners in this country can be expected to develop lung cancer by the end of 1985, according to a study now receiving highly restricted circulation in the Executive Branch of Government," Reistrup reported in the *Washington Post* on April 14th.

"Another secret document charges that the radiation standards now under study for the mines are 'significantly inadequate' to keep the number from rising," he added.

Reistrup went on to say that an underground briefing paper within the Department of Health, Education and Welfare blamed the weakness of the proposed standards on "the psychology of organizations." People in the Public Health Service and other concerned branches of government hesitated to toughen the requirements even after they had learned the extent of the hazards. They were afraid that such changes "would reflect badly on this and other departments for previous failure to act, and that some mines would be shut down and some miners be put out of work."

But Reistrup discovered that there was a federal agency that did have the authority to regulate radiation in the mines. Someone told him about an old depression-era law on the books that gave the Labor Department jurisdiction over miner safety. Under the Public Contracts provision of the Walsh-Healey Act passed in the 1930s, the secretary of Labor had authority to set safety standards for mines whose production was purchased by the federal government. The stipulation stated that a federal contract would be cancelled if there were working conditions that were "hazardous or dangerous to the health and safety of employees. . . ."

Reistrup phoned Labor Secretary Willard Wirtz to ask him about the statute. The secretary admitted that he was unfamiliar with the act. When, in his article of April 14th, Reistrup reported that "Secretary Willard W. Wirtz was unaware of this [the Act] until last month," Wirtz was furious and embarrassed.[4]

"It didn't make us look very good," Wirtz said later. "We didn't get any Brownie points. I was terribly conscience-stricken."[5]

Wirtz met with his assistant secretary Esther Peterson on the morning that Reistrup's statement appeared. He asked her to keep a close eye on the inter-agency Federal Radiation Council meetings that she attended in his place. They must push hard to get a safe radiation standard established, he told her.

When the Federal Radiation Council convened on May 4, 1967, Peterson alerted Wirtz to attend the session personally. Reistrup's articles, which had been nationally syndicated, had been stirring up public protest over the plight of uranium miners. Newspaper editorials had criticized the council for its two years of inaction in setting maximum levels for radon in the uranium mines. The committee members had decided that they must bring the matter to a head. They would take a vote that day.

So Secretary Wirtz attended the meeting. But the usual lengthy discussions pitting miner safety against the economic concerns of maintaining a viable uranium industry wore on. When time for the "ayes" and "nays" arrived, the committee members still failed to agree.

Secretary Wirtz abruptly left the room. He was weary of the "hundreds of efforts, studies, meetings, conferences and telephone calls—each of them leading only to another—most of them containing a sufficient reason for not doing anything then—but adding up over the years to totally unjustifiable lack of needed consummative action."[6]

He was disgusted with the endless divergent viewpoints between state and federal agencies, state health and mining authorities, mining associations or companies and governmental agencies, the Department of Labor and the AEC, the Department of Labor and the Interior Department, the Labor Standards Bureau and the Wage, Hour and Public Contracts Division of the Labor Department, over "who was responsible for doing what, and authorized to do it."

The next day, Wirtz decided to proceed independently, rather than through the joint inter-agency committee.

"We clearly have the statutory responsibility," he told Esther Peterson.[7]

Exercising his regulatory rights as secretary of Labor, Wirtz invoked his powers under the Walsh-Healey Act and set his own maximum standard of radiation. He stated that in the future, three-tenths of a working level of radon daughters was "the maximum level of radioactive material to which a person can be exposed without . . . increased hazard of lung cancer. . . ."[8]

Washington's reaction to Wirtz's extreme standard was heated. Representative Chet Holifield, of California, challenged the scientific basis of the radon level and deemed it "arbitrary." Wayne Aspinall of Colorado objected because the mining states had not been consulted prior to Wirtz's announcement. Langan W. Swent, of the American Mining Congress, charged that the standard was impossible to impose under present technology and that "insurmountable problems" would be encountered in endeavoring to comply with such a rigid regulation.

Four days later Wirtz defended his controversial stand before the Joint Subcommittee on Atomic Energy. Then, publicly admitting that his own Labor Department had been uninformed and neglected its responsibility, he said, "Since so much of the uranium ore mined in this country is used by mills which have contracts with the Atomic Energy Commission, the Public Contracts Act authority has clear applicability to the uranium mining situation."[9]

"Yet," Wirtz continued, "despite this recognized coverage—and in contrast to the exercise of standard setting and inspection functions by the Department of Labor in almost all other areas of production and commerce—including the processing of uranium ores by AEC contractors—no safety and health standards covering uranium mining have ever been set by the Secretary of Labor—until last week. . . ."

J. V. Reistrup of the *Washington Post* relayed the good news to Eola Garner.

"You asked me to write you if anything new came up concerning uranium miners, and it has," he said in a letter on May 11, 1967.

Reistrup told Eola about the joint committee investigation "into why it was taking Federal agencies so long to take any action." He said that Wirtz had announced that he was setting very strict standards, and better still, "Secretary Wirtz and the Atomic Energy Commission admitted that the Federal Government has a responsibility to be sure that medical care and compensation is provided for those miners who have already contracted lung cancer or will get it from radiation they have received."

But the encouraging news was short-lived. After three months of congressional hearings, Secretary Wirtz had been unable to prevail over the industrialists, the AEC and state and federal politicians. Finally, he was forced to back down on his strict standards. On August 8th, he amended his order to allow one working level of radon, instead of his earlier edict of three-tenths of one working level of radon per liter of air. It was the same amount that Duncan Holaday had recommended ten years before. However, Wirtz added, there would be the provision for "further review within one year and with the injunction that lower levels be maintained so far as is practicable."

Shortly after Wirtz's partial retreat, Eola learned that the small breakthrough in Washington had had little effect in Utah. On August 24, 1967, in a landmark decision, the Utah Supreme Court stated, "We cannot confirm that the lung carcinoma was caused by exposure to uranium miner occupation." The Utah State Industrial Commission's prior decision was confirmed. The Garner claim was denied. [10] No costs were awarded.

In desperation, Eola sent photostats of the Supreme Court decision to Utah's U.S. senators Wallace Bennett and Frank Moss. She appealed to

Secretary Wirtz and Assistant Secretary Peterson. All responded sympathetically. None promised immediate solutions.

"I realize it may be of small comfort to you to know that your husband's death, and the difficulties you have had in trying to get compensation, have aroused much concern in the Department and elsewhere," Esther Peterson wrote on behalf of herself and Secretary Wirtz.

"It is a long and hard problem," she added in a hand-written note, "but please be assured that we *are* working hard at it."[11]

In the meantime, Charles Traylor moved the Garner case to the Colorado courts. On November 16, 1967, he appeared before the Colorado Industrial Commission to plead for *Eola Garner v Vanadium Corporation of America and the Colorado State Compensation Insurance Fund.* Although it was not a Utah case, the hearing was conducted in the State Capitol Building in Salt Lake City, so that some of Garner's co-workers and expert witnesses such as Dr. Victor Archer would be readily available for testimony.

No longer was the emphasis on Utah and Garner's last exposure to radiation. Now Traylor emphasized the state in which Garner had worked the *longest,* which was Colorado. He had worked in various mines from 1948 until 1956. Dr. Archer testified that Tex Garner had been exposed to excessive radiation in all of them. This amounted to approximately eighty-four percent of his total working level months.

" . . . since the exposure in Colorado was the larger one," Archer said, "I would have to say that the odds are that the initiation of his tumor was more likely to have occurred during the Colorado exposure."[12]

The Colorado Industrial Commission didn't see it that way. They denied the claim on grounds of the Statute of Limitations. Too many years had elapsed since Garner's Colorado employment.

Traylor notified Eola Garner that they would file an appeal. Wearily, she anticipated the lengthy, exhausting replay of the drawn-out proceedings in Utah.

"My relatives are really mad I have spent so much money and received nothing," she wrote Dr. Archer. "I have spent $2,000, which I couldn't afford to, but I know I should and deserve to get it [compensation]."[13]

Equally frustrating was the fact that other widows of uranium miners failed to join her protests.

"I would ask them to write a letter about their case and they said, 'I have to work,' and, 'I don't have time,' and so forth. That's when I became so bitter at so many different people."[14]

Garner's case was not helped by the prevailing attitude of the active miners, either. With the recent resurgence of the uranium industry, the pay was good. Jobs were plentiful, and living costs in the small mining towns were minimal.

"Would you gamble your life for a good salary, low taxes and cheap housing?" Guy Halverson, staffer on the *Wall Street Journal* asked.

"Around these parts [Nucla, Colorado], the answer is yes," Halverson wrote. "Uranium miners—gamblers all—are steadfastly ignoring Government warnings that they face a deadly hazard in the underground mines here and in nearby hamlets such as Uravan, Vancorum and Naturita. The hazard: Radon daughters."

Halverson went on to report that six Nucla miners had died within the past seven years. Yet the surviving workers objected to the federal orders for ventilation to reduce the radiation, because they feared the expensive installations might put the uranium mines out of business.

"Those fellows who died with cancer probably had weak lungs and shouldn't have been in the mines in the first place," Halverson quoted forty-year mining veteran Clarence L. Neilson, "who likes to prove his point by chewing on a chunk of uranium ore."

"Those damn eggheads in Washington just don't know what they're talking about," an area newspaper editor told Halverson. "If I'm going to get cancer, I'm going to get it, and that's all there is to it."[15]

An anonymous poet, penning a rhyme about Uravan's plight, wrote:

> So if by chance the AEC and Willard Wirtz
> Would close us down
> Then everyone in Uravan
> Would leave this dismal mining town.
>
> I wonder how our cause would go
> If to Montrose we fled
> And sent this note to Welfare
> Thank you for our daily bread.'

It was Tuesday, November 19, 1968, when Eola Garner learned that Secretary Wirtz had not given up on his fight for lower radiation standards in the mines. She was leaving her mobile home to go to the post office about eleven o'clock, when the telephone rang.

"Mrs. Garner?" someone asked.

"Yes," she replied.

"This is the airport," the voice said. "You're to leave on the two o'clock plane this afternoon for Washington, D.C. You have a round trip ticket with all expenses paid."[16]

"Who sent this ticket?" Eola asked in amazement.

"It doesn't have any name on it," was the reply. "Could you be here a little before two?"

Eola couldn't imagine who had sent the ticket. The only person she could think of was Leo Goodman of the AFL/CIO. He was in Washington; she had talked with him recently.

When she reached him half an hour later, Goodman told her that he had sent her the ticket. She was to appear before a public hearing that Willard Wirtz had scheduled for Wednesday and Thursday, November 20th and 21st. Wirtz was inviting persons involved in the uranium mining industry from all over the world to "show cause" why his previously announced radiation standards of three-tenths of a working level should not become fully implemented on January 1, 1969. Eola would be a key witness on the side of labor. Goodman had sent a ticket to Eola's sister, Gladys Ellison, as well. She and her daughter Pat would meet Eola in Washington. They would appear at the hearing on Thursday.

Eola's brother-in-law, a miner named William Ellison, had died of lung cancer in 1958, but at the time no one suspected any relation between the man's twenty years of uranium mining in Colorado and his death. The idea of having an autopsy didn't occur to Ellison's wife, Gladys. It was only after Eola Garner had started her fight for compensation that Gladys wondered if she might be eligible, too. By then, it was too late. Without an autopsy, there was no proof of cause of death. In addition, the Statute of Limitations had run out.

However, Eola had mentioned her sister's story to Leo Goodman. When Wirtz called the hearing, Goodman thought it was significant that two sisters, whose husbands—both uranium miners, but who had worked in different states—had died five years apart from the same cause. He figured testimonies from both women would strengthen his position. When Pat Ellison asked to join her mother on the trip, Goodman sought authorization to bring the three women to Washington.

As Eola tried to digest the fact that she had less than two hours to get ready for her first trip to the nation's capital, her friend, Ruth Jones, knocked at the door.

"I have the most wonderful news to tell you," Eola said.

Then, as she moved around the trailer, pulling clothes from drawers and closets and hurriedly pressing the garments she would take with her, she told Ruth about the call from Goodman.

The two women jumped into Ruth's car and sped around Moab on last-minute errands. When they arrived at the windy bluff eighteen miles north of town, where a small plane waited on the dirt airstrip, it was almost two o'clock.

Eola grabbed her suitcase and overnight bag and dashed for the runway. "Call the kids," she shouted over her shoulder. "Tell them I didn't have time to write them a note."

It was a long trip, with several plane changes. First there was a puddle-jump to Grand Junction. There Eola boarded another plane for Denver. After another lay-over and change of aircraft, came the lengthy flight across the country. She didn't arrive at the airport in Baltimore until two-thirty Wednesday morning. Leo Goodman and his secretary were waiting.

"You'll be staying at the Jefferson Hotel," Goodman told her, as they walked to the car. "It's just a couple of blocks from the White House."

He handed her a map of the city. "You and your sister are welcome to stay all week to see the sights. We'll furnish you a driver, and pay all expenses."

As they drove to the hotel, Goodman said that he would like to brief Eola about Thursday's session of the hearing after she had rested. Perhaps they could have breakfast. Eola assured him that she was too keyed up to do much sleeping.

At breakfast, Goodman told her that people from all over the world would be at the hearing, and they would be interested in her story.

"People know so much about your case," he said. "They will know you."

He told her to give her testimony in the same way she had testified in her Utah cases.

When they had finished eating, they headed for Room 102-A in the Department of Labor building at 14th Street and Constitution Avenue, where the hearing would take place. Gladys Ellison and her daughter would not arrive until late Wednesday afternoon, so the three women would not make their official appearance until Thursday. But Goodman felt it would be helpful and informative for Eola to attend the opening session.[17]

The hearing chamber was packed with about four hundred people—mine operators, AEC officials, industrial hygienists, federal and state representatives, congressmen. The spectators and participants were seated in rows that faced a long, rectangular conference table. Eola recognized Duncan Holaday and Dr. Geno Saccamanno among the crowd of strangers. Goodman introduced her to Esther Peterson, one of the few women in attendance.

The opening session rehashed the old arguments on both sides of the issue. The Labor Department and the AEC, joined by Kerr-McGee Industries and other uranium producers, locked horns over Wirtz's tough proposed limits on radiation in the mines.

Dr. Robley Evans, professor of physics at Massachusetts Institute of Technology and a consultant for the AEC and Kerr-McGee Industries, argued for retention of the present limit of one working level. He said that a threshold figure at about nine hundred months of mine work at that exposure would be "far more than the average miner would receive in thirty years."

In direct conflict with Evans' statement, Karl Z. Morgan, director of the Health Physics Division of Oak Ridge National Laboratory in Tennessee, strongly supported Wirtz's standard, and told of lung cancer cases resulting from cumulative exposures well below Evans' threshold figures.

A lineup of spokesmen for various mining companies countered that lowering of the standard below one working level was not only unnecessary, but technologically impossible. Small operators insisted that the added costs of ventilation equipment and fans would put them out of business.

When the session ended, Goodman took Eola Garner to the *Washington Post* office, where she finally met J.V. Reistrup. Reistrup invited her to dinner, along with her sister and niece.

Later, at the restaurant, Reistrup told Eola that her story, and those of John Morrill and other uranium miners, had been the catalyst for Secretary Wirtz's long-delayed regulation of radiation standards. Now her appearance at the hearing was bound to be of further help.

At nine o'clock on Thursday morning, November 21, 1968, Eola Garner and Gladys and Pat Ellison were ushered to front-row seats in the hearing room. Eola was surprised to realize that she didn't feel especially nervous.

Leo Goodman took the floor first and voiced Labor's support of Wirtz's three-fourths working level of radon. Anything over that would mean an increase in lung cancers, he said. Goodman charged that government agencies and management no longer adhered to the philosophy of safety-first. Now, he said, it was a philosophy of benefit versus risk—the miners take the risk and management reaps the benefits.

Keith Schieger, a health physicist engaged in research at Colorado State University, objected. He submitted that miners had refused to use respirators, which would protect them from radon in the underground environment. The masks inhibited their work efficiency, thus reducing their productivity, and consequently, their income. Contract miners, Schieger stated, "are not willing to reduce the risk if it means a proportionate reduction in individual benefits."

Goodman returned to the table and concluded his remarks by quoting a statement made by President Lyndon B. Johnson at a Consumer Conference in February of 1968:

It has been said that each civilization creates its own hazards. Ours is no exception. While modern technology has enriched our daily lives, it has sometimes yielded unexpected and unfortunate effects.

Goodman then said that, to illustrate these effects, he would like to ask Eola Garner, and Gladys and Pat Ellison a few questions. The three women took their seats at the hearing table. Goodman first asked Gladys to tell about her husband's experience.

"He worked in uranium mines for twenty years, fifteen years underground and five years in the uranium mills," Gladys said. "He was one of the first to die of lung cancer in 1958."

"Has anybody from the AEC ever been in touch with you about his death?" Goodman asked.

"No one."

"Have you sought any assistance from the State of Colorado, following his death?"

"Yes, I did," she answered, "but the Statute of Limitation had run out because we didn't know anything about it at that time, in '58, or lung cancer."

"Did any of the newspapers tell you anything about lung cancer before that?"

"No."

"Do you know if the company he worked for ever advised him of the danger of radon gas?"

"No."

"Did he have any training or knowledge about radiation?"

"Not to my knowledge," Gladys said.

"Now would you tell us in your own words what his experience in the mines was," Goodman said.

"Well, the first mines he worked in had no ventilation at all because he would come home saying that they were dusty, it was hot in there and he would come home with a headache a lot of times from not having enough ventilation," Gladys said.

"I have readings from where he worked that ran from 4.4 up to 162 working levels, and during all the times he did work in these mines he didn't know that it was hazardous to his health, for if he did, he would not have worked in the mines. He would have left."

She described his lingering illness and painful death that had never been adequately explained.

"We feel the AEC knew the danger of the radiation from the uranium mining but never let the miners know what the expulsion of this radiation would cause," she said.

"Then I think one of the most important things is to prevent any more exposure from happening to any other miners. If it isn't done there will be a lot more deaths from uranium mining."

Goodman asked her if she thought such prevention was the duty of the state of Colorado.

"I think the U.S. Labor Department should do it, through the Federal Government," she replied.

"Why do you think it is the responsibility of the Federal Government?" he asked.

"Because they don't have enough money in the States to take care of all of it," she answered.[18]

Next, Goodman called on Pat Ellison.

"Miss Ellison," he said, "would you tell us the last time you saw your father?"

"I was sixteen," the girl said.

Goodman asked her to tell the hearing about it.

"Well, I would just like to ask some of these men here, particularly the ones from AEC, if they ever saw anyone die because their bones were rotting away. My father went very slowly. . . . The doctor in El Paso told us his bones had completely rotted away."

Goodman asked what Pat did after her father's death.

"I stayed with my mother," she said. "She just got a business, which took all of the money we had, of course. I stayed with her and tried to help her. I didn't have enough money to go to school. I did start school [college] a year after my mother got a little bit more money, when I was twenty-four years old."

"Had your father promised to send you to college?" Goodman asked.

"Yes," she replied.

"Then he died from this lung cancer and could not send you?"

"Yes."

Goodman then called on Eola Garner. She told about marrying her young baseball player, how active he was, how healthy until the very last. She told about Tex being drafted during World War II, and being sent back to Uravan with seven other former millers to work in the Manhattan Project's first uranium mill.

"The Government kept a complete check on these seven service boys . . . to see what effect uranium would have on these men," she said.

"Do you know if he was ever aware of the danger of radon gas?" Goodman asked.

"He was never aware of the danger of it," Eola replied. Then she testified that Garner's first suspicion was when he read the article in the *Denver Post* while he was in the hospital.

"This is the first time we ever knew about it and he told me at the time, 'I will never go back into another uranium mine. . . .' Now, had he known the danger involved in this, had they told him, he would have never, never gone into a uranium mine, and I don't think most of these men sitting here would either, knowing the danger of it," she said.

Eola added that Dr. Geno Saccamanno had told her only the day before that "had the working levels been as low as they are getting them down now, that my husband would be alive right today."

Next, Goodman asked Eola if she had sought assistance from Utah or Colorado after Garner's death. She told him about the Utah Supreme Court's denial of her claims.

"The results came back carcinoma of the lung, one hundred percent; occupational disease, none," she said.

"Do you think there should be a work standard set up for the men who are now in the mines?" Goodman continued.

"I definitely think there should be because, after all, if anyone could see anyone die so suddenly or so fast from this, they would never want any of their relatives to work at a high-radiation level."

"Did you tell me you came here because you thought other women were going to go through the same thing?" Goodman asked her.

"I have no doubt that there will be many more who will die of this," Eola answered. "The early miners which my husband and brother-in-law were, I think were hit the hardest from radiation, but I think they [miners] will be dying up to 1985, those who have been over-exposed to this radiation at the present time."

"Do you think they should be compensated once they lose the services of the breadwinner in the family?" Goodman said.

"I certainly think they should be compensated." Eola went on to tell about her own hardships while getting her case through the courts. She had spent over two thousand dollars for travel, food and lodging of witnesses during the hearings. Her two older boys had been denied college educations.

"Mrs. Garner, do you think that these other women who might have the same problem should rely on the State of Colorado and the State of Utah and the State of New Mexico?"

Eola Garner at Arlington Cemetery during her trip to Washington, D.C.. *Eola Garner.*

"No," she said, "I do not think so because . . . this uranium was mined for the Federal Government, for the Atomic Bomb for World War II, and I am aware of the fact that the AEC knew about this. . . . Had they followed through with this [regulations of standards], there would not have been as many deaths from this uranium mining as there have, and will be."

Goodman then turned to Gladys Ellison.

"Mrs. Ellison, is there anything further you would care to say?" he asked.

"Well, I really feel as though my husband gave his life for the protection of his country, as did the soldiers who died in combat," she said.

"Mr. Examiner, I think that closes our presentation," Goodman concluded.

When the hearing ended, many of the spectators introduced themselves to the women and thanked them for appearing. Esther Peterson invited Eola Garner and the Ellisons to her office, and they were driven back to the hotel in a White House staff car. Leo Goodman and his wife invited the threesome to dinner, and made arrangements for them to tour the White House, visit the John F. Kennedy Memorial and other Washington sights, and even spend a day in New York.

"It was like a fantasy world," Eola said later, "something that wasn't true."[19]

When it was all over, Eola felt a sense of elation, importance. She had delivered her testimony easily, and without nervousness. Some of the opposition had even accused Goodman of coaching her or giving her written statements. Many well-wishers said that she should be present when the bill was signed and receive the official pen from the President. Her experience in the Utah courts had stood her well.

Though it was too late for Tex, she felt that her efforts might benefit other miners and their families. Perhaps by telling her story, she had played a part in establishing Secretary Wirtz's standard of three-fourths working level of radon in the mines as law.

But when she returned home, she still lived in a mobile home. Her savings account showed only eight dollars and fifty cents, and progress on the Colorado case was not going well.

In October 1969, Charles Traylor notified her that her hearing would be delayed. The Colorado State Insurance Fund had announced that all bills regarding compensation that had been introduced the year before had been killed when Congress adjourned. With a change of administration, new legislation would have to be proposed.

A year later, Traylor informed her of another postponement. There were possible problems with the Statute of Limitations. And as time passed, Eola begrudged the thousands of dollars she had spent on her landmark case all the more. The Utah Supreme Court decision had been written into the United States *Congressional Record*. But all of the money she had poured into the Utah hearings could just as well have gone down the drain, she thought.

"I could have bought a dress shop and had plenty of money to retire on," she remarked, with considerable bitterness.[20]

As it was, in 1970, she borrowed funds from her children and other relatives so that she could buy a Mode O'Day shop in Silver City, New Mexico. There was a university there. She thought it was a good payroll town.

"So far I have done good and I must say that it isn't much harder work than I did working on my industrial case on my husband," she wrote Traylor the first year.[21]

But the long hours and seven-day work weeks finally took their toll. Eola suffered from high blood pressure and arthritis. And she was continually frustrated when she read newspaper articles about other uranium widows who had received compensation. These women, in her opinion, had sat back while she went to bat.

"Now I certainly am not mad because Mrs. Smith received compensation, because I think every miner's widow should receive it," she wrote Traylor after reading one report. But, she added, "I did so much to see that others would live. Spent all my time and insurance money my husband left me. It seems like some of the weaker cases are getting through and these widows have done *nothing*."[22]

Gradually, Eola's disillusionment turned into dissatisfaction with her attorney's performance. She couldn't understand why her case was taking so long, why other women were winning theirs. Her letters to Traylor became caustic.

On July 13, 1970, she wrote him that a reporter from the *Atlantic* had called to discuss the possibility of writing a story about her.

"He asked if I would like to sue the AEC," she said, "as one of the wives whose husband was on the plane that dropped the bomb over Hiroshima died of leukemia, and she sued the AEC for $100,000, and collected out of court."

"Don't fool around with those small town lawyers, get Marvin Belli," the journalist had told her. "He'll get you some big money."

Eola considered getting in touch with the famous attorney.

Writing Traylor, she warned, "If I have to get another attorney, I intend to get money out of this. I will keep you, but *no mistakes.*"

Traylor tried to assure Eola that his firm was doing everything in its power to win the case.

"I appreciate the fact that you want me to exert all influence to win your case," he wrote on September 17, 1973, "and all I can tell you is that it is our policy, no matter what type of case we have or who we represent, we always try our best, and in your particular case I would say that the time we have spent amounts to at least $4,000 or $5,000, and we have spent money of the firm in order to carry on this litigation. There is no one in the world that can say I am not interested in winning. . . . if I could bring this litigation to a conclusion in your favor, I would almost be willing to give my left little finger."

But by 1975, there was still no settlement. The economy in Silver City had weakened. Eola sold her shop and moved to Albuquerque. She and her sister, Gladys, bought a four-plex apartment, where they lived in adjacent units.

Eola's compensation case in Utah and Colorado spanned fourteen years. The struggling widow spent close to eight thousand dollars on her battle. After the drawn-out march from Industrial Commission hearings to the Supreme Court in Utah, she repeated the process in Colorado. Finally on December 15, 1977, Traylor advised her that a decision had been made.

"It seems that there is a real Santa Claus after all," he wrote. "We enclose herein a copy of the Colorado Supreme Court's decision in your case. . . . It appears that by holding on for about fourteen years with our teeth, we have finally succeeded."

Eola Garner was awarded $13,948.93. Her minor son James received $701.25, and daughter Pamela got $1,173.75.

It was a token reward. Eola remained bitter.

"I have often thought I worked my fool head off on that and I could have been making good money," she said.

But she retained satisfaction knowing that her efforts did pay off for others. "I did have a lot of the men come into the stores in Moab and shake my hand and say, "You did more for the uranium miners than anyone ever has. We're much safer now."[23]

19 COMPASSIONATE COMPENSATION

THE MARKS ON THE MAP were within a two block area of her home in St. George. Irma Thomas made a dot for every friend or neighbor who had contracted a radiation-induced disease or died. By the late 1970s there were twenty victims, fourteen deaths.. Wilford—cancer. His wife—stomach cancer. Carl—throat cancer. The boy across the street—leukemia. Irma's sister—breast cancer. Her sister-in-law, Hattie Nelson—dead at the age of forty-seven from a brain tumor.

From all outward signs, Irma was a typical Mormon housewife. She and her husband, Hyrum, manager of the local J.C. Penney store, lived in a Victorian-style home on Tabernacle Street, "with a gorgeous garden and the sweetest grapes in back."[1] A mother who often made elaborate costumes for her seven children, Irma was also an accomplished potter whose artwork was collected in America and abroad.

Irma was not afraid to speak her mind. She read voraciously and kept up-to-date on local and world affairs. When fallout clouds started drifting over St. George she launched into a one-woman crusade by stacking the dining-room table with books, articles, scientific journals—everything she could find about atom bombs—and then peppering local newspapers with letters to the editor lambasting nuclear activities. Her super-patriotic neighbors eyed her actions with suspicion, even insinuating that she was a Communist.

The fact that Irma wore rubber gloves and surgical masks when she was outdoors was also a topic of gossip. Her practice of re-washing her laundry if it was hanging on the clothesline at the time of a bomb test provoked further back fence talk.

Irma's nephew, Dennis Nelson (the boy that used to watch wash water from cars driven through fallout run into the garden), remembers visiting the big Victorian home.

"She was an activist from the time I can remember," he said. "Every time we would go to visit her she would come up with some new article in a newspaper or some foreign journalist who had come over and interviewed

284

Activist Irma Thomas. *Lolly Seal.*

her. She did not fit into the role of 'obedient patriot' or 'non-questioning Mormon.'"[2]

Irma might have been looked upon askance by her conservative neighbors, but her reputation spread outside of Utah borders as nuclear concerns increased. Reporters came to see her from all over the world. She would welcome them into her living room and then sit on the piano bench, with photographs of her seven children on the wall behind her, and answer their questions. Then, adjusting her owlish, shell-rimmed glasses she would show them her clippings and the home-made map, pointing out the markers depicting incidences of Hodgkins disease, leukemia, thyroid cancer and other radiation-induced ailments.

"I'm mad as hell and I'm not going to take it any more," said a placard on her piano. She did her best to live up to the motto.

In the late sixties, Irma's crusade grew more personal when her teenage daughter, Michele, became ill. Voted student body vice president of Dixie High School, a dancer and member of the Jetettes drill team, Michelle was suddenly stricken with ovarian cysts. A short time later, she became extremely fatigued, started to fall down and have trouble keeping her balance. She was diagnosed with polymyositis, a rare auto-immune disease that has a degenerative effect on muscles.

Michele was born in 1952, the second year of Nevada testing. Irma was certain her illness was caused by radioactive fallout. Little did she know that her suspicions would reach further into her family.

Michelle, partially crippled by paralyzed muscle groups, would suffer a mastectomy, six months of chemotherapy, radiation, and later cancer of her salivary glands. Irma's other children were plagued with miscarriages, stillbirths, and sterility, and both she and Hyram had bouts with malignancies.

Sister Hattie's family was plagued as well. In addition to her own cancerous death, her son, Dennis, fought basal cell-carcinoma twice, another son had leukemia and bladder cancer, her husband died of lung and bone cancer, and daughter Margaret succumbed to colon cancer at the age of forty.

Cancer statistics worsened during the sixties and seventies. Dr. Arthur Bruhn, the popular geology professor (and subsequent president of Dixie College) passed away in his early forties of acute leukemia. His student, Joanne Workman, who lost her hair after Shot Harry, underwent a four-year experimental treatment at Stanford University for malignant polyps that had penetrated into her colon, kidneys, bladder and stomach, as well as a brain tumor and polycythemia, a rare blood disorder. Miraculously, she survived the rigorous program of remission, although she retained a number of unusual side effects until she died at fifty-three.

Gloria Gregerson, who related memories of playing in radioactive "snow" under the oleander trees in Bunkerville, was diagnosed with ovarian cancer at the age of seventeen (1958.) During the next twenty years, she underwent thirteen operations for stomach and intestinal cancer, vaginal skin cancer and leukemia. She died at forty-two in 1983.

Claudia Boshell Peterson was an eleven-year-old in Cedar City when fifth grader Darwin Hoyt died of leukemia in 1961. Two years later her friend Bruce Stone developed bone cancer and had his leg amputated, then her best friend's mother died of colon cancer. Twenty years later, Claudia's own sister died of melanoma and a few weeks after that she lost her six-year-old daughter, Bethany, to acute monoblastic leukemia.

The statistics could not be ignored. Despite the AEC's tunnel vision about "maintaining nuclear supremacy" and "saving the free world," a few scientific minds were tackling the troubling evidence of radioactive injuries to people living downwind of the tests.

In 1960, Dr. Harold A. Knapp, a researcher with the newly-formed Fallout Studies Branch of the Division of Biology and Medicine, commenced a three-year study to evaluate fallout from nuclear weapons tests and possible warfare. Knapp's investigations centered on Shot Harry and other operations during the early 1950s, which were mostly high-yield bombs detonated from 300-foot towers that brought the explosions closer to the ground. He was interested in defining overall doses to people "around the test site and around the country as a whole, from external radiation sources, from internal radiation sources—from iodine in milk, and so forth."

". . . the entire focus up to that point had been on external gamma radiation, and there was neither data being collected or any particular interest in the amount of radiation doses which might have come from internal sources, particularly Iodine-131 in humans via the cow—the dairy cow or foraging animals," he later testified.

"It came as as much of a surprise to me as the other people;" he continued. "Namely, that we had vastly under-estimated the potential doses to the thyroids of children who drank fresh milk from dairy cows who were grazing in the direct path of the fallout from some of the shots—the "Harry" shot in particular, which I had estimated that as a result of the deposition alone, children that drank a liter of fresh milk from the dairy cow over in the vicinity of St. George might have gotten in the range of 120 rads to 140 doses of rads to their thyroid."[3]

Knapp's findings did not sit well with the AEC. Gordon Dunning, a fallout analyst and staunch defender of the test site's safety record, Charles Dunham, the new director of the Division of Biology and Medicine, and other scientists claimed the allegations were not sufficiently documented. They also feared that inferences of an AEC coverup of radioactive contamination could fester negative public opinion, engender lawsuits, and endanger continuation of the testing program.

Knapp, disgruntled by the vitriolic debates his work engendered, resigned in the midst of the flak. His paper, edited to exclude the troubling statistics, was finally published with the committee's adverse comments added.

Dr. Edward Weiss, of the Division of Radiological Health of the PHS, was the next scientist to raise questions. In 1961, he became troubled by the clusters of deaths from acute leukemia in the vicinity of St. George and Cedar City, a region in which the incidence of that disease was historically far below the national average. He undertook an epidemiological study and found that there had been twenty-eight leukemia deaths between 1950 and 1964.[4] A peak of six deaths occurred in 1959, the expected time lapse for incidences to appear following Shot Harry. But when the doctor revealed his findings, the AEC stepped in once again. Weiss's study was not published until much later.

The Centers for Disease Control published its findings on leukemia clusters in 1965.[5] They discovered a higher than expected incidence of the disease in the small towns of Fredonia, Arizona, and Monticello, Parowan and Paragonah, Utah, all in the area downwind of the Nevada Test Site. The government agency offered no speculation as to the cause, however.

In 1969 Dr. John Gofman and Arthur Tamplin co-authored a study disputing the AEC's theory of a radiation threshold.[6] Their research proved that low-level doses of fallout could cause cancer. In addition, they postulated that radiation released into the atmosphere could generate cancers with a death rate ten times worse than AEC predictions. The commission summarily dismissed the report and re-affirmed its own safety record.

The studies multiplied. Dr. Marvin Rallison, a University of Utah medical researcher, focused his work from 1965 to 1971 on the increased thyroid cancers of downwind children.[7] In 1979, Dr. Joseph L. Lyon published his findings about "Childhood Leukemias Associated with Fallout from the Nuclear Testing" in the *New England Journal of Medicine*. Dr. Glyn Caldwell at the Centers for Disease Control investigated leukemia in military personnel who witnessed Shot Smoky or other tests.[8] Dr. Carl Johnson studied cancer incidence in Utah.[9]

Nine research projects were undertaken. Eight of the studies either found quite strong linkage between fallout and the respective diseases or concluded it was still too early to make a definitive judgment. Government scientists continued to discount most of these deductions, citing insufficient data or flawed epidemiological processes while staunchly defending continuing operations at the Nevada Test Site.

The incriminating reports disappeared into government files until rumors of their existence circulated in 1978. Utah Governor Scott Matheson was furious when he saw stories in the media and then learned there were cartons of forgotten documents hidden in the state archives. During a visit to the nation's capitol he wrote a memo to President Jimmy Carter suggesting he look into the unusual preponderance of cancers in the downwind region. He had lived in Cedar City as a boy and remembered the atmospheric testing all too well.

"You could see the blasts just flash across the entire sky and feel the earth shake," he said. "And then sometimes later, you could feel the particles settling down on you like fine dust."[10] (Matheson, who considered himself a downwinder, died of cancer.)

President Carter directed Joseph Califano, secretary of health, education and welfare, to investigate the situation. Meanwhile, demands from the media to see the reports forced the federal agencies to open the AEC files and make all of their documents available to the American people through the Freedom of Information Act.

On November 13, 1978, *Deseret News* reporter Gordon Eliot White, wrote an expository piece entitled, "Leukemia Reports Never Published." The story was followed by the *Washington Post* in July when reporter Bill Curry wrote an article entitled, "The Clouds of Doubt Haunt the Mesas," questioning whether the cluster of cancer deaths striking communities downwind of the Nevada Test Site was coincidence or the result of the atomic weapons testing. In another piece on January 8, 1979, Curry wrote that "federal health officials had evidence as early as 1965 that excessive leukemia deaths were occurring among Utah residents exposed to radioactive fallout from the U.S. atomic bomb tests."

Opening up of the files in 1979 also caused the United States Senate Subcommittees on Health and Scientific Research of the Committee on Labor and Human Resources and the Committee on the Judiciary to meet in joint session to investigate the "Health Impact of Low-Level Radiation." Chaired by Senator Edward M. Kennedy, the hearings dealt with the effects of radioactive fallout on people living downwind of the Nevada Test Site and radiation exposures in underground mines. They wanted to get to the bottom of the heated accusations that hundreds of illnesses and deaths were the fault of the federal government.

"In the 1950s and early 1960s, because of the compelling needs of national security, the U.S. Government conducted an extensive series of atmospheric nuclear tests," Kennedy said in an opening statement on June 19. "During that time it was widely believed that nothing should be allowed to stand in the way of completing those tests. As a result, the Federal Government consistently minimized the effect of fallout, and failed to walk the last mile to protect the health of the people living in those communities."

Kennedy went on to discuss the "Americans who went down into the uranium mines day after day, year after year, to mine the ore to keep the testing program going. We will learn," he said, "what the Government knew about the health risks to those miners and when they knew it. What we will learn today is an American tragedy—a tragedy which could have been and should have been prevented."

The senators listened to testimony from widows and relatives about those who had died of radiation and from AEC representatives and health specialists such as Duncan Holaday, Victor Archer and Merril Eisenbud. The oft-repeated story of how responsibility for safety was passed from department to department in the federal and state governments and private mining companies was told again. As before, no one accepted the blame.

"Senator," Duncan Holaday remarked, "there is enough blame in this whole episode lying around so that every agency of the Government, the mine operators, the mine companies that owned the mines, and the individual workers can reach out and pick out what he thinks is his rightful share, and there will be plenty to go around for everybody."[11]

The lengthy congressional hearing accomplished little but put the question of compensation on the shoulders of Congress, as it was deemed too expensive, unpredictable and time consuming to decide the matter in the courts. The joint subcommittee agreed that the government should "accept at least 'compassionate responsibility,' if not legal responsibility, for the injuries sustained as a result of the nuclear weapons testing program."

However, a resultant report, "The Forgotten Guinea Pigs," might have laid the basis for a number of lawsuits against the federal government. Committee members concurred that "the government failed to provide adequate protection for the residents of this area. . . . to inform them of the exact time and place of each test and necessary precautions . . . the residents of this area merely became guinea pigs in a deadly experiment."

"The greatest irony of our atmospheric testing program is that the only victims of U.S. arms since World War II have been our own people," the report concluded.

As for the sheep incidents, they agreed that "Despite evidence indicating a causal relationship between the sheep deaths and exposure to nuclear fallout, the Subcommittee found that these sheep ranchers remain uncompensated to this day because of the government's suppression and disregard of such data. . . ."

The downwinders, whose only effort to connect radioactive fallout to the sudden onslaught of disease was the Bulloch sheep case—which failed—were heartened. Drawn together by their common suffering, they started to fight back.

That summer, Dan Bushnell and his wife were returning to Salt Lake City from a trip to Los Angeles, when they stopped in Las Vegas. Bushnell knew that government documents recently declassified through the Freedom of Information Act were stored there, and he wondered if the archives would contain undiscovered materials relating to the Bulloch case. He made a short stop at the library to satisfy his curiosity.

"I was absolutely ecstatic with what I found," he said later. "I found reference to the attorney's work papers. The result was that we were able to establish that the government had defrauded us and the court. They had intimidated witnesses, covered up documents, caused witnesses to lie, and it was just one big cover up."[12]

Normally, a case cannot be appealed after so many years have gone by. But Bushnell found one exception: Rule 60B of the Federal Rules of Civil Procedure. Under this rule, fraud on the court may be rectified at any time. The young lawyer filed a petition to set aside the decision in the 1956 hearing and reopen the case. Armed with new information from the declassified documents, he took depositions and marshaled enough evidence to warrant a new appearance before Judge Sherman Christensen (by coincidence the same judge that had decided against the sheepmen in the first proceeding).

The four-day trial commenced in May 1982. Bushnell focused on three key pieces of new evidence: veterinarians Lt. Col. Veenstra and R. E. Thompsett, who originally examined the Utah sheep, had been pressured to change their reports; the government had not disclosed results of radiation experiments on sheep conducted in 1952 by Dr. Leo Bustad at the Hanford laboratory; and the AEC had not been forthcoming in their testimonies at the first trial.

Bushnell exposed incriminating correspondence between army veterinarian Veenstra, and Lt. Col. Bernard F. Trum, of the Oak Ridge Agricultural Research Program, who co-chaired the final investigation of the Bulloch sheep. After examining the animals, Veenstra had judged that radiation was "at least a contributing factor." On March 25, 1955, Trum wrote his associate a thinly disguised admonition that he might be better off changing his mind.

"In the last report I had a chance to see that you rendered, I got the distinct feeling that you felt there was a chance that radiation could have caused the death of some of the sheep at least. In the interim substantial work has been done which may have caused you to change your mind. Beta burns have been produced on sheep at Los Alamos, Hanford and Oak Ridge. They have a very typical formation period, pathology and repair. . . . It took more than 350 r [roentgen] to kill the sheep. Those given 550 r died about three weeks later. The integrated dose in the area where the sheep were near the test site were less than 5 r."

The letter concluded with a strong hint.

"Because you didn't have this information available when you were asked to make your statement I've been wondering if you might not have changed your mind about these things. If you haven't changed your mind I'd like to know what you are basing your opinion on for I shouldn't like to go into this thing divided within our own Corps if we can avoid it."

Veenstra replied, making the points that despite revelation of the experiments on sheep at the Hanford facility, "We feel that our position has not been materially changed: basically we are still of the opinion that radiation could have contributed to the deaths of the animals. Although it is of course difficult to decide what degree of damage would have followed from a given exposure of this type, we are aware of the additional studies that have been carried out and we are convinced of their validity, but not completely of their relevance."

Veenstra noted that the Hanford studies had been on healthy animals while the Utah sheep were "in poor condition and suffering from malnutrition, 'disease,' exposure and whatever other hardships existed with range animals." His letter on April 7, 1955 was never sent or published and was marked "confidential to the Department."

Trum turned the screws a little harder with a personal visit to Veenstra, after which Veenstra wrote him a hand-written note stating, "In view of all your data and lack of ours the laboratory has decided not to make any official statement. . . . If called upon for a statement I will just report what we found and say we felt it a possibility that should be printed out for consideration."

Trum then turned his attention to Thompsett. On May 9 he sent the contract veterinarian a model letter with the suggestion, "If the letter is not exactly to your liking or not your style, you make the changes as you see fit."

Referring to Thompsett's original report, Trum wrote: ". . . I was of the opinion that radiation caused the deaths of the sheep or at least contributed to them and due to the presence of affected horses and cattle, too, the situation appeared to me to be very serious.

"Subsequently I've re-evaluated my position as more information became available.

"I believe it is reasonable to assume sheep could tolerate with impunity the maximum radiation dose possible under the range condition in 1953."

Thompsett's own undated letter indicated his judgment was unchanged.

"A thorough study of sheep diseases as well as a survey of endemic diseases at this time makes the syndrome of our sheep sickness aim toward a diagnosis of radio-active damage," he wrote.

When Dr. Bustad's report was finally revealed at the Senate subcommittee hearing, the evidence withheld at the first Bulloch trial demonstrated the similarities between the two examinations. The majority of lambs were stillborn. Full-term lambs were approximately half the normal birth size. Most newborns were weak in appearance and actions and unable to stand and nurse. The ewes had little or no milk. The hidden information could have supported the plaintiff's case.

More artifice by the AEC surfaced when Peter Libassi, attorney for the Department of Health, Education and Welfare, told the subcommittee that, "There is no question that the dose levels were almost 1,000 times the permissible count for human beings in the thyroid, and 50 percent higher than the permissible levels in the bone marrow." Trum's letter to Veenstra stated that the Utah sheep received a dosage of only 5 r.

Libassi revealed that hidden memos had also shown that Public Health Service officials met with the AEC in 1953 and stated they were convinced that the Utah sheep and animals radiated in the laboratories exhibited the same signs. Their opinions were not recorded in the minutes, however. AEC members summarily voted to only acknowledge that the dissenters were present at the meeting. In January 1954, an AEC press release stated that *all* branches of government concurred that radiation had not caused the animal deaths.

Judge Christensen was appalled at Bushnell's revelations. He ruled that, indeed, a fraud had been committed on the court. He listed "false and deceptive representations of government conduct," "improper but successful attempts to pressure witnesses," "A vital report intentionally withheld," "Interrogatories deceptively answered," and "deliberate concealment of significant facts with reference to possible effects of radiation upon the plaintiff's sheep."[13]

The judge overturned his own 1956 judgment.

But higher courts prevailed. The case was appealed before the U.S. Court of Appeals, Tenth Circuit, on November 23, 1983. The three judges seated reversed Christensen's decision, concluding that "the showing made by plaintiffs in the Bulloch II hearings falls short of proof of fraud on the court or any other kind of fraud."

However, the Tenth Circuit agreed to review the case *en banc* (with the full court seated) in May 1985. Christensen's decision was again overturned.

"We thought we had a chance," Bushnell remembered, "but we lost again. The same judge that wrote the decision the first time in Circuit Court assigned himself as chief judge. He was from Los Alamos, New

Mexico, and had affiliation back in those times. That's the only justification
I can give for him reversing."[14]

A final plea before the U.S. Supreme Court failed.

"The damages that my sheep owners had were about $250,000, docu-
mented, so many lambs, so many ewes," Bushnell said later. "And the government could have settled for that, or less, way back then. But they were
afraid of settling with us because of the admission of liability that they had-
n't been as cautious as they should have been. . . . They were afraid of cre-
ating liability as a precedent, even though they'd spent millions of dollars
trying to cover up this fraud."[15]

Stewart Udall, a personal injury lawyer from Phoenix, Arizona, was
another citizen appalled at the inferences of government coverup appearing in the press following passage of the Freedom of Information Act. One
of six children of a Mormon family in the rural southwest, Udall never lost
touch with his humble beginnings and felt a kinship with the people living
in small towns downwind of the Nevada Test Site who were suffering from
radiation-induced illnesses.

Udall started a private practice in Tucson and later served three terms in
the United States Congress. He was appointed secretary of the interior and
served in that capacity for eight years under Presidents Kennedy and
Johnson. After he left the cabinet and resumed law practice as a lobbyist in
Washington, he became weary of "being hired to influence the creation of
new laws" and returned to Arizona to be a practitioner of "real law," dedicat-
ing himself to environmental and conservation causes. Working alone in an
unpretentious one-man office, he became a pioneer "in the most provocative
and fast-moving area of the law—radiation and toxic chemical injuries."[16]

"Where you grow up is almost as important as your parents," he once
told a *Rocky Mountain News* reporter. "You live close to the land . . . it was
almost a 19th Century upbringing. But I was very lucky to have it. It gave
me an understanding of the land and our place in it."[17]

In the late 1970s as the controversial medical research surfaced and the
media took up the cause of the downwinders, Udall did some checking of
his own and attended a meeting with relatives of eight radiation victims at
the home of Irma Thomas in St. George. The stories were moving and the
numbers frightening. Widows and relatives claimed that their loved ones
had sickened or died from cancer as a result of radioactive fallout from over
one hundred atmospheric atomic weapons tests between 1951 and 1962.
Cancer rates were three to four times the normal incidence and involved
young children as well as adults.

Udall agreed to help the women institute a lawsuit against the federal government. Within three months, he had interviewed over 125 people and joined forces with other interested attorneys, Dale Haralson from Tucson and J. MacArthur Wright of St. George. The lawyers devised a screening criterion for potential plaintiffs. A victim must have lived in a downwind region where there was heavy fallout (approximately an arc north to southeast about 250 miles from the Nevada proving grounds) during the period of atmospheric tests. They must have medical records or death certificates to prove cancers were of the type known to be caused by radiation.

The first one hundred claims were filed in December. The lawyers contended that the Department of Energy was liable for approximately $100,000,000 in payments due to negligence in providing adequate monitoring, warnings to civilians downwind of the blasts, evacuation procedures or medical care. They claimed the agents had lied and falsely mitigated the dangers.

The defendants had six months to process the claims. They did not act in time, so the downwinders' only recourse was to sue under the provisions of the Federal Tort Claims Act, which essentially negates the doctrine of sovereign immunity and allows citizens to sue the federal government. Plaintiffs must prove that the government's actions were not discretionary, or according to personal judgment.

While the attorneys were preparing the case, the legislative arm of government was also at work. Both Senators Edward Kennedy and Orrin Hatch proposed bills to compensate radiation victims, but both efforts failed.

Finally, on September 14, 1982, Federal Judge Bruce S. Jenkins called for order in the oak-paneled courtroom above the United States Post Office in Salt Lake City. After several failed attempts at a settlement, the case of *Irene Allen et al v. The United States* commenced. (Mrs. Allen, first plaintiff alphabetically, was from Hurricane, Utah. She lost two husbands: the first died of leukemia in 1965, the second from cancer of the pancreas in 1978.)

The room was packed with spectators there to hear testimonies from victims and their relatives, test site workers, AEC officials, and scientific researchers. For two months the highly emotional trial was filled with stories of personal suffering, patriotic defenses and scientific statistics.

The plaintiff's attorneys adopted the premise that the tests themselves were not at fault, but the fact that they were carried out negligently and without considering harm to off-site citizens was. The defense lawyers claimed absolute immunity from tort liability and insisted there was no evidence of negligent acts by government officials.

Federal Judge Bruce S. Jenkins.
Bruce S. Jenkins.

Twenty-four plaintiffs represented nearly 1,200 persons who blamed the nuclear tests for their leukemia, Hodgkin's disease, lymphoma or cancers involving breasts, brain, stomach, pancreas, bladder and other organs susceptible to radiation.

In moving testimony Karlene Hafen's mother told of losing her fourteen-year-old daughter, a talented singer and dancer who had been voted to be harvest queen at a dance she didn't live to attend. She died of acute myelogenous leukemia.

Willard Bowler's wife related how one spring, while her husband was rounding up the cattle, he saw that a fingernail had turned black. He finally went to the doctor and had the nail removed. There was still no relief from pain so the doctor ordered a biopsy. Four months later Willard was dead at fifty-seven from malignant melanoma.

Melvin Orton was thirty-two when he died of stomach cancer. Geraldine Thompson was a victim of ovarian cancer. Kent Whipple, 38, succumbed to lung cancer.

The results were painfully evident. Causation was harder to prove.

Dr. Joseph Lyon took the stand. His studies showed that childhood leukemia deaths in Utah increased forty percent in the 1960s, and deaths from the disease in the downwind regions occurred 2.4 times higher than the rest of the state. Worse, the five counties closest to the test site increased by 3.4 times. Lyon concluded that there was a seventy-one percent probability that the illnesses had been a result of atomic tests.

Former employees of the AEC and PHS testified that test officials had not given warnings prior to detonations and had falsified records by minimizing radiation standards when they were excessive.

The government witnesses defended the AEC guidelines and denied that the cancers were caused by fallout. They brought their own experts to testify on dosimetry and medical issues. Summing up, the defendants cited the statute of limitations issue and claimed AEC negligence had not been proven.

Judge Jenkins took the matter under advisement for seventeen months. In addition to studying the mountains of transcripts and exhibits, he learned all he could about physics and other scientific data. In what was considered a landmark opinion, he ruled that the AEC failed to warn off-site citizens, educate them about the hazards and preventative measures, or adequately measure the radiation. Because these operational functions were not performed, the AEC was not immune from tort liability.

Secondly, he refuted the fed's claim for the statute of limitations. "Knowledge starts the two-year clock running," he said. Due to the lengthy latency period of cancer, it was even difficult for scientists to prove causation.

That the government had responsibility for the health and welfare of the off-site citizens was written in the Atomic Energy Act of 1946, he said. He questioned why the Nevada Test Site did not receive the same standard of care exercised at nuclear laboratories.

Jenkins ruled in favor of ten of the twenty-four plaintiffs.

The decision was reversed by the Tenth Circuit Court of Appeals in April 1987. The atomic-testing program's public information plans could not be held liable under the tort claims law, the court decreed. The plans "clearly fall within the discretionary function exception."

"The primary purpose of the 'discretionary function' exception is to maintain the constitutional separation of powers and to prevent the judiciary from evaluating or interfering with discretionary actions of the U.S. Government's agents, especially in the areas of national defense and security," Howard Ball explained in his book, *Justice Downwind: America's Atomic Testing Program of the 1950s.*

"The downwinders can overcome the government's defense only by showing that it was a *ministerial,* that is, a nondiscretionary activity, that caused the injury or death," Ball stated, "and that the Atomic Energy Commission ignored standards of 'ordinary care' that the FTCA [Federal Tort Claims Act] requires of the government."

Udall and his colleagues attempted to do just that. They appealed to the United States Supreme Court to reopen the case. But on January 11, 1988, the justices refused to grant a hearing.

"I never thought I'd say shame on the United States of America," Udall told *Salt Lake Tribune* writer Conrad Walters after learning of the denial.[18]

Udall's distress was compounded by the fact that he had experienced similar disappointments with two other extended court battles. One case was on behalf of the families of twenty underground workers from Marysvale, Utah. *Deseret News* writers Robert D. Mullins and Joe Costanzo claimed that Marysvale was where "one of the deadliest battles on the Cold War was waged. . . ."[19] Of one hundred and eleven miners, thirty-one had succumbed to lung cancer.

According to Duncan Holaday's records, the Vanadium Corporation of America's mines there registered some of the highest radon levels of any uranium works. Fifteen samples taken in 1950 averaged 120 times the safe working level recommended by the USPHS. A year later, three readings taken in the productive Freedom shaft registered radon concentrations of 2,750, 1,300, and 2,200 times the safe level. Despite construction of a cross-cut tunnel in 1955 that connected all of the VCA shafts, radon gas concentrations remained well above the safe standard. It was estimated that more than half of the Marysvale uranium miners who had worked underground for over five years had developed cancer or respiratory disease.

The cumulative working level months (wlm) of exposure to radon by miners in Marysvale was astounding. Thirty-four of those represented by Udall averaged 6,182 wlm; none was below 300. Yet the established mine safety standard was 120 wlm.

Still, widows of the cancer-ridden miners struggled vainly to get worker's compensation. Reasons for denial ran the gamut of questionable causation, latency periods, attending small town physicians' lack of experience or advanced medical studies, unavailability of scientific testing apparatus, the fact that the miner smoked, or even that the appellant lacked legal counsel to counter rigorous applications of the statutory requirements of medical records, autopsies, filing dates, etc.

Byron Anderson, who died in 1972, had accumulated an estimated 22,000 wlm of radiation. His claim was denied because of questions regarding the primary site and cell type of his cancer. His death certificate stated that cancer of the lung and liver were the cause of death, but an autopsy reported that the cancer originated in the colon.

Dr. Victor Archer wrote in a report for the USPHS, "When cancer is widespread throughout the body at death it is often difficult for the pathologist to be certain as to where it originated. The site he chooses is often a guess, guided somewhat by histological type of the cancer and the relative size of the tumor in different organs. . . . This type does, however, originate in the lung and has been demonstrated to be increased among uranium miners."[20]

Eva Dean Hanson, who was crippled with arthritis and had two minor children to support, was denied compensation after her husband's death from cancer because she narrowly missed the filing deadline and had not acquired medical proof of causation.

It wasn't until 1979 that the Utah Industrial Commission began to honor some of the Marysvale miner claims. By 1982, five awards had been made. But the process was not easy. Vivian Peterson Howes, whose husband had been exposed to approximately 13,670 wlm, petitioned courts and congressmen for fifteen years before finally receiving an award of $10,780. The widow of Alvin Christensen, who had received one of the highest exposures of any miner in the USPHS study (16,000 cumulative wlm), was required to authorize a second autopsy two years after the first and undergo three years of medical review before her $15,600 in compensation was granted.

"Fairness and equality demand a reconsideration of the other earlier cases which have never been finally determined," Kenley Brunsdale, an associate of Stewart Udall, wrote in a summary of the Marysvale tragedy. "Further, in light of current medical and scientific evidence, claims for deceased Marysvale uranium mining victims should be settled in summary fashion, not requiring exhaustive medical inquiries, not exhumation of bodies for second autopsies. These people should not be required to prove over and over again what has been obvious for some time."[21]

Udall's other legal fight was a thirty-million-dollar suit involving eighty-five Navajo Indians who died after working in uranium mines on their reservation in Monument Valley, on the Arizona-Utah border.

Prior to the uranium boom, the Navajo Nation was virtually free of lung cancer. In one study involving fifty thousand X-rays, not a single case of the disease was discovered. Then, in the 1970s, many of the men who had worked in the uranium mines for a number of years began to wheeze, develop severe coughs, and spit up blood. Unprecedented numbers of young men died. The cause of death was lung cancer.

Newspaper and magazine articles began to appear. "Full-Circle. Navajo U-Mine Cancer Probe: Nobody's Responsible!" Amanda Spake wrote for the *Washington Post Service,* June 9, 1974. The story revealed that of one hundred miners in the Mesa Mines, eighteen were dead and twenty-one ill.

"Bury My Lungs At Red Rock," was a headline in *The Progressive,* February 26, 1979. Author Barry told of Native Americans who lived in Monument Valley, Arizona, building their hogans (houses) out of the gray, cement-like materials scavangered from tailings piles at the mines. Children played in the dumps. No one told them there was any danger from the radioactive waste.

In 1974, Betty Jo Yazzie, who lost two husbands to lung cancer, joined with twenty-five other widows to sue for workman's compensation. Betty

Stewart Udall and Melton Martinez. *Melton Martinez.*

Jo's first husband, Peter, a miner for twenty years, was the first Navajo uranium worker to die.

"Those mines had 100 times the levels of radioactivity allowed today,"
says LaVerne Husen, director of the Public Health Service in Shiprock.
"They weren't really mines, just holes and tunnels dug outside into the
cliffs. Inside the mines were like radiation chambers, giving off unmeasured
and unregulated amounts of radon."[22]

The *New York Times News Service* reported more horror stories. Harry
Tome, Red Rock representative on the Navajo Tribal Council, told
reporter Molly Ivins, "No one ever told us of the dangers in it. We were
not educated."[23]

Most of the miners and their wives had grown up in the 1920s or '30s,
when there were few educational opportunities. Their lives revolved
around herding sheep and goats, gathering wood, living a nomadic life in
the barren redrock country. The Native American language had no word
for radiation or lung cancer.

According to Navajo miner Leo Goodman, "Here, there was never any
talk of safety. They didn't provide drinking water so people drank from

water in the mine. I knew that was bad, but there was no water. . . . There were no showers."

Besides the deplorable conditions, the mining companies lowered the pay scale for Navajos, with hourly wages as much as $1.16 less than regular rates.

In the fall of 1979, Udall and his associate Bill Mahoney, troubled by the published accounts of an epidemic of lung cancer among Native American miners and the plight of their widows and children, drove to the Navajo Reservation near Shiprock, New Mexico, to see for themselves. They visited Betty Jo Yazzie and a number of widows and, with the help of an interpreter, discovered that the reported stories were all too true. Husbands died as young as thirty-three. Women, some left with as many as twelve children to raise, struggled to keep their families together and provide high-school educations for their young. None received workman's compensation checks because it was necessary to file for the payments within a year of onset of the disease, but the latency period for cancer was much longer.

Udall saw it as a mission. He completed his investigation and then returned to Phoenix and filed the case *Begay v. United States* in federal court. The suit, naming ten representative plaintiffs, was based on the presumption that since the federal government was responsible for the Navajo's health care, it owed them trust responsibilities. In contrast to existing laws that controlled radiation and recommended wearing respirators and taking other precautions, the uranium miners of the 1950s were virtually unprotected. The attorneys claimed that the dangers to miners were known by the AEC but concealed. Not only had mining companies and Public Health Service experts failed to warn the underground workers, they had used them in the uranium miner study to experiment on radiation effects.

The feds made no concessions and based their case on "sovereign immunity."

Because the Navajos had no financial backing, Udall 's wife, Lee, and their children sold lithographs donated by Navajo artist R. C. Gorman to finance the court appearance. Doctors Victor Archer, Merril Eisenbud and Duncan Holaday waived their fees to appear as witnesses.

The trial finally convened in August 1983. For two weeks a federal judge listened to the arguments. The doctors reviewed the epidemiological studies they had undertaken and described instances of official negligence in warning the miners and protecting them from radiation.

"Did you ever in your seventeen or eighteen years with this program . . . warn . . . a Navajo miner at any time?" Udall asked Duncan Holaday.

Holaday said he had not warned any individual miner.

"Did you on behalf of the PHS [Public Health Service] ever set up briefings where Indian miners were to be told, with an interpreter, what the dangers were, what the facts were?"

"I do not recall any such situation occurring," Holaday replied.

"Well," Udall continued, "do you know of any instance where any of your employees working under you . . . where Indian miners were told what the dangers were and how to protect themselves?"

"No, I don't."

"Was that a byproduct of this agreement that you made with the mining companies? [The Public Health Service had to promise not to alarm the miners in order to have access to the mines for their studies.]

"Only in part," Holaday answered. "The difficulties of communication and of attempting to get an understanding of what the situation was across to a group of people through an interpreter were really pretty insurmountable."

Then Udall went after his point.

"Looking back on this, Mr. Holaday, was this good industrial hygiene practice to not tell these miners?"

"Have to look at that two ways, Mr. Udall," Holaday said. "One of them is it certainly was not good industrial hygiene practice. The second way, if you hadn't been able to get in and do a survey, the only time you would have known there was a problem was when you started to count the bodies."[24]

But in a hearing before the Tenth Circuit Court of Appeals in 1986, Udall accused Holaday of failing to warn the individual miners of radiation dangers. Udall claimed that Holaday's "temporary decision made in the field" was not influenced by government policy or national security considerations. Therefore, the tort claims lawsuit was not subject to the discretionary function exception, which would absolve the government from blame.

The government again claimed the discretionary function, invoked the statute of limitations and attempted to blame the lung cancers on smoking.

The judge ruled in favor of the United States. He cited the significance of national security and found "that those who carried out the uranium miner study were not 'experimenting on human beings' but rather were 'gathering data' so they could establish safety standards in the uranium mines."[25]

Udall filed a brief to the U.S. Court of Appeals in San Francisco. The three-man panel of judges upheld the lower court's decision.

And on October 13, 1987, the United States Supreme Court upheld the lower court's ruling and refused Udall's petition for them to hear the case.

Duncan Holaday, retired in Guilford, Connecticut, was supported in his decision to conceal information about the health risks from uranium miners. His judgment was declared a discretionary action. The case was dismissed.

The idealistic Phoenix lawyer was devastated. Ten years of work had resulted in failure. He tried to write a letter to the Navajo plaintiffs, but couldn't bring himself to do it. When the Red Valley Chapter later invited him to a meeting, he sent a representative.

"I did not go because I was humiliated and sick at heart," Udall wrote in his book, *The Myths of August.*

"I did not go because for so many years, and on so many occasions, I had urged the Navajos to be patient and to have faith in their country's system of justice.

"I did not go because I was ashamed of the outcome of their lawsuit and could not think of a convincing way to explain to them such concepts as 'national security' and 'government immunity.'

"And I was ashamed to go because I didn't know how to explain to friends who had trusted me that the government in Washington that had betrayed them—and had needlessly sacrificed the lives of their husbands in the name of national security—could, under the law I had urged them to respect, avoid responsibility for the tragedies that had engulfed their lives."

20   AFTERMATH

"PEOPLE HAVE GOT TO LEARN to live with the facts of life, and part of the facts of life are fallout," said AEC commissioner and noted scientist, Willard F. Libby, at a commission meeting on February 23, 1955.

"I was a teacher in Panguitch in '51 when the testing started," said Irene McEwen of St. George. "One day after a blast it was my turn to take my phys ed class out on the field, and I didn't want to because I'd read about Hiroshima. But the principal told me I had to. Later he died of cancer. So did three of my students. So did the other phys ed teacher."[1]

"Of course, we want to keep the fallout in our tests to the absolute minimum, and we are learning to do just that," AEC Chairman Lewis L. Strauss wrote November 21, 1957, in response to a petition from Martha Bordoli Laird. "But the dangers that might occur from the fallout involve a small sacrifice when compared to the infinitely greater evil of the use of nuclear bombs in war."[2]

"I do not consider my son's death a small sacrifice," Mrs. Bordoli responded.[3]

Circulating a petition to get signatures from seventy-five complainants was not a common undertaking for an isolated Mormon ranch wife who spent her days milking cows, churning butter and tending a vegetable garden. The Bordoli Ranch was on the flank of the Humboldt National Forest, seventy miles from the Nevada Test Site and sixty-five miles from the nearest town. Martha had few neighbors close by, but following a severe gastrointestinal hemorrhage, her seven-year-old son Martin died of stem cell leukemia on October 24, 1956, and Martha began spreading the word that something should be done to prevent more illnesses and deaths. Her grief had turned to anger and frustration. She felt betrayed by a government she had revered and respected, and refused to stand by doing nothing.

No one had ever warned her and her family to stay out of the clouds enveloping their land or told them about the effects of radiation. "I feel like we were used more or less as guinea pigs," she later told a congressional hearing. "The forgotten guinea pigs. Because guinea pigs, they will come to the cage and check, which they never have. To this day, they have never checked anyone in my family or anyone that I know of from the fallout of these bombs."[4]

Martha's efforts to enlist the help of congressmen and other government officials to get the tests suspended or see that "some equally positive action be taken to safeguard us and our families," were unsuccessful. The federal powerhouse stood firm. Martha's argument was deemed minuscule in the larger picture of things. Gordon Dunning of the AEC even minimized estimates of Martin's exposure to radiation.

But with passage of the Freedom of Information Act, the situation changed. Downwind citizens discovered official reports of agencies like the AEC and PHS had been doctored, facts withheld. Their absolute, unquestioning faith in the federal government was shaken.

"It was in the late 1970s when my people finally began to realize what had happened to us and we began to organize in response to the betrayal and anger we felt," environmentalist Janet Gordon told a conference of radiation survivors commemorating the fiftieth anniversary of the end of World War II in Hiroshima, Japan.

Janet had already suffered from the effects of radioactive fallout. Her brother Kent was, with his father and other ranch hands, bringing sheep back to Cedar City for lambing when Shot Harry emitted its dirty cloud. At the time, Kent felt stickiness and burning on his skin and got horrible headaches. Nine years later he was dead from cancer of the pancreas. Of the eight cowboys with him that spring, six later died of cancer.

In 1978, while working on a strip mining issue, Janet attended a uranium mining conference in Laramie at the University of Wyoming. She met an investigative reporter from San Francisco who had worked with Nobel laureate Linus Pauling, one of the first scientists to detect the harmful effects of fallout. The reporter showed a film he had made about the Nevada Test Site and radiation victims downwind.

Coming on the heels of the congressional hearings that exposed injustices perpetrated by federal agencies, the documentary made a profound impression on Janet. She and Elisabeth Bruhn Wright, daughter of Dixie College President Arthur Bruhn, joined forces with Jay Preston Truman, head of a young environmental group called the Brineshrimp Alliance, and Mary Lou Milburg. Elisabeth had been writing informative letters to radiation victims under the letterhead "Citizens Call," so the foursome took that as the name of their new organization.

Activist Janet Gordon. *James Lerager.*

"We started with four goals," Janet said. "1) Educate ourselves about what had happened; 2) organize; 3) stop additional exposure to radiation; and 4) get help for victims.

They began to call friends and neighbors who had been stricken by malignancies and other illnesses, to disseminate information, provide mobile cancer screening centers and arrange transportation to medical facilities for those in need. Included in their mission was continuous monitoring of the underground testing program and active campaigning for a comprehensive test ban treaty between the United States and the Soviets.

On January 27, 1980, the first annual candlelight vigil was held at the Nevada Test Site. It was the twenty-ninth anniversary of the first atmospheric shot, and the purpose of the gathering was "to remember and honor victims of radiation from nuclear weapons testing programs, as well as the victims. . . . along the nuclear weapons cycle. . . ."

"I liken it to a deadly nuclear chain," Janet said later. "Each aspect of the process, the mining, milling, processing, manufacture of weapons or power plants, the operation of testing or use of the weapons or power, the experiments, and the waste are all linked in this chain, and the chain is capable of killing us all. . . . [the] only way to stop it is to break the chain, and the only way to accomplish that is to work together to pull it apart."[5]

Jay Preston Truman soon left Citizen's Call to organize the Downwinders, with the goal of ending worldwide atomic testing and

Activists stage candlelight vigil at Nevada Test Site. *Janet Gordon.*

opposing all proposed nuclear weapons. Truman had spent his childhood in Enterprise, Utah, and was a radiation victim, himself. At fifteen, he developed Walderstrom's disease, a type of lymphoma, and after recovering from that he suffered thyroid problems, an illness that eventually killed his mother. But it was when he was twenty-eight and attended his high school reunion that the severity of nuclear testing hit hard. A segment of the commemoration was held in the city cemetery. Of nine teenage buddies, he was the only one left.

Still, by 1986, the underground tests in Nevada were continuing. According to Janet Gordon, the Department of Energy admitted approximately twenty percent of these detonations leaked, or vented.

One of the worst incidents was the 10-kiloton bomb Baneberry. On December 18, 1970, the bomb was placed in a vertical hole 910 feet deep. The device was detonated at 7:30 A.M., and three-and-a-half minutes later a fissure 315-feet long split the ground and belched a veil of black smoke and particulate matter. A vapor of gaseous nuclides continued to escape for twenty-four hours.

Prevailing winds took the fallout to the north and northeast, but at higher elevations the effluent drifted over Nevada, Utah, Wyoming and into California. Approximately three million curies of radioactivity were released, and presence of radioactivity was detected over most of the western United States.

Nine hundred civilian workers were based at Camp 12, about four miles from ground zero. Personnel were evacuated at 9:05 A.M., and 475 vehicles were monitored for radiation and decontaminated. Persons with detectable contamination were ordered to shower and given clean clothing. Sixty-eight were given additional decontamination and direct measurement of thyroid radioactivity.

"After evaluating all film badge results, thyroid counts, whole-body counts and urine sample results, it was concluded that no individual received external or internal dose in excess of AEC exposure guides in effect at that time," Donald M. Kerr, acting assistant secretary for defense programs of the DOE later testified.[6]

But by then, three of the test site workers had already died of leukemia.

Employees at the Nevada proving grounds emulated the sheepmen and downwinders in attempts to sue the government or convince legislators to compensate radiation victims. In 1969 Keith Prescott, who developed multiple myeloma after digging tunnels and mucking out debris after the blasts, attempted a landmark litigation against the DOE and the Nevada Test Site.

Bennie Levy was an iron worker whose job was to recover instruments from ground zero shortly after a detonation so they could be analyzed by scientists. Assured it was safe, he wore no protective clothing. In 1980, suffering from cancer and leukemia, he organized the Nevada Test Site Workers Victims Association on behalf of one hundred co-workers who had died. Levy's goal was to inform Washington legislators and federal judges of the situation and urge them to take actions to furnish relief for the widows and families.

It appeared that remedies through the courts had been exhausted. When the *Irene Allen et al vs. United States* completed its eight-year spiral into defeat, attorney Stewart Udall told a *Salt Lake Tribune* reporter, "Write the obit. The only avenue remaining open for compensation to families of radiation victims is the United States Congress"[7]

Senator Edward Kennedy rallied to the cause and proposed the Radiation Exposure Compensation Act of 1979 on behalf of downwinders, uranium miners and the sheep herders. But the 96th Congress adjourned without taking action.

Utah Senator Orrin Hatch introduced compensation bills in 1981 and 1983. Both failed. He tried again in 1984, this time attempting to amend the *Compact of Free Association,* which recognized the Marshall Islands as an independent nation and created a $150 million health care trust fund for residents injured by radiation from U.S. nuclear tests. Hatch proposed tacking on another $150 million fund for downwinders. He was again unsuccessful.

Finally in 1990, the Radiation Exposure Compensation Act was passed with provision of a $100,000,000 "Atmospheric Nuclear Testing Compensation Trust Fund."

"Compassionate payments" would be made for deaths or injuries due to exposure to radiation resulting from an above-ground nuclear test at the Nevada Test Site, or as a result of employment as a uranium miner. Damages of $50,000 would be paid to individuals, or their heirs, who lived in an area affected by fallout for one year between January 2, 1951, and October 31, 1958, or from June 30, 1962, through July 31, 1962. Specified diseases to be awarded compensation were leukemia (other than chronic lymphatic leukemia), multiple myeloma, and cancers of the thyroid, lung, breast, stomach, colon, esophagus and urinary tract.

Miners who had worked in Colorado, New Mexico, Arizona or Utah from January 1, 1947, through December 31, 1961, having been exposed to one hundred or more working level months (wlm) if a non-smoker and two hundred-fifty wlm if a smoker, would be awarded $100,000 upon proof of lung cancer or other respiratory disease. Heirs of deceased miners were eligible for payments.

The act included the statement: "Congress recognizes that the lives and health of uranium miners and of innocent citizens who lived downwind of the Nevada tests were sacrificed to serve the national security interests of the United States, and Congress apologizes to these citizens and their families on behalf of the Nation."

On October 15, 1990, aboard Air Force One, President George Bush signed the Radiation Exposure Compensation Act (RECA) into law.

But not everyone was satisfied.

Claudia Peterson of St. George wasn't. There had never been a history of cancer in her family. Then her father died of a brain tumor. Her sister passed away at thirty-six from myeloma skin cancer. And, after three years of weekly drives all the way to Salt Lake City for chemotherapy and radiation, her six-year-old daughter Bethany succumbed to monoblastic leukemia.

"At the time my daughter was diagnosed, four others of my high school class of thirty had children with cancer. We've got two with stomach cancer, one that died of liver cancer, and one that has a teenager now diagnosed with cancer. The odds of that happening are astronomical," she said."[8]

Claudia was asked to speak at the First Global Radiation Victims Conference in New York in August 1987. It was the year Bethany died.

Activist Claudia Peterson.

"I looked at it as a free trip," she told reporter Lucinda Dillon. "but when I got there, there were people from all over the world who had the same stories. There was a guy from Hanford, some people from Savannah River where Three Mile Island happened, from Rocky Flats . . . I was blown away, and I thought I am never going to go home and be a guinea pig or be silent again."9

Since that time Claudia has worked with a number of groups promoting medical care and compensation for radiation victims.

"I think the RECA bill is basically a slap in the face to most of us," she said. "I've heard it been called 'The Utah Lottery.' If you're lucky enough to get one of those cancers [designated in the Act] you'll be compensated.

"Is $50,000 any kind of compensation when you were used as a guinea pig and they knew exactly what they were doing? Test site workers were eligible for $75,000, and uranium miners were worth $100,000. So we just keep feeling like we are the low use segment of the population."10

American Indians had other problems with RECA. The act imposed stringent standards of proof for individuals or their families who became ill or died due to radiation. Underground miners had to give written evidence of working level months of exposure. This task was made more difficult because mine companies in the 1940s and '50s knew little about radiation levels and their measurement, and accurate employee records were not always kept. When testing facilities were available, they were usually several hours' drive from the reservation. Worse still, few Navajos spoke English and many communicated through interpreters. As there were no words in their language for nuclear terminology, communication was very difficult.

Miners had to be diagnosed with a specific lung disease to qualify for payment. Those who smoked or developed cancers of the liver, kidneys and other internal organs were excluded from eligibility. Many American Indians smoked only when using special tobaccos during ceremonial rites.

Proof of the disease required two of the following: a pathology report of tissue biopsy; autopsy report; X-rays, with reports by two "B" readers showing presence of fibrosis; a physician summary report; hospital admitting and discharge reports; death certificate.

But Navajos did not always get a diagnosis. Medical facilities were miles away and not always equipped with diagnostic instruments. Besides, many retained traditional beliefs:

"The first thing we do when we're bothered by an illness is go to see the star gazer," said Melton Martinez, a representative of Diné Citizens Against Ruining Our Environment, and grandson of Paddy Martinez, who discovered the first uranium in New Mexico in the 1950s. "They can look into our body and see our past history and the future of what's going on. They'll try to pinpoint what's bothering your body, then they'll do a ceremonial on it. They say it is 'the wind way,' meaning all living beings, the environment, animal life, have to have air. And they do this thing to see if they can restore their lungs or what they're breathing in. But it didn't work."

Claimants were required to verify that they lived in an affected area or worked for a certain mine at a given time and for a specified number of years. The Navajos frequently change residences and miners are prone to move from job to job, depending on work opportunities. Stricken miners or their widows rarely had such records or documents from business dealings, medical histories or church affiliation to corroborate their claims. Marriage licenses were rare because no certificates were issued for traditional weddings and there were no records in state or tribal files.

By the spring of 1994, of 324 claims by Navajo workers or their families only 155 had received the $100,000 compensation, according to the Department of Justice. There were 97 persons awaiting a decision, 12 filing an appeal, and 72 claims denied.

Written medical opinions by local doctors and specialists did not always produce acceptance of claims. The final judgment was made by laboratories operated by the federal government.

Navajo miner Billy Chee worked underground from 1955 to 1961, until he broke his back in a mine accident. A plaintiff in the fated *Begay vs. United States* case, who vowed, "I have never smoked in my life," he died in a nursing home of pneumoconiosis (a lung disease caused by inhalation of mineral or metallic dust) in 1994. His widow Anna hired Stewart Udall to file a RECA claim.

Udall retained Dr. Victor Archer, an expert in the PHS uranium miner study, to tabulate Chee's wlm record and asked two radiologists from Albuquerque to read the chest X-rays. Archer determined Chee had a cumulative exposure of approximately 458 wlm. His X-rays, taken less than three months before his death, showed fibrosis in RECA-qualifying profusions. Chee's claims were denied. "There is not a preponderance of evidence that the decedent contracted a compensable disease," the official letter reported.[11]

Udall was incensed. He had received another claim denial the same day. He felt the Navajos were not getting a fair shake and was convinced that the RECA bill should be amended and broaden its scope to include more of the radiation victims. He dashed a handwritten letter off to Gerard Fischer at the Department of Justice.

"Two 'deficiency letters' which arrived today arrive at such outlandish results that I am sending them, with my comments, to the three U.S. Senators who are working with me re: amending the law to demonstrate the strained results under the existing program," he wrote.

"Navajo widow Anna Chee. . . . Dead husband was a non-smoker, had 458 wlm's of exposure. The CXR's [chest X-rays] were taken three months before Billy Chee died, and the Albuquerque experts found qualifying profusions. . . . Again your super-experts disagree . . . a result that slams the door in the face of this widow.

"With all due respect," he concluded, "these results leave me wondering whether a decision was made somewhere to have a Screw-Navajo-Widows week in the Justice Department."

Signing, "With a fierce sincerity," he postscripted, "These results cry out for a change in the RECA."

Udall was not alone. A number of congressional bills for reform of the act were introduced. Scores of organizations involved with radiation exposure survivors lobbied and testified at hearings in Washington. Finally, on July 10, 2000, the Radiation Exposure Compensation Act Amendment of 2000 was passed.

The revised bill provided for payment to above- and below-ground miners, uranium millers, and those who hauled ore for at least one year. The geographical area covering downwinders was expanded and the numbers of diseases and cancers eligible for compensation were increased to include those affecting the salivary glands, bladder, brain, and ovary. Congress also significantly lowered the radioactivity exposure thresholds and simplified the application process.

A year later, the Department of Labor kicked in when the Energy Employees Occupational Illness Compensation Program Act honored claims made by test site workers, lab technicians and other Department of Energy employees in the nuclear weapons industry. That ruling provided for lump-sum payments of $150,000 for compensation and medical expenses to those seriously ill from exposure to beryllium, silica or radiation. For those already awarded $100,000 through RECA, an extra $50,000 was added.

It appeared that the prayers had been answered. Then the RECA fund ran out of money. An appropriation of only $10.8 million had been earmarked while $20 million in compensation had already been approved, but not paid, in the year 2000.

"The RECA Amendment of 2000 law required payment by the federal government within 6 weeks," said a report of the Western States RECA Reform Coalition dated February 22, 2001. "However, uranium workers who qualify for compensation are receiving 'IOU's' from the Federal Government. There is no money in the RECA Trust Fund to pay those who qualify."[12]

"The Justice Department admits it made a mistake by not requesting enough money," Senator Hatch told the *Deseret News*.[13]

It was estimated that the budget request was $76 million short. Approximately 355 persons were issued notes and many were dying "with 'IOU's' in their pockets," while they waited,

"I've lost 10 of my IOU holders since October," said Rebecca Rockwell, a private investigator from Durango, Colorado, who assists uranium miners applying for compensation. "The problem is people are dying. I've gone to about as many funerals as I can take."[14]

Lori Goodman, of the RECA Reform Coalition, told the *Denver Post*, "Since the coalition was founded in January 1999, we've lost four of our leaders."

One of the deceased was a downwinder who went on a weekend camping trip with her pregnant cousin during an atmospheric bomb test. Her cousin later delivered a child with a serious birth defect, and she died at 52 of a rare thyroid cancer.

In 2001, several bills were introduced to remedy the situation. Senators Hatch and Pete Domenici proposed an emergency appropriation of $84 million to stabilize the RECA fund and increase downwinders' benefits to match those of uranium miners and nuclear weapons workers. Representatives Tom and Mark Udall and Jim Matheson moved to secure payment of existing IOUs. Senator Hatch proposed permanent, indefinite appropriations to the RECA Trust Fund to avoid its getting into the red again in future years.

On July 24, in front of cheering American troops in Kosovo, President George W. Bush signed a $6.5 million supplemental spending bill to make good on the embarrassing IOUs and provide funding for the next ten years. By December 12, payments had been made to 54.8 percent of childhood leukemia victims, 68.2 percent of other downwinders, 27.4 percent of onsite participants, 54.8 percent of uranium miners, 98.1 percent of uranium millers and 100 percent of ore transporters, for a total of $363,485,905.

"This is truly a major victory," said Senator Hatch. "It's about time. . . . all eligible individuals will receive their compensation in a timely matter."[15]

But there was one segment of the population affected by radioactive fallout that didn't even receive IOUs.

"The only people left out of the compensation act were the sheepherders," Dan Bushnell said. "When they were considering amending RECA, I contacted them and said, 'Look, we've got documented evidence. It's not a big item of damage, we can show you what they were, how much, etc. They said, 'Well, we'll put it in the bill.' The bill came out and it wasn't there. It never had any consideration by Congress."[16]

At the same time activists were waging the lengthy campaign for radiation exposure compensation, another aftermath of the atomic age threatened the environment. Nuclear waste.

The uranium boom got its second wind in the 1970s when there was a growing awareness of radiation and of incidences of cancer among uranium miners and mill workers. The U.S. Congress realized that mine dumps and mill tailings piles were poisoning the air and waterways. Accordingly, they passed the Uranium Mill Tailings Radiation Control Act, to legislate construction of new facilities and reclaim those that had been abandoned. The act put closed mills under the jurisdiction of the DOE, with their clean-up financed by the government and supervised by the new Nuclear Regulatory Commission (NRC). Mills still operating in 1978 had to be maintained according to NRC guidelines, but the owning companies had to pay all costs of bringing them up to code.

Uranium mills crush huge amounts of ore to extract and process the final product of yellowcake. Most of the remaining pulverized rock is discharged into a pond as a slurry and is contained by dikes until the water dries up or seeps away. The dry tailings remain. Along with the noxious elements.

"The greatest hazard from mill tailings is associated with radon-222, a short-lived daughter of radium-226 found near the end of a chain of long-lived radio-nuclides that begins with uranium (half-life, 4.5 billion years) and its daughter thorium-230 (half-life, 80,000 years). Unless covered deeply with clay and other material [or relocated and buried in a safer environment], a tailings pile may exhale radon gas at up to 500 times the natural background rate," Luther J. Carter reported in *Science* magazine.[17]

The refuse also contains heavy metals, arsenic and chemical reagents.

There were twenty-four abandoned mills at the time of the new law, and many of them were on the banks of rivers. The DOE proceeded to reclaim the workings in situ, but an outcry from environmentalists forced them to move those beside rivers to safer locations. All but the Atlas Mill in Moab, on the banks of the Colorado River.

Being an operating mill, Atlas was classified Title II, and required to do an Environmental Impact Study to determine the best way to decommission its mill and clean up its site. Atlas investigated nine alternatives involving moving the pile or capping it in place. The NRC had no reclamation standards at the time, so the company was allowed to select the method that was most cost efficient. Since capping was the least expensive option, Atlas posted a $6.5 million bond to dredge sand and gravel out of the Colorado River for a covering.

Ten years later the NRC established reclamation standards with the provision that tailings must be isolated from the environment and not require continuing maintenance, preferably with below-ground disposal at a stable and remote site. The agency gave Atlas an out, however, by adding the provision that companies could get around the requirements if they got the job done "to the extent practicable." As Atlas deemed the new standard too expensive and difficult, they declared compliance would be *impractical* for them.

"Through this loophole Atlas is pouring 30,000 gallons of toxic sludge into the aquifer of the Colorado every day," wrote Bill Hedden, a leader in the fight to clean up the site. "NRC accommodates situations like this by issuing what it calls 'Alternate Concentration Limits,' and Atlas' whole reclamation plan is built around 'ACLs.'"[18]

Atlas succeeded again in 1993 when it claimed that, if the tailings adhered to existing regulations for height, they would block two adjacent

Atlas Mill tailings pile was covered with grass seed.

highways and reach the river. It would be necessary to make the pile three times higher than the ordinance allowed. The NRC approved the change.

"It was at this point that the Grand County Council intervened," former councilman Hedden wrote. "In strongly worded comments, local government asked how a reclamation plan that left 10.5 million tons of tailings towering 100 feet high on unstable alluvium in the floodplain of the southwest's largest river, adjacent to a town and the entrance to a national park, blocking the mouth of a major wash, at the confluence of two fault systems and directly across the river from the largest wetland on the upper Colorado could possibly meet the objectives of below grade disposal at a suitable site?"[19]

The local *Times-Independent* also fanned community interest in requiring Atlas to move the pile. Senator Hatch agreed and requested an NRC environmental impact statement and technical evaluation report. The two-year study concluded that moving the pile to Klondike Flats next to the county landfill would be preferable, but when Atlas complained that that would cost about $150 million as opposed to $20 million for capping it, NRC gave in again.

In 1998, Atlas filed for bankruptcy. Their funds for capping the pile had diminished to $9 million. To make matters worse, the Oak Ridge National Laboratory revealed their estimates that, if no cleanup of the groundwater was done, approximately 9,648 gallons of contaminants, including one-half pound of radioactive uranium, would leak into the river every day for 270 years. And the U.S. Fish and Wildlife Service discovered the pile was leaking ammonia and killing endangered fish.

Pricewaterhouse-Coopers was named trustee and directed to dry out the soggy mountain of waste, study the entire site and initiate the NRC groundwater treatment program.

Moisture was removed from the tailings by transferring 800,000 yards of treated soil from the sides of the pile to the top so that the pressure would force out water trapped inside through a vacuum-assisted wick drainage system.

"It works kind of like a straw," said Diane Nielson, executive director of the Utah Division of Environmental Quality. "It's like putting celery into colored water and the colored water moves up the veins of the celery. Basically, it works through capillary action because it's under pressure and water moves up these straws, or wicks, to the surface, and then we pull the moisture into the evaporation pond. That helps get water out of the pile and into a position where it can be moved more easily."

The *Times-Independent* and Moab activists campaigned in 1998 to transfer oversight of the Atlas reclamation to the DOE, which would be in a better position than the NRC to plan mitigation and funding for removal of the pile. The Utah congressional delegation helped the cause by convincing their California colleagues that contamination of the Colorado River by the tailings would eventually pollute the Golden State's drinking water. President Bill Clinton signed the resultant legislation effecting the transfer on October 30, 2000.

Three months later Pricewaterhouse-Coopers ran out of money. Mother Nature made things worse that spring with a ferocious windstorm that tore at the tailings and pelted Moab with dust and particulates.

"The spring winds are blowing with a force so fierce, there are days we cannot even see the pile, the red dust is swirling so violently in town," Bill Heddon and Terry Tempest Williams wrote in an article, "Bound By the Wind." "We watch not only our town blanketed with uranium dust, but our own skin. You cannot wash it off. Our eyes and throats are irritated. It is a story we know too well."

Authorities were forced to take action. In May, the trustee negotiated with the NRC and the Utah Department of Environmental Quality to release $400,000 owed the trustee, and DOE added $1 million to complete dust control measures. A Grand Junction contractor scattered native grass seed over the tailings to prevent further erosion until the pile can be relocated.

Still, despite a deadline of November 2002, by June of that year there was no definite decision on the ultimate fate of the Atlas waste—moving or capping in place. A panel of eminent scientists assembled by the National

Research Council concluded that more answers to technical questions were needed. The group called for a new, "focused" study to make a final determination on the matter.

In the meantime, activists are fighting on other fronts: a mill recycling uranium from scrap ore near Blanding, approximately eighty miles from Moab, and dumping grounds for casks of high-level radioactive refuse proposed for Yucca Mountain, about ninety miles northwest of Las Vegas, and in Skull Valley on the Goshute Indian reservation, approximately fifty miles west of Salt Lake City.

The White Mesa Mill, Utah's only operating uranium mill, is a facility of the International Uranium Corporation (IUC). The plant receives shipments of alternate feeds from former Manhattan Project operations and other nuclear facilities in the East and California and then re-processes the material to recover uranium.

"The piles of feed material are no different than ore, in my opinion," said Bill Deal, general manager. "It's been processed somewhat in the past and there's still residual, recoverable uranium in it, and it is our intent to process this material and recover more uranium."[20]

Once the uranium has been extracted and processed into yellowcake, it is sold to make fuel rods for nuclear power plants.

Southern Utah activists fear White Mesa is becoming another Atlas. They consider the alternate feed as toxic waste containing radioactive lead and other potent toxins. They worry about pollution of the aquifer, radioactive dust, and accidents to trucks rolling through the center of town loaded with uranium leavings. The facility, close to the Ute Indian Reservation, poses a threat to the water supply.

Sam Taylor, of Moab's *Times-Independent,* thinks "it's a sham and a subterfuge. They're getting upwards of $100 a ton for material from New York and California, it's coming in from all directions. There's not enough uranium in that stuff to really recover. And even if there was, there's so much cheaper uranium available through South Africa, Australia and Canada."

Many environmentalists, burned by the drawn-out Atlas mess, fear that the IUC, whose annual report showed substantial losses in recent years, "may be intentionally stockpiling large amounts of radioactive and toxic metals and other wastes, with the intention of declaring bankruptcy and closing the facility."[21]

Bill Deal denies the allegation. "Presently, we're in stand-by while we're receiving alternate feeds from various sites around the country," he said. "We plan to start up and process those materials for uranium recovery

The White Mesa Mill.

starting in the spring of 2002. . . . We have a vested interest in taking care of this community."[22]

In Nevada and northern Utah, people are worried about deposition of over 40,000 tons of spent nuclear material being stored in 131 above-ground sites in 39 states. A permanent graveyard beneath Yucca Mountain for as much as 77,000 tons of the refuse was approved by President George W. Bush on February 15, 2002. The State of Nevada immediately challenged this plan with a lawsuit claiming the decision of the Bush administration was based on flawed guidelines. The House of Representatives countered by approving the Yucca Mountain project by a vote of 306 to 117.

In his story, "House Okays N-Storage Proposal," *Washington Post* correspondent Eric Pianin reported that Energy Secretary Spencer Abraham claimed the Nevada site was "scientifically sound and suitable" and would provide protection against terrorist attacks by consolidating all of the waste underground in hardened steel alloy casks that would provide safe storage for 10,000 years.

Opposing scientists argue that Yucca Mountain's volcanic rock is to porous, the containers' dependability over the long haul is not proven, and cross-country shipping by hundreds of train and truck convoys would be "a mobile Chernobyl," subject to accidents and acts of terrorism. A 4.4 earthquake near the site on June 14, 2002, added to fears of geological instability in the region.

Piles of seed materials stand beside the White Mesa Mill.

Utah Representative Jim Matheson added his voice to the opposition with a negative vote in the House. On May 8, 2002, he told *Salt Lake Tribune* reporter Dawn House, "Utah and Nevada produce no nuclear waste. We have no nuclear power plants. We have no proprietary stake in the waste that will be coming to our states."

Reflecting on past mistakes of the U.S. government, he said, "I can tell you, as a son of a downwinder and a congressman who represents thousands of sick, dying and widowed victims of our nuclear testing, that the federal responsibility on this issue has been appalling. Enough is enough."

A final vote on Yucca Mountain by the U.S. Senate is expected in late 2002.

One of the greatest concerns of Utahns is that the Yucca Mountain waste site, scheduled to open in 2010, will not be big enough. Approximately 45,000 tons of waste is currently stored around the country and another 20,000 tons is expected to accumulate before 2010. Projections call for 3,000 more tons of refuse per year for 23 years.

In the meantime, Private Fuel Storage (PFS), a consortium of eight electric utilities with nuclear power plants, proposes to pack their spent nuclear fuel rods into approximately 4,000 steel and concrete canisters and store them on 820 acres of Skull Valley at the Goshute Indian reservation. The depository is proposed as a temporary facility, but Utah citizens fear the "temporary" status of the dump is unlikely.

"Over my dead body," Utah Governor Mike Leavitt said when the project was proposed.

Opponents to the plan point out that completion of the Yucca Mountain underground storage site by the year 2010 is debatable. Like many Utahns, the people of Nevada don't want to become "the nation's nuclear dumping grounds" and are fighting the proposal with additional lawsuits.

Spent nuclear fuel is "hot" for 10,000 to a million years. A person exposed to the pellets without protection would die almost instantly. Utahns fear other hazards such as accidents in transporting the casks, natural disasters such as earthquakes, off-target missiles from the nearby Utah Test and Training Range crashing on the storage pad, and, since 9/11/2001, the possibility of terrorist attacks.

PFS's proposal is that the storage would be safe and temporary, only remaining in Utah until a permanent site at Yucca Mountain, is ready. The depository would open in 2004 with an initial twenty-year lease that could be extended to forty years. The depleted coffers of the destitute Goshutes and Tooele County would benefit by $3 billion. Besides, the state has no direct regulatory control over a project on tribal lands.

Besides, "there's an easier solution," a *Salt Lake Tribune* editorial commented on June 12, 2001. "Store the stuff the same way near the nuclear reactors that produce it. That is already happening at 16 of the nation's 103 reactor sites. If the dry-cask storage method is safe, as the nuclear industry claims, why move the stuff to Utah? Leave it where it is, and don't risk accidents during transport."

Meanwhile, as the Atomic Safety and Licensing Board of the U.S. Nuclear Regulatory Commission conducted hearings on the issue, Nevada politicos campaigned with environmentalists to oppose the Yucca Mountain repository, and Utah's governor told crowds of rallying protestors, "We do not want high-level nuclear waste in our state—period." A decision is expected in the fall of 2002.

As for the position of the uranium industry itself, the jury is still out. Early expectations of having some two-hundred-thirty atomic power plants in the United States did not materialize. The nation relied on other energy sources, such as oil, coal, and better conservation practices. But, due to recent power shortages, the picture is likely to change in the future.

The mining industry has slumped accordingly. In addition to slackened demand for uranium ore, the price per pound has plummeted from forty-three dollars to about eleven dollars. In 1999, only three underground uranium mines were operating and some other production came from in situ leaching, mine water, and mill tailings. Seventy-six percent of uranium used in U.S. utilities came from foreign sources.

So the uranium boom is essentially over. Except for the aftermath. On January 27, 2001, the fiftieth anniversary of the first atomic bomb test in Nevada, Governor Mike Leavitt proclaimed Downwinders Day to honor those who had suffered from radioactive fallout. The year 2002 marks fifty years since Charlie Steen struck ore at the Mi Vida and triggered uranium frenzy in a little Mormon farm town that is little no more.

Moab is now an energetic, modern community. The Atlas Mill stands no longer. The Uranium King's money-green mansion on the clifftop is now a restaurant, and neighborhoods of neat, brick-and-frame ramblers have replaced the trailer village of Steenville. Art galleries and espresso shops mingle with banks and real estate offices housed in historic buildings of native rock along Moab's Main Street, and water slides and sightseeing tramways clutter the beautiful redrock hills.

It was the uranium boom that transformed Moab from a little Mormon farm town to an energetic, modern community. It was the uranium boom that unwittingly fostered the area's second expansion after the mining bust by opening up the previously inaccessible desert with miles of improved roads. Today, the former "Uranium Capital of the World" is a destination for retirees and international tourists, the fulcrum for two national parks—Arches and Canyonlands—and the center of some of the most spectacular river scenery in the world.

The perspective of fifty years casts fragmented lights and shadows on the *people* who played a part in the fabled days of the uranium boom. Clear-cut observations don't apply so easily to human beings. Men and women can't be gauged like the infrastructure and tax coffers of a town. More often, they finally emerge as "victims" or "survivors" of a period. Perhaps those who lived through those crazy times on the Colorado Plateau were a little of each.

Sheldon Wimpfen, past manager of the AEC Grand Junction office (now DOE) was satisfied with the outcome.

"I'm gratified about the era," he said. "I think we were given a simple mission to get more uranium—like put a man on the moon—and we did it."[23]

But in contemplating the atomic era Stewart Udall points out what might be the most tragic episode and most important lesson of the times.

"Like gold, the uranium craze produced its share of big winners and big losers. What is more poignant is that the losers were innocent victims—

and the Atomic Energy big shots were 'patriots' who lied to protect what they conceived as the 'national interest.'"[24]

"For most of these people," said former stockbroker Dick Muir, "whether their subsequent lives have been fair or foul, even if in 1954 they had been a net loser, or they'd been a loser since, I tell you it was worth paying for to have experienced that year. Where else do you get the Yukon in Utah? And it was an equalizer, where things were possible for people for whom, under the normal course of life, things weren't so possible."[25]

NOTES

LETTERS, DOCUMENTS, AND TRANSCRIPTS of court cases used in the text were from the files of Charles and M. L. Steen, Stewart Udall, Kenley Brunsdale, Dr. Victor Archer, Mitchell Melich, Eola Garner, Judge Bruce S. Jenkins and Dan Bushnell. While some of the records from the Atomic Energy Commission (now the Department of Energy) and United States Public Health Service may be acquired by requesting copies from the appropriate offices through the Freedom of Information Act, many of the documents have been destroyed. Personal medical records may only be obtained by written permission of the family. Transcripts of trial proceedings and depositions may be obtained through the clerk of the U.S. District Court. Theses and doctoral dissertations listed in the notes are located in the Utah State Historical Society and the University of Utah Library Special Collections Room, where my personal files are also catalogued.

The most complete oral history collection concerning the uranium boom is at the California State University at Fullerton. A smaller oral history collection is at the Grand Junction, Colorado, Public Library. The Museum of Western History in Grand Junction has a file of clippings and photographs concerning uranium mining, the Moab Public Library contains films about the Mi Vida Mine, and the Dan O'Laurie Museum in Moab and the Helper Mining and Railroad Museum have photographs of mining and Charlie Steen.

The following books contain information about various facets of the uranium era: John D. Fuller, *The Day We Bombed Utah: America's Most Lethal Secret* (New York, New American Books, 1984); Philip L. Fradkin, *Fallout: An American Nuclear Tragedy* (University of Arizona Press, 1989); Peter H. Eichstaedt, *If You Poison Us* (Santa Fe, N.M., Red Crane Books, 1994); Stewart L. Udall, *The Myths of August: A Personal Exploration of Our Tragic Cold War Affair with the Atom* (New York, Pantheon Books, 1994); Howard Ball, *Justice Downwind: America's Atomic Testing Program in the 1950s* (New York, Oxford University Press, 1986); Carole Gallagher, Keith Schneider,

America Ground Zero: The Secret Nuclear War (MIT Press, 1993);Terrence R.
Fehlner, F.G. Gosling, *Origins of the Nevada Test Site* (U.S.D.O.E., 2000);
U.S.D.O.E., *United States Nuclear Tests July 1945 through September 1992;*
Raymond W.Taylor and Samuel W.Taylor, "The Ten–Million Dollar Puzzle
of Vernon J. Pick," *Uranium Fever, or No Talk Under $1 Million* (New York,
Macmillan Company, 1970.). All books, articles, and manuscripts used in the
text are documented in the notes.

1 THE SIREN CALL

1. "rebellion against authority": Maxine Newell, *Charlie Steen's Mi Vida* (privately published, 1976).
2. "...risks are being":A.W. Knoerr,"Can Uranium Mining Pay?" *Engineering and Mining Journal,* December 1949.
3. "barely lifted his eyes": Newell, *Charlie Steen's Mi Vida.*
4. "if it takes me": Ibid.
5. "In retrospect": Letter from Charlie Steen to Bill Hudson, September 4, 1952.
6. "To encourage prospectors": Paul Leach, Jr.,"Uranium Ore: How to go About Finding and Mining It," *Engineering and Mining Journal,* Sept. 1948.
7. "We got to jagging": Interview with Bob Barrett, 1968.
8. "We had some discussion": Deposition of William R. McCormick, from Official Report of Court Proceedings in United States District Court In and For the District of Utah, Central Division. *Roselie Shumaker, plaintiff, v. Utex Exploration Company, a corporation, et al defendants.*
9. "happier than I have been": Letter from Charlie Steen to Edna W. Miner, of Houston. Written from Dove Creek, Colorado, probably 1950.
10. "The scenery is magnificent": Letter from M. L. Steen to her friend "Beverly," March 13, 1951.
11. "Hardships never seem to cease";"I spent more time hunting";"Charles finally found himself": Ibid, April 1951.

2 THE EUROPEAN EXPERIENCE

1. Starting with "I think we might have a problem" through "At some of the more dusty operations": Ralph Batie's actual words come from Report of Court Proceedings in United States District Court In and For the District of Arizona. Civil No. 80–982. *John N. Begay, et al, plaintiff, v. the United States of America, defendant. August 3, 1983.* Deposition of Ralph Batie, from Report of Court Proceedings in the United States District Court In and For the District of Utah, Central Division. Civil No. 81005. *Sylvia Barnson, et al plaintiff, v. Foote Mineral Co., a Pennsylvania corp., Vanadium Corporation of America, a Delaware corp., and the United States of America, defendants,* a letter from Ralph Batie to Stewart Udall, December 26, 1979; interview with author, April 1986.
2. "After that we ran": Ibid.
3. "The mining was": Ibid.
4. "I took Eisenbud": Ibid.
5. "At some of": Ibid.
6. "one that you wouldn't": comment in telephone conversation with Dr.Victor Archer, 1987.
7. "Eisenbud recommended": Derived from Ralph Batie's depositions in the court proceedings listed above.

8. The story about George Gallagher "tak(ing) over" was derived from an interview with Phil Leahy in Grand Junction, Colorado, September 1986.
9. "My job was to teach": Interview with Ralph Batie, April 1986.
10. "Mr. Jacoe, if you ever": Interview with Henry Doyle, May 1986.
11. "The mining company": U.S. Senate Committee on Indian Affairs, "Leasing of Indian Lands," Hearings on Senate Bill 145, 57th Congress, Jan. 16–23, 1902.
12. "Survey of Uranium Mines in Navajo Reservation," Henry N. Doyle, Nov. 14–17, 1949; Jan. 11–12, 1950.
13. "While I rather anticipated": Letter from Duncan Holaday to Henry Doyle, February 21, 1950.

3 THE DAWN'S EARLY LIGHT

1. "I remember the day": Testimony of Gloria Gregerson, Radiation Compensation Act of 1981, Hearings.
2. "It was apparent" Terrence R. Fehner and F.G. Gosling, *Origins of the Nevada Test Site*, United States Department of Energy, December 2000.
3. "became radioactive stoves": Meeting of Joint Chief of Staff's evaluation board following Shot Baker, July 25, 1946, Ibid.
4. "policy and psychological": Ibid.
5. "result in no harm": Howard B. Hutchinson, "Project Nutmeg," Armed Forces Special Weapons Project, January 28, 1948. Ibid.
6. Quotes and descriptions of the proceedings August 1, 1950 meeting are based on "Discussions of Radiological Hazards Associated with a Continental Test Site for Atomic Bombs," notes taken by Frederick Reines, *Deseret News*, February 3, 1979.
7. "most nearly satisfies": Origins of the Nevada Test Site.
8. "that it might": Harry S. Truman, *Memoirs of Harry S. Truman: Times of Trial and Hope*.
9. "Two aspects came": Discussion of representatives of the Department of State and Defense and the AEC, December 19, 1950, *Origins of the Nevada Test Site*.
10. "every effort to educate": Ibid.
11. "But Dr. Shields Warren": Conversation with Dr. Shields Warren, July, 1979, Stewart Udall, *The Myths of August*.
12. "shown to be": *Origins of the Nevada Test Site*.
13. "less spectacular": Ibid.
14. "as anticipated": "No Radiation Damage From Blasts, AEC Claims," *Albuquerque Journal*, January 31, 1951
15. "a point": *Origins of the Nevada Test Site*.
16. "All tests": Hank Greenspan, "Where I Stand," *Las Vegas Morning Sun*, January 30, 1951.
17. "Children played": "Atomic Snow," *Las Vegas Morning Sun*, February 3, 1951.

4 DEADLY DAUGHTERS

1. "The mining companies": Interview with Duncan Holaday, 1986.
2. "There's a lot of country": Interview with Duncan Holaday, May 1986.
3. "It's ridiculous to me": Interview with Mark Shipman, Oral History Collection, California State University at Fullerton.
4. "Most of the miners thought": Interview with G.A. Franz, Sr., 1986.
5. "That's supposed to be"; "see what was there": Interview with Duncan Holaday, 1986.
6. "That radiation is capable": "The Development and Progress of a Health Study of Uranium Miners with Consideration of Some of the Problems Involved," report by Duncan Holaday. (No date.)
7. Scene taken from actual quotes in "Summary: Minutes of First Meeting of Uranium Study"; memorandum from Dr. William F. Bale to the Public Health

Service Files, entitled, "Hazards Associated with Radon and Thoron," March 14, 1951; and Duncan Holaday's testimonies in previously mentioned court cases involving miner radiation.

5 BONANZA AT BIG INDIAN

1. "More uranium was": *Times-Independent*, January 3, 1952.
2. "We arrived looking like": Newell, *Charlie Steen's Mi Vida*.
3. "Have found no ore": Letter from M. L. Steen to her friend "Beverly," March 13, 1951.
4. "Charles A. Steen": *Times-Independent*, September 4, 1952.
5. "Nobody would believe": Burt Meyers and Robert Grant, "Uranium Jackpot," *Engineering and Mining Journal*, September 1953.
6. "Well, you surely fooled": Letter from Bill Hudson to Charlie Steen, September 10, 1952.
7. "Spread a layer": Collie Small, "For 'Gold' in the Atomic Age the Forty–Niners Listen to the Hills," *Collier's*, September 10, 1949.
8. "It was bad geology": Grand Junction *Daily Sentinel*, December 12, 1952.
9. "Maybe it was a good": Interview with Mitchell Melich, 1968.
10. "I can remember that night": Interview with Charlie Steen, 1968.

6 URANIUM FRENZY

1. "There were dozens": Murray Teigh Bloom, "Charlie Steen, Ten Years Later," *Cosmopolitan*, April 1963.
2. "had a knack for reading": Robert Coughlan, "Vernon Pick's $10 Million Ordeal," *Life*, November 1, 1954.
3. "You can go look": Interview with Sheldon Wimpfen, September 1986.
4. "The trouble with most people": Coughlan, "Vernon Pick's $10 Million Ordeal."
5. "It would indicate": Peter Wyden, "Greenhorn's Trail to Uranium Riches," Part 2, *St. Louis Post-Dispatch*, June 30, 1953.
6. "he hinted that he": Raymond W. Taylor and Samuel W. Taylor, "The Ten–Million–Dollar Puzzle of Vernon J. Pick," in *Uranium Fever, or No Talk Under $1 Million* (New York: Macmillan Company, 1970).
7. George Morehouse: Interview with George Morehouse, 1968.
8. "He was 'shocked'": Letter to author from Vernon Pick, from Switzerland, 1969.
9. Letters courtesy Sheldon Wimpfen.
10. "A long, lean, lanky": *Times-Independent*, January 28, 1954.
11. "We've got twenty–one feet": *Times-Independent*, February 2, 1953.
12. "Poverty and I have": Newell, *Charlie Steen's Mi Vida*.
13. Utex party story derived from interview with Mitchell Melich, Sam Taylor, and articles in *Times-Independent*, May 23, June 11, 1953.
14. "All of a sudden": Interview with Mitch Melich, 1968.
15. "I was stone blind": Interview with Bob Barrett, 1968.
16. "This settles an irreconcilable": *Times-Independent*, December 17, 1953.

7 DIRTY HARRY

1. "absolutely no": From testimony of Mrs. J. T. Workman, *Irene Allen et al. v. The United States of America*
2. "emphasize too strongly": *Deseret News*, May 8, 1952.
3. "the amount of dust": "Eye Witness Account Given of Yucca Flat's 'Atomic Device' Explosion," *Iron County Record*.
4. "I don't believe": *Deseret News*, March 26, 1953.

5. "There was dust": Author interview with Dennis Nelson, June 8, 2001.
6. "Fairly large amounts" Philip L. Fradkin, "The Tests," *Fallout: An American Nuclear Tragedy.*
7. "A 7,000 square mile": Ibid. From AEC, "Radioactive Debris from Operations Upshot and Knothole," Health and Safety Laboratory, New York Operations Office, June 25, 1954.
8. Description of Shot Harry. "The Crime," Ibid.
9. "You saw the flare" Testimony of Mrs. J. T. Workman, *Irene Allen et al. v. U.S.*
10. "It appeared": Testimony of Dan Sheahan, *Bulloch v. United States, JH,* vol. 1.
11. "the fallout was": Testimony before the Committee on Labor and Human Resources, *Radiation Compensation Act of 1981,* Hearings. 97th Congress, 2nd Session, April 8, 1982.
12. "the teeth": Testimony of Mrs. Workman, *Allen v. U.S.*
13. "He said it was": Author interview with Janet Gordon.

8 THE BURDEN OF PROOF

1. "The only way we": Interview with Duncan Holaday, August 1987.
2. "I went to talk": Deposition of Duncan Holaday. Transcript of Trial Proceedings in United States District Court, In and For the District of Utah, Central Division, Civil No. 5-80-0119-A, *Rowena Anderson, plaintiff, v. Foote Mineral Company, a Pennsylvania corp., defendant.* November 12, 1981.
3. "they were using": Deposition of Duncan Holaday, in *Barnson v. Foote.*
4. "our conclusions": *Anderson v. Foote.*
5. "The Division of Raw Materials": "Radiation Problem in Uranium Mines of the Colorado Plateau," report by the Director of Raw Materials, 1951.
6. "Attempting to drill": Gary Lee Shumway, "A History of the Uranium Industry on the Colorado Plateau," (Dissertation, University of Southern California, 1970).
7. "The Biology and Medicine Boys": Testimony of Duncan Holaday, in *Begay v. USA.*
8. "when they [Anaconda] found": Ibid.
9. "You know who": *Annual American Conference of Industrial Hygiene,* Vol. 7, 1964. Interview with Duncan Holaday by Charles D. Yaffe.
10. "Any ruling they may set"; Letter from Denny Viles to Alan Look, chief Arizona Section Bureau of Mines, August 5, 1953.
11. "If the initial census": "Statement of General Plan of Operation and Cost Estimates for Field Medical Survey in the Colorado Plateau During the Summer of 1954" (memo), March 31, 1954.
12. All the quotes were taken from the transcripts of the Seven-State Uranium Mining Conference on Health Hazards, Salt Lake City, February 22–23, 1955.

9 "THE FUTURE OF AMERICA"

1. "As the conversation": Letter from Jack Coombs to author, April 13, 1968.
2. "Uranium is the nucleus": Interview with J. Walters, Jr., *Western Mining Survey,* June 4, 1954.
3. "The first issue": Letter from Jack Coombs to author, 1968.
4. "We'll look at a deal": Interview with Wallace R. Bennett.
5. "So we had to offer": Letter from Jack Coombs to author, 1968.
6. "Talk to Lyman Cromer": Letter from Jack Coombs to author, April 13, 1968.
7. Segment on Whitney Coffee Shop derived from interviews with Jack Coombs, Dick and Frank Whitney, April 1968.
8. "Walters came in one day": Interview with Plato Christopulos, April 1968.
9. "He could maneuver": Interview with Robert Cranmer, May 1968.
10. "One day that fall": Story and quotes from interview with Dewey Anderson, May 1968.
11. "The weather was so mild": Derived from interviews with Jack Coombs, Ralph and Ray Bowman, April–May 1968; 1986.
12. Information on Standard Uranium Corporation from the *Prospectus,* January 11, 1954.

13. "Zeke Dumke": Interview with Ezekial Dumke, May 1968.
14. "The pattern of property": Richard W. Muir, "Uranium Boom in Utah," *Harvard Business School Bulletin*, Spring 1956.
15. "had a face": Interview with Dick Muir, May 1968.
16. "Now that we": *Time*, October 10, 1955.
17. "All you have to do":"Big Boys Give Uranium a Rush," *Business Week*, April 3, 1954.
18. "Boom! . . . Boom! . . . Boom!": Jack Coombs, "Uranium Topics," *Salt Lake Tribune*, May 25, 1954.
19. "It was a get-rich-quick": Interview with Ray Bowman, May 1968.

10 THE COLOSSUS OF CASH

1. "That chair's too big": Interview with Pete Byrd, May 1968.
2. "With Hartford of": Martin L. Gross, "Floyd Odlum: Colossus of Cash," *True*, September 1954.
3. "So partly because": Interview with Floyd Odlum, February, 1968.
4. Winchell tip information in letter from Jack Coombs to author, January 18, 1969.
5. "One of these": Bob Bernick, "Up and Down the Street," *Salt Lake Tribune, May* 1954.
6. "Well, if the current": Coombs, "Uranium Topics," May 7, 1954.
7. "During the week": Bernick, "Up and Down the Street," July 24, 1954.
8. "By 1975, I believe": Speech by Floyd Odlum at Colorado Mining Association quoted in Bernick's article, "Odlum Activities Hailed as 'Spark of Life' for Small Uranium Firms," *Salt Lake Tribune*, February 20, 1955.
9. "Mr. Odlum, being": Interview with A. Payne Kibbe, April 1968.
10. "My representative": Interview with Floyd Odlum, February 1968.
11. "Charlie Steen later": Gordon G. Gauss, "Charlie Steen Found Headache Went with Wealth,*" Associated Press,* June 13, 1954.
12. "The one thing": Peter Wyden, "Greenhorn's Trail to Uranium Riches," *St. Louis Post-Dispatch,* June 29, 1953.
13. "Pick had had": Interview with Nels Stalheim, May 1968.
14. "it was because": Address of Floyd B. Odlum, "Uranium and Its Commercial Future." National Western Mining Conference, February 4, 1955.
15. "They had nothing but": Interview with Floyd B. Odlum, February 1968.
16. "There were fellows": Interview with Reed Brinton, 1987.
17. "The arbitrage opportunities": Letter from Jack Coombs to author, January 18, 1969 and interviews 1986.

11 SUCCESS AND SUBPOENAS

1. "Some sixty–five million": Newell, *Charlie Steen's Mi Vida.*
2. Poker game story from interview with Mitch Melich, 1968, and Charles A. Steen's speech, "Rendering Unto Caesar," at Uranium Reduction Company Dedication, September 14, 1957.
3. Story about Queen Mary cruise from: Bloom, "Charlie Steen, Ten Years Later."
4. "Moab was being touted": Elizabeth Pope, "The Richest Town in the U.S.A.," *McCall's,* December 1956.
5. "When I hit Moab": Interview with Charles A. Steen, May 1968.
6. "I love the rings": Bloom, "Charlie Steen, Ten Years Later."
7. "We sued him": Interview with Charles Traylor, April 1986.
8. "We have prepared a map": Scene based on testimony of Archie Garwood, from Report of Court Proceedings in United States Court of Appeals, Tenth Circuit, No. 5541. *Utex Exploration Co., a corporation, plaintiff, v. Archie Garwood, R. C. Gerlach, and W.E. Bozman, defendants.* December 20, 1956.

9. The source for the Bilyeu confrontation came from testimonies in court proceedings listed in preceding note.
10. "One courtroom battle": Interview with Pete Byrd, May 1968.
11. "I've only been a plaintiff": Interview with Charles Steen, 1968.
12. "He went around town": Bloom, "Charlie Steen, Ten Years Later."
13. "I couldn't have been": Interview with Sheldon A. Wimpfen, August 1986.
14. "both country boys": Remarks by Mitchell Melich, at the Dedication of Uranium Reduction Company Mill, September 14, 1957.
15. "Hell, that's easy": Interview with Henry Ruggeri, Oral History Collection, California State University at Fullerton.
16. "Well, he said": Deposition of Mitch Melich, in *Shumaker v. Utex.*
17. Dinner party story from interview with Mitchell Melich, 1986.
18. Charlie Steen referred to the URECO mill as a "coffee grinder" in his speech, "Rendering Unto Caesar," at the Dedication of Uranium Reduction Co.
19. "These ceremonies": Remarks by Mitchell Melich at the Dedication of Uranium Reduction Co.
20. "the Precious Few": Steen's speech, "Rendering Unto Caesar."
21. Story about Rose Shumaker based on testimonies in *Shumaker v. Utex.* Deposition of Charles A. Steen. Also interviews with Mitchell Melich, 1968, 1986.
22. "Steen was numbed": Bloom, "Charlie Steen, Ten Years Later."

12 THE BUBBLE BURSTS

1. "Buying uranium stocks": Coombs, "Uranium Topics," May 28, 1954.
2. "Don't put your": Ibid, May 31, 1954.
3. "The remainder are putting": "Colorado Plateau-Fabulous Treasure House of Energy," *Times-Independent, Special Edition:* 1954.
4. "We have options" Interview with Richard Muir, May 1968.
5. "We first permitted": "State Halves Broker Share for U Stock Sales," *Salt Lake Tribune,* May 29, 1954.
6. "A lot of people": Interview with Harold Bennett, 1968.
7. "This department considers": Letter to Licensed Securities Brokers from Milton Love, Utah Securities Commission, March 25, 1955.
8. "We do it" Bernick, "Action on Market Proves Salt Lake 'Wall Street' of U Stocks," in "Up and Down the Street," March 13, 1955.
9. "In the beginning for the full amount": Interview with Jack Coombs, 1986.
10. "The Bowmans get Mr. White": Letter from Jack Coombs to author, January 18, 1969.
11. "some of these brokers": Quote of Milton Love in Bernick, "Action on Market Proves Salt Lake 'Wall Street' of U Stocks."
12. "Because of the nature": Interview with Noland Schneider, 1968.
13. "We came to the point": Interview with Al Bain, 1968.
14. "I was one of nine men": Interview with Kenneth Cole, Oral History Collection, California State University at Fullerton, August 5, 1970.
15. "to provide closer surveillance": *Salt Lake Tribune,* August 1954.
16. "It just seemed to be": Interview with Alex Walker, Jr., 1968.
17. "The cash requirement": *Salt Lake Tribune,* September 1, 1955.
18. "Don't you think": Quotes of Representatives John B. Bennett, Arthur D. Klein, and Lyman Cromer, president of Salt Lake Stock Exchange, Reports on Senate Subcommittee Hearing, *Salt Lake Tribune,* September 14–16, 1955.
19. "The gathering and subsequent publication": Alan Pritchard, "Means to Cover Up Errors Seen in AEC Clamp on Ore Reserves Info," *Salt Lake Tribune,* March 1955.
20. "And then the 'wolves": Interview with Noland Schneider, 1968.

21. Incident about Trading and Investment Company from interview with Marilyn Coon, 1968.
22. "as an accommodation": "SEC Decision Issued in Sale of Stock," *Salt Lake Tribune*, November 17, 1955.
23. "The free world": Ralph Hendershot, *New York Herald-Tribune*.
24. "It sometimes seems": Letter to author from Richard Muir, February 20, 1988
25. "It was a fantastic": Letter from Jack Coombs to author, January18, 1969.
26. "I loved the": Ibid, March, 1927.

13 LEETSO—THE MONSTER THAT KILLS

1. "The Navajo word": Esther Yazzie and Jim Zion, "Leetso: The Powerful Yellow Monster," *Navajo Uranium Worker Oral History & Photography Project newsletter*, Albuquerque, New Mexico.
2. "Well, they shot": Author interview with McRae Bulloch, October 15, 2001.
3. "Boy, you guys": Ibid.
4. "Well, we trailed": Testimony of Kern Bulloch, "The Forgotten Guinea Pigs," Report Committee on Interstate and Foreign Commerce, United States House of Representatives and its Subcommittee on Oversight and Investigations, August, 1980.
5. "the concentration": Preliminary report by Arthur H. Wolff (then Acting Chief of the Radiological Health Training Section, U.S. Public Health Service) "The Forgotten Guinea Pigs."
6. "You know": Testimony of Dr. Stephen Brower, "Low-Level Radiation Effects on Health," Hearings before the Subcommittee on Oversight and Investigations of the Committee on Interstate and Foreign Commerce, House of Representatives, 96th Congress, April 23, May 24 and August 1, 1979.
7. "[But] they were": Testimony of Dr. Stephen Brower, "The Forgotten Guinea Pigs."
8. "Dr. Pearson told": Ibid.
9. "it is apparent": "The Forgotten Guinea Pigs."
10. "You are in": AEC Report, Atomic Tests Effects, 1955.
11. "of reduction factors": Stewart Udall, "Big Lies of Bomb Testers," *The Myths of August*.
12. "Did anyone": Documents furnished author by Dan Bushnell.
13. "I was just": Author interview with Dan Bushnell, October 8, 2001.
14. "It was a bad": Author interview with McRae Bulloch.
15. "wasn't exactly": Testimony of Martha Bordoli Laird, *Allen v. United States*.
16. "So he said": Ibid.

14 THE AMERICAN EXPERIENCE

1. Information about Van Arsdale case is from letters given to the author by Dr. Victor Archer: Dr. Archer to Dr. David J. Berman, September 28, 1956; Archer to Dr. Lynn James, September 28, 1956; Archer to Dr. Berman, December 4, 1956; Archer to Mrs. Van Arsdale, October 15, 1957; Archer to Charles Traylor, May 14, 1962, February 5, 1965.
2. "Urgent repeat sputum smears": Letter from Dr. Victor Archer to Dr. O. P. Gableman, February 18, 1958.
3. "History of chest pain": Letter from Dr. Victor Archer to Dr. David J. Berman, August 27, 1958.
4. "We both went back": Interview with Dr. Geno Saccamanno, August, 1986.
5. By the end": Joint Committee on Atomic Energy, Radiation Exposure of Uranium Miners" (report), 1967 p. 193.
6. "the Bureau of the Budget": "Origin, History and Development of the Uranium Study," prepared August 1964 from material in P.W. Jacoe's and Duncan Holaday's files by Duncan A. Holaday.

7. "They 'did their best'": Transcript of *Begay v. USA*.
8. "This was the requirement": Ibid.
9. Information and quotes on these pages taken from "Report of the Inter-Agency Committee on HealthHazards in Uranium Mining," November 12, 1959.
10. "State attempts to improve": "Uranium Miners' Cancer," *Time*, December 26, 1960.
11. "The consensus": "Killing Off Small Mines," *The Daily Sentinel* (editorial, probably 1961).
12. "This has been": Memo from G. A. Franz, December 24, 1960.
13. "It is now fourteen years": Duncan A. Holaday, "Origin, History and Development of the Uranium Study" (report), August 1964.
14. "it is no longer": AEC, "Summary History of Domestic Uranium Procurement" (Not dated).

15 Senator Steen

1. The proclamation was read by Orval Hafen in the Utah state senate, March 1, 1957.
2. "Charlie, I've been thinking": Interview with Mitchell Melich, 1968.
3. "Hell, I'm paying": Ibid.
4. "So we decided": Interview with Mitchell Melich, 1987.
5. Information and quotes derived from strategy notes made for a meeting of the Steen for Utah State Senator Election Committee, August 15, 1958. Notes furnished by Mitchell Melich.
6. "I will fight": Charles A. Steen campaign brochure, "The Man Who Didn't Move Away."
7. "I know the proper way": Speech made by Charles A. Steen, Class of '43, Texas College of Mines and Metallurgy, Ex-Students' Association Meeting, El Paso, Texas, November 1, 1958.
8. "we come to Homecoming": Letter from Paul H. Carlton to Dr. J. R. Smiley and Faculty, Texas Western College, November 2, 1958.
9. "'fess up' Paul": Letter from Charles A. Steen to Paul H. Carlton, November 7, 1958.
10. "I realize in retrospect": Letter from Charlie Steen with reprinted copies of his Texas College speech sent to friends October 22, 1959.
11. "I retract no part": Letter from Charlie Steen to John Fitch, editor of *The Prospector*, at Texas College, December 1, 1958.
12. "I said to him": Interview with Mitchell Melich, 1968.
13. "I was a wild radical": "'Hindsight in Depth!' Uranium King's Topic at U. Lecture," *Nevada State Journal*, February 27, 1964.
14. "I then thought of introducing": Ibid.
15. "but not sleep with them": George D. Lyman, *The Saga of the Comstock Lode* (N.Y.: Scribners).
16. "If I go broke": Bloom, "Charlie Steen, Ten Years Later."
17. "On May 29, 1962": Material on sale of Atlas derived from Closing Memorandum of loan to Atlas Corporation by Manufacturers Hanover Trust Company, First Security Bank of Utah, N.A., and Crocker-Anglo National Bank, August 17, 1962; document for loan to Charles A. Steen from First Security Bank.
18. "the Three B Syndrome": Carol Mathews, "'Cinderella Man' Still Thinking Big," *Herald Tribune* News Service, *Deseret News*, October 28, 1965.
19. "Ordinary income doesn't": James E. Bylin, "The World Collapses for U.S. Uranium King, But He Vows to Recoup," *Wall Street Journal*, January 17, 1969.
20. "it cost $285": Interview with M. L. Steen, 1968.
21. "They're supposed to send you": Ibid.
22. "Everybody kept advising me": Charles Hillinger, "Charlie Steen Awaits Fate Ponders Riches to Rags' Career," *Salt Lake Tribune*, November 2, 1969.
23. "I got outside": *Wall Street Journal*, January 17, 1969. (See above.)
24. "Who helps the destitute millionaire?": *Salt Lake Tribune*, December 4, 1969. Benedec's sympathy drive was in Nebraska.

25. "life isn't like that": Bloom, "Charlie Steen, Ten Years Later."
26. "we stopped at a café": "Charlie Steen's Back in Prospector Saddle Again," *Salt Lake Tribune*, February 24, 1970.
27. "I'm going to buy": "Uranium King Strikes Again," *Salt Lake Tribune*, July 14, 1970.
28. "There's uranium": "Steen Admits U-Ore Defeat," *Salt Lake Tribune*, October 25, 1970.
29. "The following year": Information about Steen's operation from, "Steen Serious in Hospital After Surgery," *Salt Lake Tribune*, September 12, 1971.
30. "When my family": Geoff Doman, UPI, "Steen Keeps in Trim—Still Battles," *Salt Lake Tribune*, July 4, 1973.
31. "all he could come out with": Interview with Charles Steen, Jr., 1988.
32. "Recovery has been": Doman, "Steen Keeps Trim—Still Battles."
33. "a contrast to": Interview with Charles Steen, 1968.
34. "You know, the government": Hillinger; "Charlie Steen Awaits Fate Ponders Riches to Rags' Career."
35. "All I want to do": Quote of Charlie Steen, Ibid.
36. "It's always been": "Steen Family Reopening Famous Cash Mine Near Boulder, Colo., "*Times-Independent,* August 4, 1983.
37. "I didn't need": Ibid.
38. "Certainly there aren't": Interview with Edward R. Farley, Jr., May 1986.

16 A Widow Fights Back

1. The story of Tex Garner is derived from interviews and correspondence with Eola Garner, 1986–1988.
2. The pathology report: Autopsy Protocol, Dr. Geno Saccamanno, Pathologist, September 16, 1963.
3. "The average range of lead–210": Letter from Dr. Victor E. Archer to Dr. Geno Saccamanno, January 20, 1965.
4. "It is our idea": Letter from Charles J. Traylor to Brigham Roberts, October 23, 1965.
5. "To begin with": "Uranium Mining Produces First Disease Death Claim," *Salt Lake Tribune,* May 2, 1964.
6. "the odds are greater": Transcript from proceedings of Utah Industrial Commission Hearing: *Eola Margaret Garner, widow, James Douglas Garner and Pamela Garner, minor children of Douglas Garner, deceased, v. Hecla Mining Company and Industrial Indemnity Company,* May 16, 1966.
7. "I wouldn't use the word": Ibid.
8. Traylor filed objection to findings of Medical Panel February 10, 1966.
9. "I really need": Letter from Eola Garner to Charles Traylor, August 21, 1966.
10. "step up to their": Letter from Leo Goodman to Charles J. Traylor, May 31, 1966.
11. "a gym teacher appointed": Interview with Esther Peterson, May 1986.
12. "That killed me": Ibid.
13. "Maybe it's my old": Ibid.
14. "We've got to do": Ibid.

17 Full Circle

1. "Nuclear Power Plant": Glynn Mapes, "Nuclear Power Plant Needs Begin to Revive Sick Uranium Industry,": *Wall Street Journal,* August 1, 1966.
2. "The nuclear power industry": Neilson B. O'Rear, "Summary: The Domestic Uranium Program, 1946–1966,": AEC, Grand Junction, 1966.
3. "to maintain a high rate": Statements by Alan Jones, manager AEC office, Grand Junction, May 24, 1956. Gary Shumway, "History of Domestic Uranium Procurement of the AEC."
4. "Throughout a world": "The Taming of the Atom for Peace, *" Unesco Journal,* July, 1968.

5. "However," Faulkner said": "A Shortage of Uranium," *Durango Herald,* February 21, 1966.

6. "Moab has been geared": Maxine Newell, "Uranium Industry Being Quietly Rejuvenated in the Moab Area," *Times-Independent,* February 23, 1967.

7. "Locating an unstaked": Ibid.

8. "A windfall discovery": "The Present Uranium Situation," speech by Elton A. Youngberg, Assistant Manager for Operations, AEC Grand Junction office, before Wyoming Mining Association, Riverton, Wyoming, May 13, 1967.

9. "In the chill": Mapes, "Nuclear Power Plant Needs Begin to Revive Sick Uranium Industry."

10. "the West is throbbing": Jerry E. Henry, "Rags to Riches Uranium Mining Now Unlikely," *Deseret News,* December 18, 1967.

11. "The more worthless": Interview with Noland Schneider, 1968.

12. "I put up about": Interview with Hal Cameron, 1986.

13. "not from local": Robert H. Woody, "Penny Stock Mart Booms, So Let th' Buyer Beware," in "Up and Down the Street," *Salt Lake Tribune,* January 11, 1968.

14. "The action by all": Letter from Jack Coombs to author, January 18, 1969.

15. "But they were smarter": Interview with Don Glenn, 1968.

16. "tout the acquisitions": Interview with G. Gail Weggeland, Oral History Collection, California State University at Fullerton.

17. "Is this resurrection real?": Woody, "Penny Stock Mart Booms, So Let th' Buyer Beware."

18. "I had one–hundred–seventy": Interview with Don Glenn, 1968.

19. "The man was quite poor": Interview with Paul Barracco, 1968.

20. "Elkton's promoters": Story derived from interview with Don Stocking, regional director SEC, August 7, 1970, Oral History Collection, California State University at Fullerton.

21. "'persuasive evidence' some stocks": Robert H. Woody, "SEC Cracks Down on Penny Stocks," *Salt Lake Tribune,* January 25, 1969.

22. "There will be": Ibid.

23. "No doubt": Robert H. Woody, "Salt Lake Penny Mart Reaches New Measure of Stability," *Salt Lake Tribune,* February 4, 1970.

24. "The market has calmed": Robert H. Woody, "Over-Counter Association Promotes Market Image," *Salt Lake Tribune,* March 3, 1970.

25. "For a number of years": Interview with Jack Coombs. 1987.

26. "a tendency toward a work": Guy Bolton, "Matheson Securities Task Force Cites Need for Better Controls," *Salt Lake Tribune,* December 15, 1984.

18 The Standard is Set

1. Information on Action Paper, "Radiation Protection Policy; Guidance for the Control of Radiation Hazards in Uranium Mining" (Report from the files of Stewart Udall); J.V. Reistrup, "50 Uranium Miners Seen Fated for Cancer," *Washington Post,* April 14, 1967.

2. "This is probably": J.V. Reistrup, "Hidden Casualties of Atomic Age Emerge," *Washington Post,* March 8, 1967.

3. "I would have": J.V. Reistrup, "Union Is Admitted to Case on Uranium Cancer Claim," *Washington Post,* July 1967.

4. "Secretary Willard Wirtz was unaware": J.V. Reistrup, "1150 Uranium Miners Seen Fated for Cancer," *Washington Post,* April 14, 1967.

5. "It didn't make": Interview with Willard Wirtz, May 1986.

6. "hundreds of efforts": "Radiation Exposure of Uranium Miners," Congressional Subcommittee on Research, Development and Radiation, Joint Committee on Atomic Energy, Statement of Hon. Willard Wirtz, Secretary of Labor, May 9, 1967.

7. "We clearly have the statutory": Interview with Esther Peterson, May 1986.

8. "the maximum level": "Radiation Exposure of Uranium Miners."

9. "Since so much": Ibid.

10. "We cannot confirm": Decision of Utah Industrial Commission upheld and stated in Decision of Utah Supreme Court case of *Eola Margaret Garner, widow, James Douglas Garner and Pamela Garner, minor children of Douglas Garner, deceased, v. Hecla Mining Company and the Utah Industrial Commission*, August 24, 1967.

11. "I realize it may be": Letter from Esther Peterson to Eola Garner, September 28, 1967.

12. "since the exposure": Dr. Victor Archer's testimony before Colorado Industrial Commission, *Eola Garner v. Vanadium Corporation of America and the Colorado State Compensation Insurance Fund*, November 26, 1967.

13. "My relatives are really mad": Letter from Eola Garner to Dr. Victor Archer, October 31, 1967.

14. "I would ask them": Interview with Eola Garner, 1986.

15. "Would you gamble": Guy Halverson, "Uranium Miners Defy Warning That Death Lurks in Form of Cancer-Causing Particles," *Wall Street Journal*, October 6, 1967.

16. Garner trip to Washington from letters and interviews with Eola Garner, 1986–1988.

17. Account of the hearing and actual quotes derived from Official Report of Proceedings before the U.S. Department of Labor, Bureau of Labor Standards, November 21, 1968.

18. "He worked in uranium": Testimony of Gladys Ellison, Official Report of Proceeding before the U.S. Department of Labor. (See above.)

19. "It was like a fantasy": Interview with Eola Garner, 1986.

20. "I could have": Ibid.

21. "So far I": Letter from Eola Garner to Charles Traylor, July 13, 1970.

22. "Now I certainly": Ibid. March 12, 1974.

23. "I have often thought": Interview with author, 1986.

19 COMPASSIONATE COMPENSATION

1. "a gorgeous garden": Author interview with Dennis Nelson.

2. "She was an activist": Ibid.

3. "the entire focus": Testimony of Harold Knapp, *Bulloch v. United States*.

4. Dr. Edward S. Weiss, "Leukemia Mortality Studies in Southwest Utah," AEC Report, 1965.

5. Shaffer Report, "Compensation", p. 27, Centers for Disease Control.

6. John Gofman, Arthur Tamplin, et al, "Low Dose Radioactivity, Chromosomes and Cancer," IEEE, October, 1969.

7. Dr. Marvin Rallison, Blown M. Dobbyns, F. Raymond Keating, Joseph E. Rall, Frank H. Tyler, "Thyroid Diseases in Children," *American Journal of Medicine* 56, April, 1974.

8. Glyn G. Caldwell, Delle Kelley, Matthew Zack, Henry Falk, Clark Heath, Jr., "Mortality and Cancer Frequency among Nuclear Test (Smoky) Participants, 1957–1979," *Journal of the American Medical Association, 250*, August 5, 1983.

9. Dr. Carl Johnson, "Cancer Incidence Downwind," *Journal of the American Medical Association,* January 1984.

10. "You could see": John G. Fuller, *The Day We Bombed Utah, America's Most Lethal Secret*.

11. "'Senator' Duncan Holaday": Duncan Holaday's testimony, *Bulloch v. United States*.

12. "I was absolutely": Author interview with Dan Bushnell, October 8, 2001.

13. "false and deceptive": *Bulloch v. United States,* 1982.

14. "We thought": Author interview with Dan Bushnell.

15. "The damages that": Ibid.

16. "being hired to influence": Zeke Scher, "Stewart Udall's 'pioneering' toxic chemical litigation," *Denver Post,* January 8, 1984.

17. "Where you grow": *Rocky Mountain News,* July 31, 1983.

18. "I never thought": Conrad Walters, "Angry Victims View U.S. As Villain Shirking Duty," *Salt Lake Tribune,* January 2, 1988.

19. "One of the deadliest": Robgert D. Mullins and Joe Costanzo, "Uranium Peril Was Concealed, Miners's Survivors Say," *Deseret News,* February 2, 1984.

20. "When cancer is": Kenley Brunsdale, "Summary of History, Law and Arguments in Support of Cited Claims for Relief Now Pending Before the Utah State Industrial Commission.
21. "Fairness and equality": bid.
22. "Those mines": Bury My Lungs At Red Rock," Barry, *The Progressive*, February 26, 1979.
23. "No one ever": Molly Ivins, "Uranium Mines falling victim to lung diseases," *New York Times Service*, June 3, 1979.
24. "Did you ever": Testimony of Duncan Holaday, *Begay v. United States.*
25. "those who carried": Stewart Udall, "Betrayal of the Uranium Miners," *The Myths of August*, Pantheon Books, 1994.

20 Aftermath

1. "I was a teacher": Anne Fadiman, "A Thousand Americans Sue for Damages Brought On By Atom Fallout, *Life*, June, 1980.
2. "Of course, we want": Ibid.
3. "I do not consider": Ibid.
4. "I feel like": Martha Bordoli Laird's testimony, *The Forgotten Guinea Pigs.*
5. "I liken it": Author interview with Janet Gordon, July 2, 2001.
6. "After evaluating all": Testimony of Donald M. Kerr, "Effect of Radiation on Human Health,"Vol. 1, *Hearings before the Subcommittee on Health and Environment, Committee on Interstate and Foreign Commerce, House of Representatives*, January 24–26, February 8, 9, 14, 28, 1978.
7. "Write the obit": Conrad Walters, "Angry Victims View U.S. As Villain Shirking Duty," *Salt Lake Tribune*, January 2, 1988.
8. "At the time": Author interview with Claudia Peterson, July 2, 2001.
9. "I looked at it": Lucinda Dillon, "Toxic Utah: Ghosts in the Wind," *Deseret News*, February 15, 2001.
10. "I think the RECA": Author interview, July 2, 2001.
11. "There is not": Decision from Gerard W. Fischer, Assistant Director Torts Branch, Civil Division, U.S. Department of Justice, February 11, 1997.
12. "The RECA Amendment": Lori Goodman, Ed Brickey and Melton Martinez, "Radiation Exposure Compensation Act of 2000," Western States RECA Reform Coalition.
13. "The Justice Department": Lee Davidson, "$76 Million foul-up hits downwinders," *Deseret News.*
14. "I've lost 10": M.E. Sprengelmeyer, "Dying for Compensation," *News Washington Bureau*, March 11, 2001.
15. "This is truly": Judy Fahys, "Utah Woman Hopeful Over the Bill to Compensate Radiation Victims," *Salt Lake Tribune*, December 16, 2001.
16. "The only people": Author interview with Dan Bushnell.
17. "The greatest hazard": Luther J. Carter, "Uranium Mill Tailings: Congress Addresses Long-Neglected Problem," *Science*, October, 1978.
18. "Through this loophole": Bill Heddon, "Atlas Politics", *The Zephyr*, September, 1998.
19. "It was at this point": Ibid.
20. "The piles of feed": Author interview with Bill Deal, June 19, 2001.
21. "may be intentionally": News release, "Groups Protest Toxic Uranium Mill Near Ute Reservation," Sierra Club, Glen Canyon Group.
22. "Presently we're in": Interview with Bill Deal.
23. "I'm gratified": Author interview with Sheldon Wimpfen.
24. "Like gold": Letter to author from Stewart Udall, November 17, 1987.
25. "For most of": Author interview with Dick Muir.

INDEX

Adams, Clarence H., 185
Adams, Don, 122
Adams, Joe, 122
Adams, T. Randall, 194
AEC (Atomic Energy Commission), 4, 8, 18, 24–29, 32, 49, 74, 88, 96, 210, 254, 257, 287, 304; assistance to prospectors, 257; Batie, Ralph, 16–27, 49; bonuses, 110, 162, 209; Brooke, Gerald, 77–78; buying schedule, 251; controls on industry, 212; disclosing ore grades, 187; Eisenbud, Merril, 21–24, 290, 301; Faulkner, Rafford L., 253; Gallagher, George, 26, 162; Groves, Leslie, 20, 31; Harley, John (New York), 98; incentive program, 74, 162; Johnson, Jesse, 26, 97, 100, 162, 168–70; Jones, Allan F., 251, 254.; Leahy, Phil, 7, 18, 26, 67; mining policy, 161–62; Morehouse, George, 78; nuclear plant safety, 51–56; O'Rear, Nielson B., 250; radon, permissible levels, 28–29; Rasor, Charles A. "Al," 74; security division, 79; Strauss, Lewis L., 106, 304; support for health studies, 100–101; Swanson, Melvin R. "Red," 77–78; Technical Committee on Miner Radiation, 247; uranium buying program, 80; Wimpfen, Sheldon A., 162–63, 187, 322; Youngberg, Elton A., 256
AEC Division of Biology and Medicine: Deal, L. Joe, 24; Dunham, Charles L. 209–10, 287; funding for mine studies, 101–2; Hardie, George A., 92
AFL/CIO, 247, 274; Goodman, Leo, 247, 274–81; Industrial Union Department, 247; air standards, 44
Alamogordo, New Mexico, 31–33, 35
Alladin Uranium Company, 116
Allen, Irene, v. United States, 295, 308
Allen, I. W., 149
American Lead, Zinc and Smelting Company, 165
American Mining Congress, 67, 270
American Smelting and Refining, 257
Anaconda Mining Company, 172; John (Jack) Warren, 103
Anderson, Dewey, 116–20, 124, 129, 266
Anderson, Gerald W. "Andy," 259–60
Apache Uranium, 138
Archer, Dr. Victor, 203–8, 213, 243, 244–45, 272, 290, 298, 301, 312
Army Corps of Engineers, 4, 6, 8
Arrow Uranium, 188, 261
Aspinall, Wayne (congressman, Colorado), 270
Associated Press, 231

Atlantic (magazine), 282
Atlas Corporation, 134, 135, 141, 142, 167, 225, 226, 229, 234; Clinchy, Walter G., 226; Farley, Edward R., Jr., 226, 234; Howard, George, 234; Odlum, Floyd, 123–37, 140–145, 165, 167–68; Stalheim, Nels, 142; Stretch, David A., 226
Atlas Mill, 315, 322
Atomic Energy Commission. *See* AEC
Auerbach, Oscar, 207

Bain, Al, 116, 183
Bale, Dr. William F., 52, 53–56, 97
Ball, Howard, 201, 297
Balsley, Howard ("Mr. Uranium"), 43, 67
Barracco, Paul, 261–63
Barrett, Bob, 9, 65, 69, 83, 84–86, 121, 125, 140, 160
Baruch, Bernard, 32, 33
Basic Chemical Corporation, 226
Batie, Ralph, 16–19, 20, 21–26, 27, 49
B.C. Life Insurance Company, 260
Begay v. United States, 301
Belli, Melvin, 282
Benedec, Dobson, 230
Bennett, Harold, 180
Bennett, John B. (congressman, Michigan), 186–87
Bennett, Wallace (senator, Utah), 271
Bentley, Jim, 14
bergkrankheit (mountain disease), 19
Berman, David J., 203
Bernick, Bob, 139, 140, 181
Bernstein, Sam, 112
beryllium, disease related to, 20–22
Beyer, Ralph, 83, 84
Big Buck claims: Adams, Don, 122; Bentley, Jim, 14; Brewer, W. Y., 14; Hayes, Dan, 14, 122, 154; Saul, Edward, 122
Big Indian Mines Incorporated, 147
Bikini, 17, 32, 33, 196
Bilyeu, Virgil, 108, 156, 158–59
Blandy, Vice Admiral W. H. P., 32
"blind pool" securities offering, 265
Bloom, Murray Teigh, 71, 154, 161, 223–24, 230
Blue Lizard, 261
Bordoli, Martin, 199
Borst, Dr. Lyle E., 90
Bowers, Dr. John Z. (dean, University of Utah Medical School), 107

337